Nonlinear Design

FETs and HEMTs

For a listing of recent titles in the
Artech House *Microwave Series*,
turn to the back of this book.

Nonlinear Design

FETs and HEMTs

Peter H. Ladbrooke

ARTECH
HOUSE

BOSTON | LONDON
artechhouse.com

Library of Congress Cataloging-in-Publication Data
A catalog record for this book is available from the U.S. Library of Congress

British Library Cataloguing in Publication Data
A catalog record for this book is available from the British Library.

ISBN 13: 978-1-63081-868-5

Cover design by Charlene Stevens

Artech House
685 Canton Street
Norwood, MA 02062

10 9 8 7 6 5 4 3 2 1

In recognition of the work and good fellowship of James Bridge, the quiet achiever without whom none of this would have been possible

Contents

Preface

"The best time to plant a tree is 20 years ago; the next best time is now."

—*Old Chinese proverb*

Everything in this book, be it a principle, method, or result, was in place 20 years ago. It represents the culmination of work done throughout the 1990s by a small, close-knit group of specialists with access to GaAs foundries, mobile phone manufacturers, defense companies, government institutions, and universities. In Europe, the effort was mainly collaborative, whereas in the United States it was commercial. The starting point was *MMIC Design: GaAs FETs and HEMTs*, published by Artech House in 1989, a book describing hitherto unpublished methods of compact device physics applied to designing MMICs for yield and manufacturability.

At that time, it was widely recognized that large-signal models were useful only up to a certain point, beyond which they lacked accuracy. We also knew that the methods described in *MMIC Design: GaAs FETs and HEMTs* should be applicable to the nonlinear case, but there were links missing. It took a year to find the missing links, but the best part of 10 years of practical work and refinement to prove them. Useful insights came from joint endeavors to design and implement a low dc-power-consumption MMIC foundry process for personal communicator devices and to design a manufacturable FET structure for a 44 GHz Monolithic Integrated Circuit demonstrator for spaceflight applications.

From 2000 to 2005, the entire resources of the vanishingly small enterprise for which I worked were committed to technology transfer to the United States of a desktop Pulsed I(V) Measurement Instrument. The instrument had been developed in the mid-1990s in support of contracts with the U.K. Ministry of Defence on the nonlinear problem. The hardware and controlling software were the result of collaboration with a similarly small company located nearby in Cambridge, whose expertise in parallel computing and digital design meant that they were naturals to create the hardware. From 2000 to 2005, work on the nonlinear question was left untouched, not only for want of manpower but because nondisclosure and right-of-first-use provisions had to be honored up to late 2006. Besides, there were many capable people active in finding better answers to the nonlinear question, so the idea that events would overtake what we had achieved was likely.

Thus, foreseeing irrelevance, it seemed that, after a working lifetime unraveling the physics of GaAs devices and their operation, the time had come to hang up the journeyman's hat and pursue other interests. Shortly afterwards, the few device specialists scattered to the four winds: for all there was a new life ahead.

Then, in mid-2020, out of the blue came a question from an engineer with decades of experience in the MMIC and power FET fields; he was still finding *MMIC Design: GaAs FETs and HEMTs* relevant and useful. By the unfathomable ways in which the world sometimes works, within a week another designer with similar experience in the United States not only said the same but hinted at a resurgence of interest in compact device physics, notably to shed light on nonlinearity in GaN FETs. Although one can never be sure to have located everything that has been published on a particular subject, such accounts as came to light indicated that no approach like that in *MMIC Design: GaAs FETs and HEMTs* had emerged to deal with the nonlinear problem, so the work that was shelved 20 years ago might still be relevant; perhaps its time had come. Fortunately, enough material had been retained at least to tell the story. These pages are the result.

To suggest that the methods herein are the answer would be ill-judged. However, what can be claimed is unparalleled accuracy (for the time) in the few practical cases investigated at length and, above all, a two-way understanding of the relationship between device fabrication technology and circuit performance. For those with the resource, enough detail has been given to implement the whole, with the prospect of a better understanding of GaN HFETs as well as replicating the kind of accuracy achieved for nonlinear circuits in GaAs. To see it used thus would be reward enough.

Acknowledgments

This book is a record of how a small group of specialists came to view a tiny niche of the world, with me as chronicler. They were James Bridge, Adrian Hill, David Bowler, and Nick Frankau.

Much is owed to the sponsors, collaborators, and adopters MA, JT, CW, RGD, AB, AJ, TB, DA and PB, together with the organizations through which they worked, principally the U.K. Ministry of Defence, the U.K. Department of Trade and Industry, the European Commission, and Plessey Research (Caswell) Ltd.

As commissioning editor, Merlin Fox suffused the project with good humor and forbearance while we worked through the challenges of an historical account with (hopefully) contemporary relevance.

Finally, if there are omissions, errors or falsehoods, I can only take comfort from the inimitable wordsmith Oscar Wilde: "I am not young enough to know everything."

Introduction

This book describes a unified approach to nonlinear analysis and design involving compound semiconductor field-effect transistors (FETs) and high electron mobility transistors (HEMTs). Although underpinned by what may be termed *compact device physics*, the method centers on an equivalent circuit of reformed topology that corrects errors in the commonplace circuit model. As the objective is one of practical application not theoretical treatise, the underlying physics is treated as a resource to be referred to rather than a prerequisite. Consistent with all circuit model approaches, the methods apply to die-sized device cells, thus providing the basis for electromagnetic (EM) simulation of large devices with cells interconnected by buses.

The book is the logical continuation of the principles laid down in *MMIC Design: GaAs FETs and HEMTs*, published by Artech House in 1989. It is the outcome of over 10 years of work by a small group of specialists working under the auspices of the U.K. Ministry of Defence to find answers to the nonlinear problem. Collaborative programs involving industry, mostly in Europe but also the United States, included successful design of the active devices for new monolithic microwave integrated circuit (MMIC) fabrication processes, thus verifying the practical utility of compact device physics.

An account is given of the rarely acknowledged errors inherent in extracting capacitive and conductance elements, as required by all circuit models, from measurements made at the terminals of a device. These limitations and restrictions can, in many cases, nevertheless yield workable results, which is demonstrated. To preserve the accuracy of the data for circuit simulation, a method of fitting sequential separated functions sympathetic to the physics is described.

A routine for recovering the dynamic I(V) characteristics from direct current (dc) biased S-parameter measurements has been developed, motivated by the need to recalculate quickly the radio frequency (RF) I(V) characteristics along which a device is working. Under large-signal conditions, the operating point may shift due to self-biasing and the circuit simulator must account for the changed RF I(V) characteristics accordingly. Examples of how these reconstructed I(V) curves compare with pulsed I(V) measurements are given.

True-to-life accuracy emerges in the few practical cases investigated at length, among them a 38 GHz frequency doubler and an MMIC 500 mW amplifier. Experience with implementation in harmonic balance and time-domain simulators is outlined, including start-up transients and a simple demonstration of log-term transients (memory effect).

The book is distinctive in that it offers a two-way understanding of the relationship between device fabrication technology and circuit performance. For those with the resource, enough detail has been given to implement the whole, with the prospect of a better understanding of GaN heterojunction FETs (HFETs) as well as replicating the kind of accuracy achieved for nonlinear circuits in GaAs.

With the exception of the review chapter, Chapter 2, all the work described in the book is original; most is hitherto unpublished.

Part I

CHAPTER 1

Introduction

At the time this work was done, there was a profusion of nonlinear models for compound semiconductor field effect transistors (FETs), mostly gallium arsenide (GaAs) FETs and high electron mobility transistors (HEMTs). All had their merits and demerits; they would work for some types of circuits, but not others. In that respect, little seems to have changed in the intervening 20 years. No consensus has emerged, no single all-encompassing approach with wide, if not universal, applicability. Could it be that no one had dug deep enough into the fundamentals of how these devices work to see how to construct a modeling scheme from the bottom up (from physical principles) rather than from the top down (from a circuit perspective)?

This book describes a complete method for constructing models of compound semiconductor HFETs. It is based on a reformulation of the standard equivalent circuit to comply with the way that such devices operate physically. As an alternative to persisting with the simpler mainstream approach with its acknowledged compromises, the method inverts the compromise in favor of fidelity to practice but at the expense of extra complexity. A guiding principle was that any extra complexity should be soundly based and manageable. The result is a model that should be applicable in a wide range of nonlinear circuits, is scalable, and is amenable to implementation as a turnkey system by a major software house. The underpinning physics corrects or replaces those deficiencies in common practice, which have meant pulsed S-parameters, large-signal network analyzer measurements, load-pull tuners, and the like when they become necessary. In their place, carefully measured small-signal S-parameters at multiple bias points suffice. An instrument for measuring the fast-pulse current-voltage characteristics of devices is an asset but is not necessary in many cases. The one reservation is the labyrinth of interacting memory effects, which remains a hurdle.

1.1 The Statement of the Problem

An acknowledged master of circuit design using nonlinear semiconductor devices is Steve Cripps, whose methods for designing power amplifiers have been widely practiced by circuit engineers for many years. Given the right test instrumentation and some experience in the art, Cripps [1] showed just how much can be done without prior characterization of the device in the form of a model. Generally, however, for different types of nonlinear circuits, a model is required. It is no surprise, therefore, to find an abundance of them, most based on an equivalent circuit. Purely measurement-based modeling schemes are fewer in number and offer less opportunity for compromising or losing essential electrical characteristics of the device, but generally are application-specific. With so many to choose from, why might a different approach be needed?

To some extent, the answer is historical. Amid the profusion of models were some configured around the devices of particular foundries or manufacturers, which they fitted quite well. That fact alone suggested a modeling scheme of wide, if not universal, applicability did not exist. Even so, claims of accuracy became numerous to the point where the word began to lose its meaning and, in some quarters, a type of fatigue set in that came to be known as the "yet another model" syndrome.

In retrospect, the assessment seems harsh: the challenge was far from simple. Nevertheless, with the proliferation of FET models discussed in Chapter 2, an effort to get to the bottom of why a procedure of wide applicability was proving elusive was justified. A thorough understanding of the internal workings of HFETs would be the place to start, and that meant device physics. In this book, therefore, physics is not an adjunct but underpins the whole. It builds upon the method published in the forerunner to the present volume, *MMIC Design: GaAs FETs and HEMTs* [2].

The emphasis is not theoretical rigor but one of finding approximations that satisfy two criteria. First, they must be plausible on physical grounds and supported by at least an element of mathematical justification. Second, they must lead to practically useful answers in a proven sense. The approach is not as academic as some, is practical in outlook, and culminates in an equivalent circuit of reformed topology. It begins with understanding how the device functions physically, bypassing the need for protracted circuit-theoretical considerations of charge conservation, symmetry, how many negative capacitances or transcapacitances there should be in the equivalent circuit model, issues of what circuit simulators of the time could cope with, and so on.

Among the many questions surrounding the use of a circuit model were those of data acquisition and how the data should be processed. Despite it being said many times that bias-dependent linear characterization is not workable as a nonlinear representation, there was no convincing evidence that the assertion had ever been tested after all potential sources of infidelity had been eliminated, including:

- Using just the direct current (dc) current-voltage characteristics rather than both the dc and instantaneous (or RF) set;
- Basing models on poorly measured and de-embedded S-parameter data;
- Accepting imprecise function fits to measured data;
- Adopting the small-signal equivalent circuit for nonlinear use.

What would happen if all such sources of inaccuracy could be corrected?

Although the outcome may be nothing more than a natural consequence of looking at the problem in a different way, the answer is a final result that appears to contest the foregoing but well-founded belief (see Section 4.2.2). Data acquired from multibias S-parameters, coupled with pulsed $I(V)$ measurements, can indeed lead to excellent nonlinear results provided that the equivalent circuit elements conform to the functional forms that physics says they should (which some published elements do not) and are then put into the form appropriate to the correct equivalent circuit. Herein, then, is evidence that:

- Having the correct equivalent circuit with physically justifiable elements outweighs any inaccuracy arising from the use of a multibias S-parameter measurement as the data source.
- Relatively commonplace microwave test equipment is sufficient after all.

1.2 Verifying the Approach in *MMIC Design: GaAs FETs and HEMTs*

This book builds on the compact model of device physics described in [2]. The formulation allowed the equations predicting device behavior from their technological makeup to be inverted, so some features of a device's makeup could be recovered from its microwave behavior. To distinguish the two procedures from one another, the terms "forward modeling" and "reverse modeling" were coined.

When weighed against known practical data, both forward and reverse modeling amount to rigorous tests of whether the underlying physics is close enough to reality to be useful for analysis and design. Although in 1989 enough practical proving had been done to justify dissemination of the work, the method had not been widely applied. Subsequently, therefore, concurrent with developments on the nonlinear problem, design and analysis of bias-dependent linear circuits were undertaken, mostly under the auspices of European collaborative ventures.

Two programs of design stand out. First, as part of a consortium on personal communicator networks (PCN), a low-voltage, low dc-power consumption, ion-implanted, 0.5 μm process was designed then implemented successfully by the commercial foundry participating as a consortium member [3]. Figure 1.1 compares the associated gain of a sample FET with the forecast made during the design phase.

Results of this kind were not selected as "best" comparisons but were as-is, attested to by Figure 1.2 for noise figure where design and practice diverge but the agreement is nevertheless respectable.

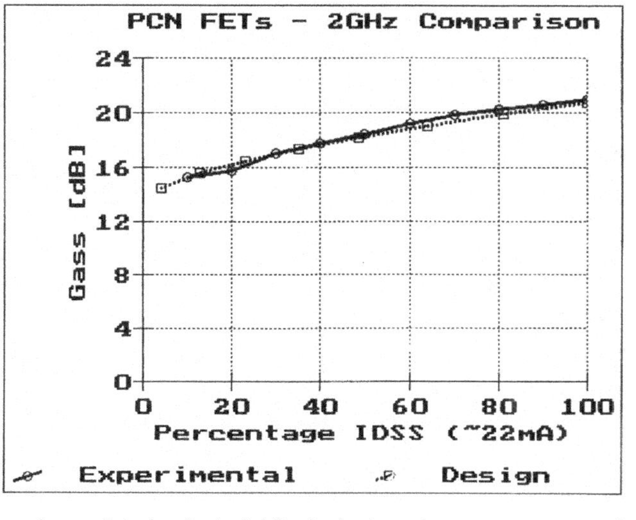

Figure 1.1 Measured associated gain (solid line) of a low-dc power consumption FET for personal communicator handsets compared with the design forecast.

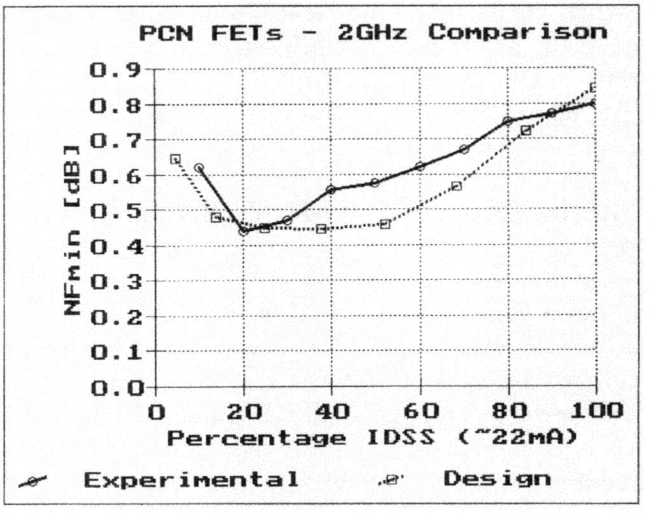

Figure 1.2 Practical and forecast noise figure for the device in Figure 1.1.

PCN was an initiative of the U.K. Department for Trade and Industry (DTI).

The second challenge of HFET process design was undertaken as part of a European consortium named MONOFAST [4]. Starting with a blank sheet the aim was to produce a millimeter-wave monolithic integrated circuit (M3IC) demonstrator at 44 GHz for spaceflight applications. Freed from the constraints of an established foundry or other fabrication process, the idea was to see what could be done at all levels of design and manufacturing to meet the demonstrator specification. A key benefit of not being subject to prior constraints was that the device and the circuit could be designed concurrently, each in concert with the other, meaning that all iteration and compromise would be resolved at the design stage without the need for costly practical trial.

Following an appraisal of the true performance differences between FETs and HEMTs [5] and with good reproducibility being an important consideration, a heterojunction field-effect transistor (HFET) structure emerged centred on a thin, highly doped active layer and an AlGaAs buffer requiring molecular beam epitaxy and electron-beam lithography.

Figure 1.3 shows what happened when sample HFETs, which were tightly toleranced, became available as a stage objective. Again, the lines through the data points are not best fits but are the forecasts from the initial design stage.
When fabricated, the M3IC worked the first time [4].

In parallel with the design activity, reverse modeling of many HFET variants was undertaken using data made available through demonstration and sale of commercial software to fit the standard equivalent circuit to measured S-parameters [6]. Measurements made at multiple bias points were processed to recover material and structural features, mainly the free-carrier profile in the channel and the metallurgical gate length [2].

Figure 1.4 shows the free-carrier profile of the MONOFAST HFET obtained by two methods: by reverse modeling measured S-parameters, and by conventional capacitance-voltage $C(V)$ profiling. In the channel layer itself, the flat region, the two

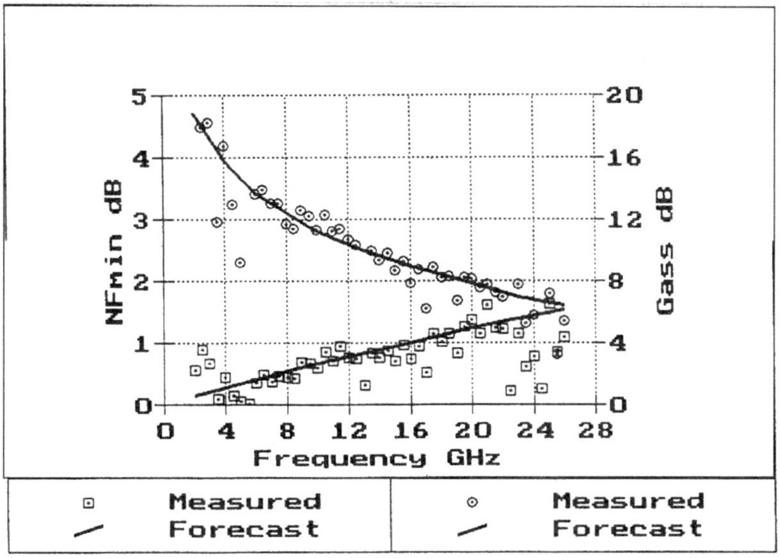

Figure 1.3 Noise figure and associated gain of a 0.2 μm gate HFET designed for a 44 GHz demonstrator M3IC. The lines through the measured points are the original design forecasts.

agree surprisingly well. There is no telling which of the profile tails is more correct, except to say the steeper tail of the reverse-modeled version seems more consistent with carriers being confined by the underlying AlGaAs buffer beyond $d \simeq 0.1$ μm.

Looking to more elaborate devices, reverse modeling enables an impression to be gained of the carrier distribution in a two-dimensional electron gas (2-DEG) (the wording is deliberately cautious). A very high-volume density of free carriers with rapid fall-off is apparent in the pseudomorphic high electron mobility transistor (PHEMT) of Figure 1.5. Reservations aside, an interesting conjecture is that

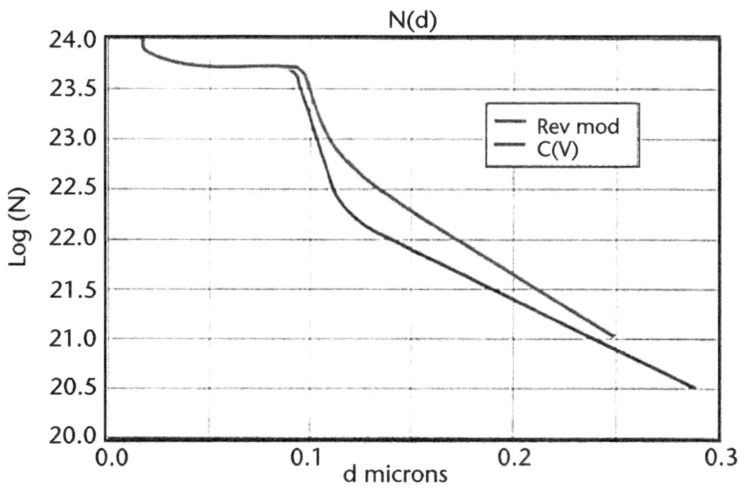

Figure 1.4 Free-electron profile recovered from measured S-parameters of a 0.2 μm gate FET (lower curve) plotted together with the *C*(*V*) profile. The numerical values on the vertical scale are the logarithm (to base 10) of the density of free carriers.

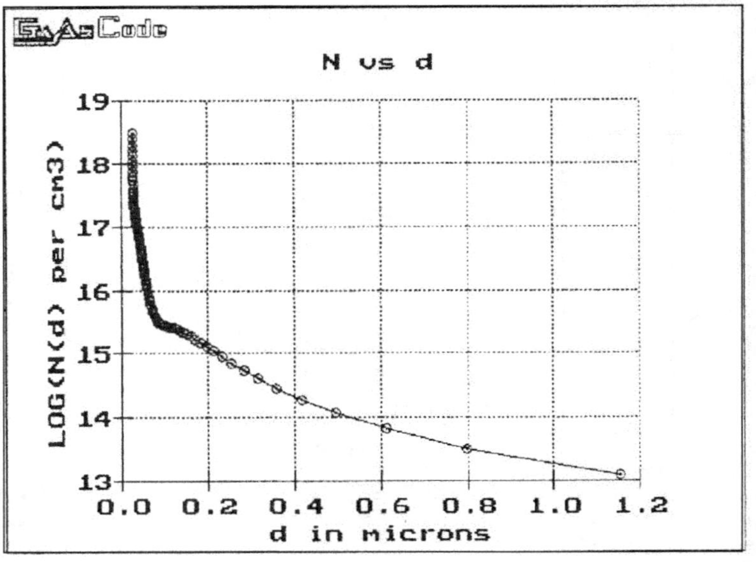

Figure 1.5 Free-electron distribution obtained by reverse modeling a 0.25 μm gate PHEMT. In this case, the divisions on the vertical scale are logarithmic, so the number 18 signifies a free-carrier density of 10^{18} cm^{-3}.

the result might even be a truer reflection of the actual carrier profile than that achieved by $C(V)$ measurement of bulk layers where the spatial resolution is limited by carrier diffusion in the profile tail: the 2-DEG confines the carriers, so there is no carrier tail in the usual sense.

Unresolved, but interesting nonetheless, is the inflection at a carrier density of around 3×10^{15} cm^{-3}. Does this mark the limit of the 2-DEG?

Figure 1.6 shows the metallurgical gate length recovered by reverse modeling for the same PHEMT, with the nominal gate length being 0.25 μm (the recovery algorithm is inapplicable below the knee in the $I(V)$ characteristics). Given fabrication tolerances, 0.23 μm agrees well enough with the nominal.

In Section 6A of [2] will be found a discussion of how the longitudinal velocity of electrons in the channel of an HFET is affected by transverse electric fields. In GaAs, the expectation is a maximum drift velocity of circa 10^7 cm sec^{-1}. Reverse-modeling algorithms can be recast to recover the velocity, again from multibias S-prameters, given known values for the technolgical features of the device. Figure 1.7 gives an example.

Finally, as outlined in [2], among other factors the extension of the space-charge layer into the gate-drain region, which we call X, can be retrieved by reverse analysis, as, for example, in Figure 1.8.

One of the tenets of device physics as applied to HFETs is that X is a key determinant of both dc and RF behavior. It gives rise to intense two-dimensional (2-D) electric fields that affect electron transport in the channel, and it is overridingly important in determining the shape of the current-voltage characteristics. Charging

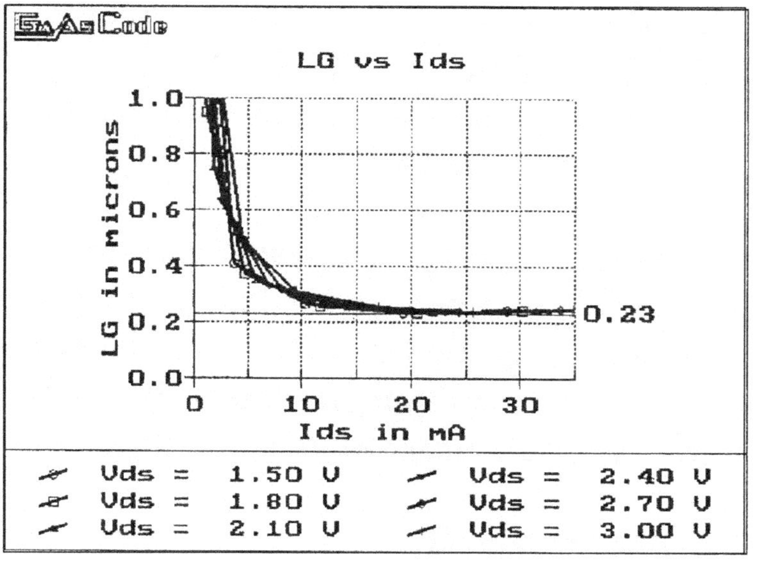

Figure 1.6 Metallurgical gate length recovered by reverse-modeling analysis of the same multibias S-parameter measurements as in Figure 1.5. The recovery algorithm is inaccurate below the knee in $I_D(V_{GS},V_{DS})$.

and discharging the X region take a time τ, which is pure delay and has ramifications to one degree or another for all the elements in the equivalent circuit.

In summary, in the majority of cases that the technological features recovered by reverse modeling agreed, within the limits of manufacturing tolerance, with the maker's numbers, where known. That fact is testimony not just to the validity of

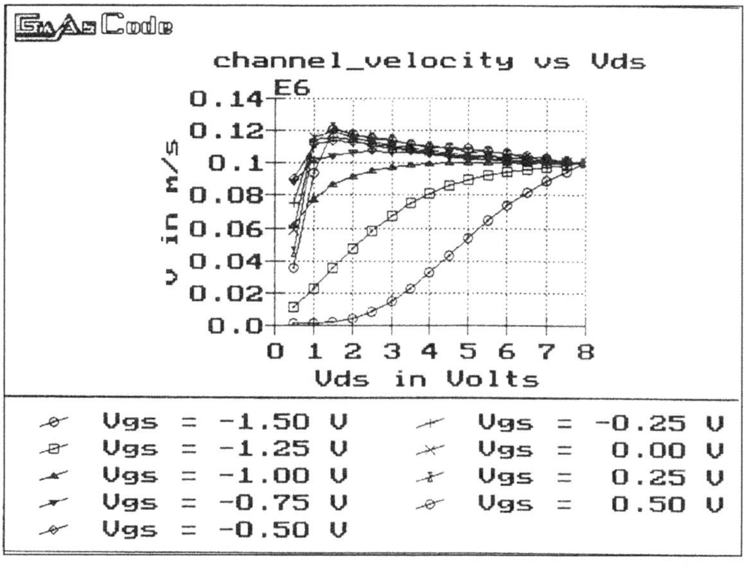

Figure 1.7 Average electron velocity at the drain end of the channel of a 0.5 μm gate FET.

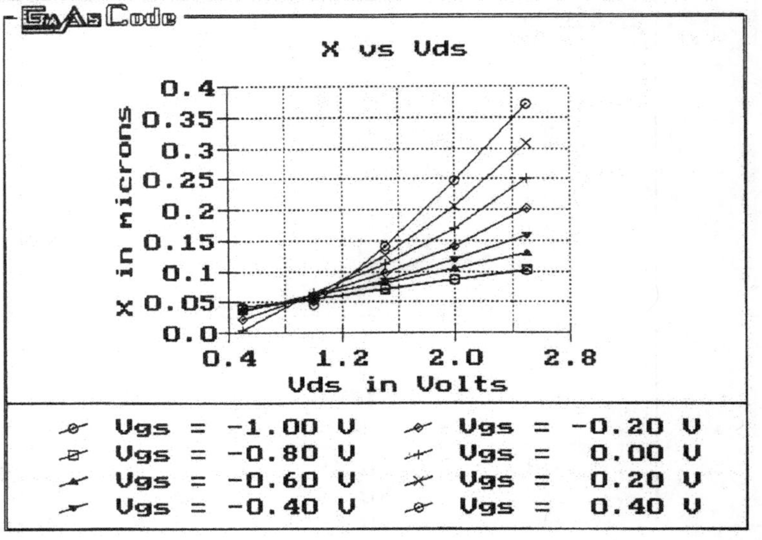

Figure 1.8 Extension X of the depletion layer into the gate-drain region deduced by reverse modeling.

the compact physical method but to the manufacturers' process control as well. All showed the approach in [2] to be fundamentally sound and therefore to be a reasonable foundation for tackling the large-signal problem.

Well before [2] was written, it was clear that the method had potential for nonlinear application, but there were links missing. It took a year to find the links and the best part of 10 years of practical work to verify them. None of the missing links required alteration of the original, so [2] stands as the essential background for the present endeavor.

To avoid duplication, reference will be made freely to the rationale, equations, and figures in [2]. As that work was primarily concerned with small-signal matters, however, some of the concepts require not only expansion but breaking down for the nonlinear case. In such instances, a full account will be given.

1.3 Aims of the Present Work

Having verified the background as in Section 1.2, the aims are:

1. To create an open large-signal model of HFET cells that allows a check to be made, at each step in the model derivation, that the results are in accordance with expected forms;
2. To give examples of these expectations at the important stages;
3. To give a full account of the reformation needed to the equivalent circuit topology to render it valid for nonlinear applications;
4. To use device physics to resolve or deduce:
   i. The extent to which charge conservation applies;
   ii. The consequences of pure delay in drain current;

   iii. Functional forms for negative capacitance or transcapacitance;
   iv. The general forms expected of the equivalent circuit elements.
5. To provide results for several nonlinear circuits;
6. To appraise forthrightly the merits and demerits of the method, including issues of implementation in a commercial circuit simulator;
7. To place all the foregoing in context with commonplace models and practices.

1.3.1 Motivation and Practical Application

It is natural to ask what motivated a fresh approach to the large-signal model problem and where the work fits in the scheme of things outlined in Section 2.1. Like others, it is empirical, centered on an equivalent circuit with parameter values derived from measurements. However, there, the commonality with established methods ends. Whereas one can get by with the commonplace equivalent circuit for linear circuits, especially if some of the minor elements are allowed to take on apparently nonphysical values, there are errors in it that render it a poor approximation for nonlinear use. The topology of the commonplace equivalent circuit needed to be reformed for it to stand a chance of being well suited to nonlinear design. It had to represent correctly the physical processes occurring in real devices, which the standard model does not.

Because, in the lesser problem of bias-dependent linear analysis and design, experience had proved the value of the link between FET technology and circuit performance (revisited in Section 1.3.2), a motivating factor was to preserve it. Through the reformed equivalent circuit the direct link is retained, but now between the nonlinear behavior of FETs in circuits and the material and structural features of which they are composed. Not only does such a facility help to judge whether the model parameter values and functions derived from measurements are physically sound, it establishes a basis for material scientists, fabrication technologists, and circuit engineers to talk constructively with one another.

Among the challenges that culminated in this book was the hope of achieving agreement with practice comparable with the best demonstrated by any modeling scheme. Insofar as it requires learning to begin with manual intervention to process the measurements and coding for commercial simulators, it fits at the slow but accurate end of the range in Section 2.1. However, specialized and expensive test instrumentation is not required; dc-biased small-signal S-parameter measurements suffice, supplemented by dynamic $I(V)$ measurements made possible by a small, inexpensive instrument.

Once derived for a specific device, the model should be applicable to the analysis and design of a wide range of nonlinear circuits. It proved manageable to implement in a major commercial simulator, although not without occasional difficulties with run time and convergence. Short of adopting the whole, some of the techniques could be adapted quickly to improve the fidelity of some circuit models in common use.

1.3.2 The Physics-to-Circuit Model Construct

To technologists, the closest model to something with which they are familiar is one based on the material and structural description of a device: in other words, a physics-based model. To circuit engineers, the closest model to something they understand is an equivalent circuit. For these reasons alone there is merit in formulating a modeling scheme around both. There are other reasons besides, notably that commercial circuit simulators since inception have made provisions for equivalent circuits. Thus, by configuring a compact physics-based model to calculate the values of the elements C_{gc}, C_{gd}, g_{m0}, g_0 ... from the technological makeup of a device, the relationship between technology and circuit performance becomes better understood.

The physics-to-circuit-model construct establishes a link between what the technologist does and what the circuit engineer has to work with (the FET), conversely between what the circuit engineer wants and what the technologist must provide—an important link available in no other method.

For nonlinear circuits, the underpinning physics is deployed differently from linear circuits. In the latter and well-rehearsed case, the standard equivalent circuit mostly works well enough for FETs in the velocity-saturated regime, so calculation of its elements from physics is practicable.

When it comes to nonlinear circuits, the first use of the underlying physics is to reform the equivalent circuit itself. Having done so, the next task is to determine the elements it requires as functions of V_{GS} and V_{DS}, in our case by network transformation from the standard equivalent circuit fitted to multibias S-parameters, with the physics relegated to conditioning the results and ensuring their efficacy. In other words, the physics provides measures and criteria for correctness at each step in the model extraction.

Continuing with the alternative of calculating the element functions directly from physics would now have to include the region below the knee in the $I(V)$ characteristics, which is problematic for two reasons: it requires a higher level of sophistication in the device simulator, and there is fundamental uncertainty in the exact form that some elements should take in the below-the-knee region. As will become apparent in Chapter 16, it is nevertheless possible to devise fast-acting simulators to calculate the forms of the elements for the standard circuit over the entire range of V_{GS} and V_{DS} (and then, as above, the elements for the revised model could be obtained by network transformation). Our simulators of this type were never tried and tested to the extent necessary to prove their worth and so remain a prospect, not a proven workable method.

1.3.3 Applicability

We have to consider applicability in two senses: device size, and type of compound semiconductor.

Beginning with the semiconductor type, as the foundation of the whole approach lies in the physics of FET operation divorced from any particular material system, the findings should apply to GaN and SiC as well as to the predominantly GaAs-based devices upon which most (but not all) of the work was done. Although no ultimate proof is available in the form of fully worked circuits, a measure of confidence may be derived from diagnostic work carried out on SiC FETs, as detailed in Section 4.3. Nothing of a fundamental nature changes in moving from one material system to another, only the material constants and the values of the parameters governing electron transport. Therefore, provided that the appropriate values are substituted, the same device behavioral principles should apply, regardless of the material system.

Moving on to the largeness question, the procedures herein are restricted to point devices, where point means on the order of one-fiftieth of a wavelength at the signal frequency or smaller. Fundamentally, there is an upper limit to the size of device that a single equivalent circuit can represent adequately, a constraint that applies to any model based on a circuit of lumped components. None of the practical circuits worked on during this study involved devices with total gate widths more than 15 mm. We are therefore talking about small FETs and the elemental FET cells of which large power transistors are composed. In the literature, the term "intrinsic" is sometimes used to describe elemental cells [7].

Generally, it is not easy to define the size of HFET for which a simple equivalent circuit model ceases to be useful no matter how faithfully it describes the dc and RF behavior of a point device. The upper bound on dimensions, which incidentally includes the width of each gate individual finger [2], depends not only upon the frequency of operation but also the specification of the circuit, for example whether phase tracking is a criterion. Accordingly, various rules apply for how large a FET cell can become as a fraction of a wavelength before a single point equivalent circuit loses validity and has to be replaced by a distributed model.

A large power FET consists of individual cells interconnected by busbars or manifolds: one to link all the gates in parallel and another to connect the drains similarly (the source electrodes typically are grounded to the backplane by via holes). Busbars resemble transmission lines that can be a large fraction of a wavelength, thus compounding the modeling problem by having to accommodate signal delay in the time domain or, in the frequency domain, having to account for different cells being driven by signals of different phases. Simply combining the output of cells with different phases can lead to suboptimal addition of signals and, in extreme cases, destructive addition [7].

To circumvent the phase problem, it may become necessary to connect each manifold to the outside world by feed points distributed along it, in which event passive phase splitters external to the device proper are required to supply and collect signals at the various ports. At this level of design, an electromagnetic (EM) simulator is indispensable. (The search string "Books on Electromagnetic Simulation" yields about 20 titles.)

None of these issues will be considered further. Good accounts can be found in the literature [7].

1.4 Preview of Results

A first impression of the relationship between this work and others can be had by abstracting the starting point and some end results. Figure 1.9 shows the core of the reformed equivalent circuit superimposed on a cross-sectional diagram of the FET structure to indicate the physical origin of the main elements. It was the starting point for referencing the circuit model.

From there, the reformulated model was expanded with branch capacitances, each having a single physical origin, two negative branch capacitances (long since referred to by the far better descriptive term of transcapacitance), each also having a single physical source, and, among other reforms, an element that resembles both a capacitance and transcapacitance, but which, as the saying goes, is perhaps both and maybe neither. Ten years later, where did it lead?

Figures 1.10, 1.11, and 1.12 are three frames from a batch of six load-pull experiments. In 2000, they were considered remarkable, the phase comparison being the most striking of all. Despite the passing of the years, there is still interest in how it was done. Extraordinarily, perhaps, only dc $I(V)$ and multibias S-parameter data were used. Which of the "3F's" was it: fluke, fake, or fact?

When theory and practice disagree, usually it is the model which first comes under suspicion, but, in this case, there was little discrepancy to explain. Either everything was correct or the practical data and the model contained exactly the same character and magnitude of error. As improbable as that may seem, the results' comparison was so unlike any previously seen that the error hypothesis seemed no more extraordinary than the results themselves. If the error hypothesis could be dismissed, however, another extraordinary factor emerged: the (again, for the time) evident accuracy and precision with which the load states on the test circuits and the FETs' S-parameters had been measured by the U.S. corporation responsible.

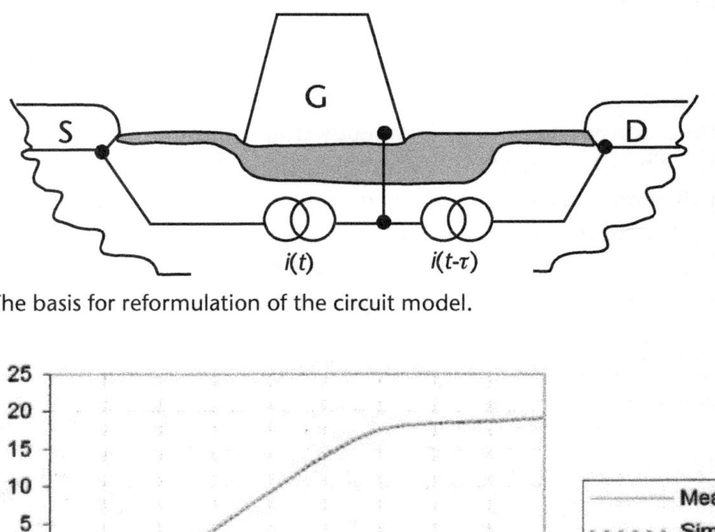

Figure 1.9 The basis for reformulation of the circuit model.

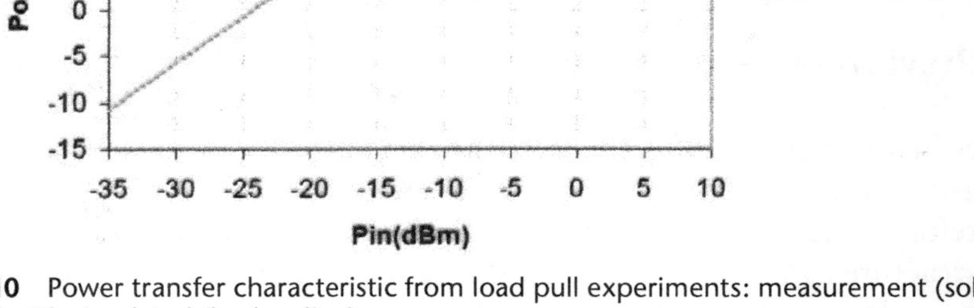

Figure 1.10 Power transfer characteristic from load pull experiments: measurement (solid line) compared with simulated (broken line).

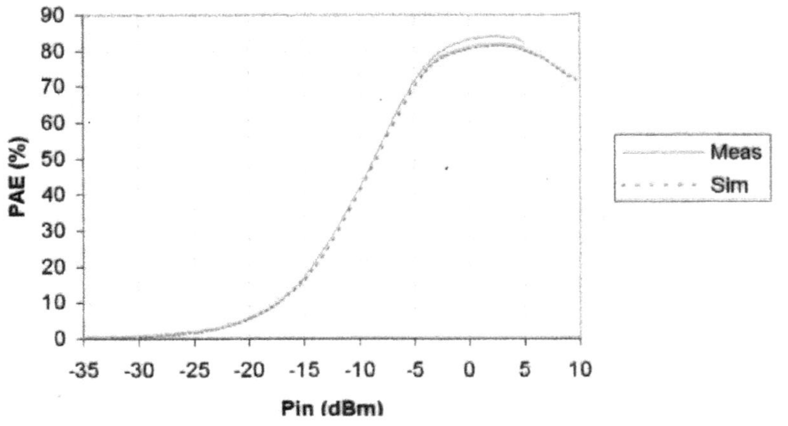

Figure 1.11 Power-added efficiency: measured (continuous line), and simulated (dotted line).

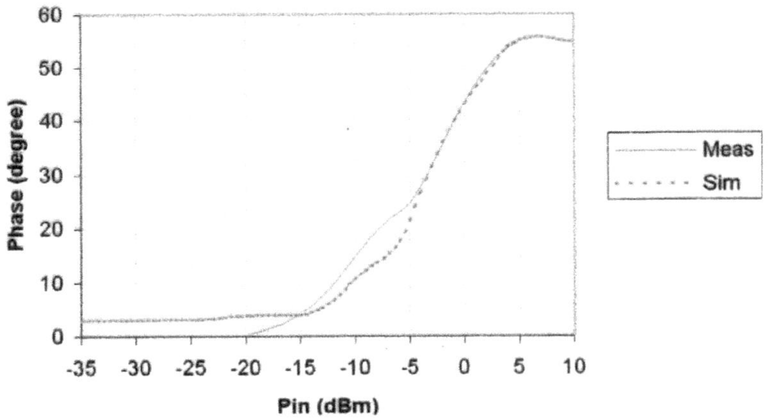

Figure 1.12 Insertion phase: measured (full line) and simulated.

1.5 Organization of the Book

The remainder of Part I is a résumé of commonplace methods and practices with emphasis, where appropriate, on where careful work can pay off and where compromise can lead to error. Chapters 2 through 6 reflect the historical preponderance of empirical methods; there is minimal physics.

In Part II, the equivalent circuit is reformulated, invoking physical arguments to reconfigure the topology and to gauge the general form of the elements within it, with practical examples. Results from implementing the entire scheme in a commercial circuit simulator are compared with the measured RF characteristics of a handful of circuits: a frequency doubler and several amplifiers. Specific challenges that had to be resolved are enumerated, followed by brief speculation as to how some might be alleviated by simulator development.

In Part III, device operating principles are dealt with roughly in the order in which they were applied in Part II. Put the other way around, some of the ideas in Part II make use of principles that are presented later in Part III. To a degree, therefore, the order is conclusion before reasoning, meaning that the essential points are brought to the fore without first having to wade through reams of foundational material with the attendant risk of losing the thread of the argument.

The word "extraction" is used to mean "fitting the circuit model and its elements to measured data," so extraction is synonymous with fitting. Commonly, other authors use the word to refer to the more complete act of not only fitting but assigning values to parameters in equations used to fit the bias dependence of the fitted elements.

There is repetition in the text which is intentional. Several concepts, some of which may be unfamiliar, are relevant to different parts of the story, so to cover the same ground at the point where it is required not only saves unnecessary turning of pages but creates scope for different wording. Sometimes an idea expressed in different words has a better chance of being understood.

1.6 A Note on Figures

Many of the figures are original experimental results reproduced exactly as they appeared on the instrument screen, meaning that the resolution is as set by the display and not subsequent copying.

Many years ago, there was a debate about the presentation of experimental data, notably the habit of sanitizing it for publication. A countervailing argument centered on the premise that the purpose of a graph of results, above all, is to convey information, not to look pretty. Importantly, if secondarily, information is contained in the scatter of experimental points, how many points there are, whether curves drawn through the points are fair, and so on.

Not only do many of the figures in this book fall into the category of unsanitized experimental plots, but also some of the computer simulation results in Chapter 4 exhibit kinks and sharp corners that have been left in for the same reason.

References

[1] Cripps, S. C., *RF Power Amplifiers for Wireless Communications*, 2nd ed., Norwood, MA: Artech House, 2006.

[2] Ladbrooke, P. H., *MMIC Design: GaAs FETs and HEMTs*, Norwood, MA: Artech House, 1989.

[3] DTI LINK Project PC012 *Evaluation of Semiconductor Technologies for the Radio Functions of Personal Communications Network Applications*, U.K. Department of Trade and Industry, 1992–1995.

[4] Ladbrooke, P. H., J. P. Bridge, and D. R. Bowler, "Forward and Reverse Engineering in the 'MONOFAST' Project," in J. Mun and A. A. M'baye, (eds.), *Gallium Arsenide Technology on Europe*, New York: Springer-Verlag, 1994.

[5] Ladbrooke, P. H., "How Different Are HEMTs from FETs Really?" *EDMO'94, 2nd International Workshop on High Performance Electron Devices for Microwave and Optoelectronic Applications Digest*, King's College, London, November 1994, pp. 55–59.

[6] Ladbrooke, P. H., "Reverse Modelling of GaAs MESFETs and HEMTs," *GEC Journal of Research*, Vol. 6, 1988, pp. 1–9.

[7] Aaen, P. H., J. A. Plá, and J. Wood, *Modelling and Characterization of RF and Microwave Power FETs*, Cambridge, U.K.: Cambridge University Press, Chapter 6, 2007.

CHAPTER 2

Summary of Approaches and Needs

2.1 Why Models Are Important

With the ever-increasing sophistication of nonlinear circuits and the demands placed upon them, particularly by digital communications systems, the role of design by computer simulation has expanded correspondingly.

It has long been true that the correctness of the algebra and numerical solution methods implemented in commercial circuit simulators can be taken for granted. Furthermore, models for passive circuit components may be taken to be dependable, often because they are the result of long and careful characterization work on behalf of either foundries, or of specialists in 3-D electromagnetic simulation. That leaves device models as the factor determining the accuracy with which a circuit can be analyzed and designed on a computer.

Herein lurks a dilemma: what is accuracy supposed to mean? Claims for accuracy have been a feature as long as the models themselves have been around. Consulting the literature evidences a history needing clarity of what is meant by accuracy. Instead, the term seems to have been invoked as a pointer to implied excellence, sometimes verging on a claim to be better than anything that has gone before. Undoubtedly, many were better in some respects than anything that had gone before, but in all respects? For the would-be users, in which of the 25 models (at least) classified as mainstream in 2013 were they supposed to believe? It is hard to resist the feeling that the wider community would have been well served by accounts of where models did not work as well as where, within limits, they did work.

In this chapter, therefore, an effort has been made to avoid repeating claims for accuracy and instead to introduce criteria based on measure.

2.2 Types of Nonlinear Models

Any involvement with nonlinear models of semiconductor devices, either as a creator or a user, leaves a lasting impression: it is a problem that has no simple solution. As a result, compromise seems inescapable. At one extreme are models that:

- Are workable and easily understood;
- Require relatively inexpensive test equipment;
- Are either supported by most circuit simulators, or are readily coded for them;
- Involve compromises and approximations.

Dominant in this sector are approaches based on an equivalent circuit, with the merit of providing basic insight into the inner workings of the FET. Further, once an equivalent circuit has been established for a specific device, it is widely (although not necessarily universally) applicable to the analysis and design of a variety of circuit types.

At the other extreme, motivated by the quest for fidelity and realism, are mostly behavioral, black box, and table-based models that require:

- Specialized and usually expensive test equipment;
- Learning advanced measurement and data manipulation techniques;
- A circuit simulator that caters for this kind of model, restricting the choice of simulator.

Arguably, behavioral models tend to be application-specific, meaning that distinct circuit applications require their own version. Also, as the label "black box" suggests, they offer no physical insight into how the device functions.

Both approaches are empirically based and, beyond repeating the exercise for maybe hundreds of devices nominally of the same type, do little or nothing to foster design accounting for production spreads.

Somewhere between these extremes, or maybe outside both, are physics-based models that are purely theoretical and necessarily computer-based. As input, they take data on the material and technological makeup of the device from which they calculate the dc $I(V)$ characteristics and, in some cases, RF properties as well.

Fully expanded physics-based models can take hours to compute a single point, let alone a full set of characteristics. Unfortunately, all the physical variables governing HFET behavior can realistically never be known, so, apart from all else, the degree of accuracy required for circuit design can rarely, if ever, be met. For these reasons their role tends to be restricted to one of investigating specific problems of electrical behavior at a generic level, for example relating nonlinearity in the $I(V)$ characteristics to the doping profile in the channel of a FET.

Retreating from the idea that physics-based models must embody the finest principles of solid-state physics, compromise is possible here too. Carefully constructed experiments, typically involving many people and spread over many years, can reveal which of the physical processes are dominant and which of those play a secondary or minor role. Basing physical models on the dominant mechanisms and omitting the secondary and minor processes leads to what are sometimes termed compact physical models.

What of the discarded physical processes? Are the inaccuracies thus built into the model worse than those stemming from the material and structural uncertainties in practical devices that limit the use of fully fledged simulators? Experience has it that the answer is no. Given acceptance of realism but not total fidelity, a compact physical model serves as a design and diagnostic tool that executes in a few seconds on a personal computer. When allied to a circuit simulator, they can produce answers close enough to reality to be of proven use in the design of practical circuits, at least in the realm of bias-dependent linear circuits, a useful step along the way to full nonlinear capability [1, 2]. By varying the material and structural values the input to the simulator, statistical analysis and yield-driven design become a reality.

All modeling approaches have their attributes and all are compromises. Some models require a minimum of instrumentation, are simple, execute rapidly in a circuit simulator but prove to be incapable of handling other than simple measures like gain and output power. Yet others require expensive instrumentation, are elaborate, are slow to execute but have a good shot at complex modulation schemes such as orthogonal frequency division multiplexing. To not put too fine a point on it, the whole field is a maze that can be hard to navigate.

Generally, accounts of models and approaches have focused on how the model does it and less on what it does, or rather what it does not do. Historically, because the emphasis has tended to be on where a model works with scant regard for what its limitations are, inordinate amounts of time have been spent uncovering the limitations of model-after-model until something that worked well enough was reached.

2.3 Desirable Attributes

Setting realism and pragmatism aside, the attributes desirable in a model make an interesting wish list.

1. Openness in a model means transparency to the user, so what it does and how it does it are demystified. It is to be contrasted with the status of being closed, where the model is opaque. Subject to (2) below, the quality of openness allows the prospective user to decide how much faith can be placed in the approach without the time and cost of having to try it.
2. The underlying principle and the requirements of practical implementation should be easily understood.
3. Constructing models for nonlinear circuit applications invariably involves measurements at dc and RF and with fast pulses. Specialist test equipment is expensive, a fact that is of paramount significance to small enterprises, contractors, and consultants. Desirably, therefore, empirical characterization should be possible with inexpensive test instrumentation.
4. Constructing the model from measurements, the procedure commonly known as extraction, should be straightforward and ideally automatic with a minimum of manual intervention, which, in turn, implies dedicated software.
5. Accepting that some organizations will not possess test instrumentation covering the desired frequency bands, a model should extrapolate to frequencies beyond the range over which measurements were made to establish it.
6. The model should exhibit proven fidelity to practice if it is to meet the demands of contemporary circuit applications.
7. It should be scalable to devices of size different from the test piece(s).
8. Implementation in a circuit simulator should be straightforward.

Unsurprisingly, roughly half are mutually exclusive with the other half, but that is not the point. The point of the list is to act as a reference against which each among the many models can be positioned.

Dealing with the issues in reverse order;

- Item (8) is a major requirement as will be well known to those who have struggled with model implementation in circuit simulators. Chapter 12 is a limited user's point of view.
- Scalability, item (7), is a distinctly useful property even given the fundamental constraints, namely the number of HFET cells that can be connected in parallel without having to account for busbars.
- Fidelity to practice, item (6), is not negotiable. It has long been understood that an apparently good fit to the measured $I(V)$ characteristics is not good enough and that high-order derivatives of the current must be correct as well [3]. In brief, the derivatives

$$\frac{\partial^n i_D}{\partial v_{GS}^n}$$

  have roles as follows: $n = 1$ governs the gain of a circuit at the fundamental frequency; $n = 2$ governs the generation of second harmonic frequencies; $n = 3$ governs the generation of the third harmonic; and so on.
- Item (5) can make the difference between circuit design being doable or not.
- It is fair to say that extraction of the model parameters from measured data, as in item (4), is a major weakness viewed overall. When it comes to equivalent circuit-based models there are two parts to the problem. First, the elements of the model, resistances, capacitances, current generators, and so forth must be found as functions of (V_{GS},V_{DS}) by fitting to the data. Second, the fitted functions must be manipulated into the particular form dictated by the model equations. The whole is far from being a straightforward process, maybe affordable by large organizations but by no means all (further comment is to be found at the end of this section). For those without the in-house resources, Modelithics [4] is a specialist in providing such services.
- So far as item (3) is concerned, the same organization, Modelithics [4], has a wide-ranging measurement capability so, together with model extraction, what is on offer amounts to a turnkey service of model generation.
- Although not essential, especially if model generation services are being outsourced, the attributes of items (1) and (2) remain desirable for building user confidence.

As far as scalability is concerned, to be properly scalable a model must be true to the manner in which the FET operates, which means that it must consist of elements that each represents a single, identifiable, physical process occurring within the device. Here we are not talking about the fundamental limit beyond which connecting FET cells in parallel requires modeling of metal interconnects, or buses having transmission line properties of signal attenuation and delay; we are talking about topological deficiencies in the basic cell model.

In small-signal applications, it is possible to get by with the customary equivalent circuit provided that some of the minor elements like R_s and R_d are permitted to take on nonphysical values [5]. Such apparently nonphysical quantities arise as compensation for topological errors in the circuit but, if accepted, lead to terminal behavior that is close to correct. Under large-signal conditions, however, the same does not apply: the topological deficiencies lead to performance predictions which can be very wide of the mark. Thus, because the standard model (and minor variants) is not compliant with the dynamics of FET operation, it is not scalable in a strict sense, although, for some applications, it may be approximately so.

Item (4) is perplexing. The tiny amount of space accorded in the literature to fitting and extraction relative to the models themselves is out of all proportion to its importance. Fitting and extraction can either make or break a model, no matter how good the model might otherwise be. Achieving passable fits to the $I(V)$ characteristics is one thing; managing the same for the capacitive, resistive, and inductance elements is quite another.

In broad terms, the $I(V)$ fit largely governs how well a model will do in designing circuits to operate at frequencies far below the current cutoff frequency f_T provided that the performance specification is limited to simple measures like gain and output power at the fundamental frequency. Under such conditions, the reactive elements of the model play a minor part so, if they are poorly modeled, it is not of much consequence. That ceases to apply when phase precision is part of the specification even for $f \ll f_T$. It similarly ceases to apply when the operational frequency is other than a small fraction of f_T. Under these circumstances, the capacitive, resistive, and inductive elements truly matter. It is here that fitting and extraction seem to have become a forsaken art.

Trawling through decades of literature reveals an extraordinary diversity of forms for $C_{gs}(V_{GS},V_{DS})$ and $C_{gd}(V_{GS},V_{DS})$. Some feature swirls, steep rises, crossovers, and even oscillations, prompting the question: Can they all be right? (The results given in this book are not beyond scrutiny either.) When scant, if any, information is provided about how the fitting and extraction were done, there is no way to judge, given that the habit of showing the quality of fit to the measured data, invariably S-parameters, fell out of fashion years ago. It is a baffling reality.

2.4 Behavioral or Black Box Characterization

In the absence of first-hand experience, the following appraisal is perforce a distillation of claim and counterclaim drawn from published accounts and from opinion of workers with at least some acquaintanceship with these techniques.

Behavioral characterization treats the device as a black box with no questions asked about what is in the box: it takes at face value how the box behaves at its terminals. Thus the device is characterized by incident and scattered waves—a generalization of small-signal S-parameters to large-signal excitation—and it requires a nonlinear vector network analyzer or equivalent to measure them. To make the measurements, the device is loaded at input and output with impedances that are adjustable and, in particular, that are tuneable to mimic the impedances with which the FET will ultimately be presented in the final circuit application. Looked at another way, it is a load-pull characterization,

Black box models have several positive attributes that are hard to contest. First, they cannot be anything but an unreservedly correct representation of what the device does in the circumstances under which it was measured. Second, preservation of the efficacy of the measurements stems from the fact that there is no intermediate step of model fitting and extraction (as in equivalent circuit models) with the potential to compromise the measurements. Third, although the no-questions-asked standpoint implies that the model is closed, it is not opaque. It works like this: the model, a large-signal wave characterization, reduces to the small-signal S-parameters when excited with small signals. The small-signal S-parameters can then be analyzed using tried-and-tested methods of reverse modeling (examples of which appear throughout the book) to reveal material and structural features of the device from which the data came. Along the way, the standard equivalent circuit model is fitted, yielding capacitance, resistance, and transconductance information for those who feel more comfortable with a circuit representation.

While conforming to the same basic principle, there are variations of approach to black-box models [6, 7]. All require more than simple stock microwave laboratory test instrumentation. For some, the necessity for specialized equipment (a relative term) might be an impediment, although, once again, full services for generating behavioral models are on offer [8].

Black box models have a limitation which is not easy to articulate. In essence, it is that their validity extends only to circuit environments closely resembling those when the measurements were made, where by circuit environments is meant the impedances presented to the device at input and output. To see why, it is necessary to call upon some fundamental aspects of nonlinear systems. It begins with a question: Is it possible to characterize a nonlinear system by terminal measurements alone to create a model with universal applicability?

If the transfer function is an instantaneous function of the input stimulus, as in a simple clipping or limiting device, then the answer is that terminal measurements alone do indeed constitute a model with universal applicability. The same applies if the transfer function includes a fixed delay that does not depend on any past condition or state of the system. At the other extreme, however, are systems exhibiting hysteresis, for example in magnetic materials. Here, the present state of the device or system depends upon the history of what has gone before. Terminal-only measurements cannot produce a model of universal applicability because they are unable to account for the history of any particular set of operating conditions. In other words, memory defeats the aim. The delay function itself may be nonlinear and dependent upon interactions with other nonlinearities occurring within the device, in turn, dependent upon the impedances with which the device is loaded and upon the magnitude and spectral content of the signals. Without bringing into play the inner workings of the device on some basis it is difficult to conceive of how a delay function that is universally applicable could be determined.

FETs and HEMTs exhibit memory in the form of charge exchange with deep levels or self-heating (or both). Further, even if the device is dispersionless (as in Figure 3.5) there is still delay, notably transconductance with current generator delay, τ_{gm}. To what extent does even short delay affect the way a device is behaving at the present instant? The obvious answer is that it does if the timescale in question is comparable with the delay. For a total delay of, say, 10 ps, the black-box model should therefore be useful for frequencies of up to 16 GHz. Whether that is true depends on the fraction of the RF cycle (i.e., phase) which matters to the circuit specification.

We are thus left with the idea that any delay phenomenon in the device frustrates the possibility of establishing by terminal measurement a black box model with wide applicability. Complex interactions between the various nonlinear processes operating in the device may interrupt or reverse delay-producing events, depending upon waveform distortions, which are, in turn, governed by the terminating loads or impedances. By implication, a black box model ideally has to be based on measurements made with terminating impedances akin to those in the intended circuit application. That apart, a behavioral model should have wider applicability quasistatically; that is where only conduction current matters.

Among the desirable attributes of Section 2.3, item (8), ease of implementation in circuit simulators, behavioral models are well served among others by Microwave Office [9], ADS, and Genesys [10]. Having had to work through the limitations imposed by circuit simulators on equivalent circuit models and then convergence failures, ease of assimilation of black box models by circuit simulators promises to be a strong argument in their favor [8].

Finally, there is an apparent conundrum to resolve. The starting point for empirically based equivalent circuit models is a set of measured small-signal S-parameters that is just as closed or opaque as the large-signal set constituting a behavioral model. It demands an explanation of what it is we do to turn the one into a model with wide, if not universal, applicability, while the other remains restricted.

If a complete physical model can be developed from the measurements, every possible operational state is accounted for at a fundamental level. However, the inference is that empirical data spanning a limited range is a sufficient resource from which to generate a generalized description of the device. When reconstruction of a complete physical model proves elusive and we have to settle for a model consisting of discrete elements, with each element representing a clearly defined individual physical process, then the explanation is much the same. An approach of this kind is an abstraction: what in nature tends to be distributed processes ends up being represented by point or discrete elements. If done properly, all physical processes should be accounted for and we retain the expectation of being able to generate something of wide applicability from a restricted base of data.

Rarely is the validity of this premise questioned in detail, which is unfortunate given the considerations of Section 4.2.2. Basing a nonlinear large-signal model on linear (but multibias) measurements has no right to work as well as it does. Ideally, the kind of model we desire should be constructed from a database is as close as possible to the operational conditions the device will ultimately experience in a circuit application. Feasibility aside, it raises the specter that instead of fitting to multibias small-signal S-parameters a circuit model should be fitted to the data that we have called a behavioral model.

2.5 Properties of Large-Signal Models in More Detail

Except for minor updates, the remainder of this chapter replicates notes from a 2-day course on large-signal modeling that was given in the late 1990s to early 2000s. Through specific models, it traces the evolution of the art from the early days, illustrating how fidelity to practice improved as deficiencies became apparent.

With one exception, all are based on the commonplace or standard equivalent circuit model.

2.5.1 List of Properties

The properties desirable in a large-signal model ultimately have to be judged from the correspondence between predictions of a FET's electrical behavior and the results of practice. Conjecturing that the best chance of achieving good correspondence follows from the model satisfying all known conditions of physical validity, in this section a physically oriented stance is adopted in reviewing the properties desirable in a large-signal model of FETs and HEMTs, an approach that, albeit hopefully sound in itself, does not answer for the extent to which each individual physical criterion is important to achieving a high degree of correspondence between theory and practice at the behavioral (electrical) level. An assessment of the practical importance of each physical criterion would likely occupy years of a systematic trial and might prove of no engineering value anyway.

Most large-signal models are based on the equivalent circuit of Figure 2.1(a), which has the following shortcomings:

1. Transreactance currents are ignored, as a result of which the minor elements must take on apparently physically invalid bias dependences as a "make-good" to achieve good fits to measured S-parameters. A good model should produce excellent fits to measured S-parameters with bias-independent parasitic elements, without reservation and excuses, for all FET variants and styles.
2. The circuit does not possess the same symmetry at V_{DS} = 0V as does the real device.
3. The provision made for the RF current-voltage characteristics to be different from the dc characteristics is valid only for small signals. A typical example of what the makeshift arrangement of C_{LARGE} and ($g_{0RF} - g_{0dc}$) in Figure 2.1(b) leads to under large-signal conditions is given in Figure 2.2.

   A serious shortcoming is that the drain current goes negative. Current flows in the direction opposite to that implied by the polarity of the applied drain-source voltage. In the model, this arises from the discharge of the capacitor C_{LARGE}, which, together with ($g_{0RF} - g_{0dc}$), emulates the long time constant associated with the dispersive effects of deep levels (and maybe self-heating). In the real FET, the current never behaves in this manner.
4. The current generator forces delayed source current in the model, which is not a feature of the real FET [5].
5. The processes of charge flow (current) and charge storage are divorced from one another in the model, whereas in the real FET the second largely occurs as a result of the first. There should be at least a built-in (if not explicit) relationship between the charge and current functions, an expectation that arises from the fact that in the real FET much of the charge storage in the main depletion edge occurs by the transport of electrons in the channel as channel current.

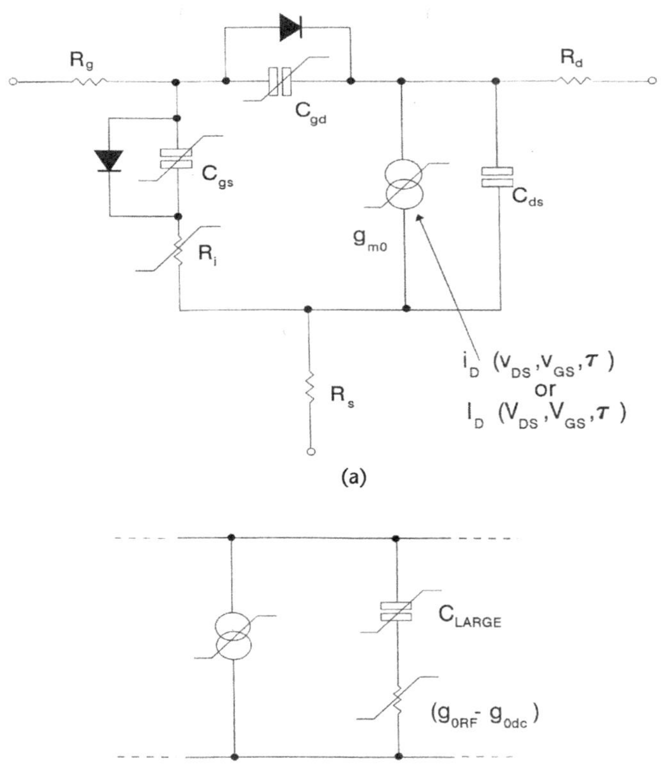

(b)

Figure 2.1 (a) Commonplace circuit model assumed as the basis of most large-signal models of the GaAs MESFET and HEMT. (b) Addition to the model in an effort to allow for the fact that the RF I(V) characteristics are different from the dc set.

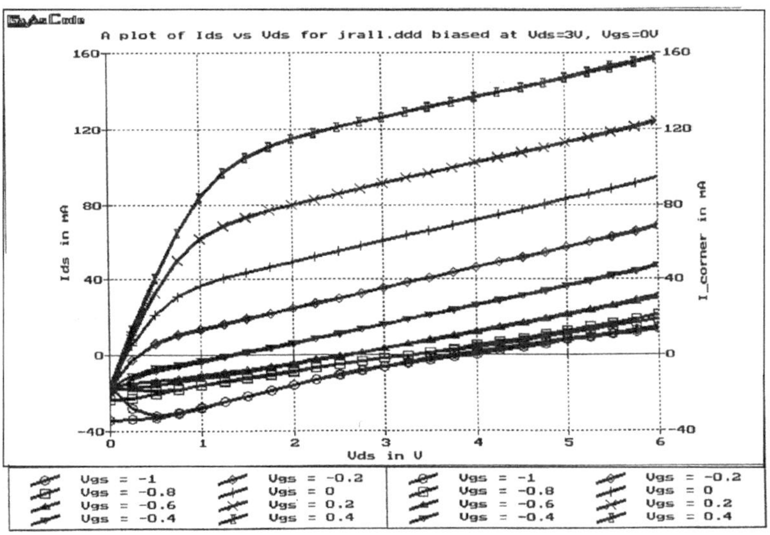

Figure 2.2 Nonphysical current flow ($I_D < 0$) which the contrivance of Figure 2.1 (b) gives rise to under large-signal drive.

In addition, there are several desirable features that have nothing to do with a specific configuration of circuit model:

1. Provision should be made for the large-signal RF drain current characteristics to be different from the dc characteristics (see note below).
2. The large-signal RF drain current characteristics of the model should be dependent upon the bias point, just as they are in many practical FETs and HEMTs.
3. The relaxation time with which the characteristics make the transition from the dc set to the RF set, typically tens of microseconds to several milliseconds, must be included if bias or pulse transients are to be modeled.
4. Although the total charge stored in the depletion region is a conservative function of V_{GS} and V_{DS}, to the extent that this charge can be viewed as being stored partly in the gate-source region (or branch) and partly in the gate-drain branch, the branch charges are not individually conservative functions of V_{GS} and V_{DS}. Preferably, therefore, charge storage in the branches should be represented by nonlinear capacitances rather than charge functions. (For a more detailed comment on this topic, refer to Section 2.5.9.)
5. The delay in the current generator of typically a few picoseconds should be included.

Finally, there are several practical attributes that are desirable in a model:

a. The equations it uses to represent the voltage dependences of all elements should extrapolate well for voltages outside the range of the measurements [11].
b. The model should be comprehensive in that it should represent the FET properly in all operational modes, including reducing to the switch FET model at $V_{DS} = 0V$, and for all operational modes it should be scalable, within limits, to other numbers of fingers and total gate widths.
c. The model should not require special measurement instrumentation to construct, beyond that found in any laboratory well equipped with general-purpose test equipment, notably a vector network analyzer.

Note: In connection with this, the set of $i_D(v_{GS},v_{DS})$ characteristics along which a FET will work under large-signal conditions cannot be gauged by, or cannot be represented by, the $g_0(V_{GS},V_{DS})$ function extracted from multibias, small-signal S-parameter measurements. This extracted function is a composite (or distortion), consisting of an assembly of points where, at each bias point, the value of g_0 comes from a set of RF $i_D(v_{GS},v_{DS})$ characteristics valid for (and only for) that bias point. That is, in a practical FET, there is no unique set of RF $i_D(v_{GS},v_{DS})$ characteristics: there are as many sets as points at which the FET can be stably biased, in other words, an infinity of sets. Valid RF representation therefore consists of knowing the bias point about which the device is operating, and knowing also the particular set of RF characteristics for that bias point. For this reason, any circuit method of attempting to construct the RF $i_D(v_{GS},v_{DS})$ from the dc $I_D(V_{GS},V_{DS})$ curves, as, for example, by the addition to the circuit model of the elements of Figure 2.1(b), at best can only be a rough approximation.

2.5.2 The Subthreshold Region

The term "subthreshold" refers to the region close to pinch-off. Taking the formal statement of pinch-off, or threshold, V_P, to be the gate-source voltage that must be applied at $V_{DS} = 10$ mV, say, to reduce I_D to zero, the drain current exhibits types of behavior that may be summarized as (i) if the FET is biased at V_P and V_{DS} is increased from zero, drain current will generally start to flow such that (ii) the dependence upon V_{DS} is initially exponential-like and (iii) if V_{DS} is held fixed once it has been increased enough for I_D to start to flow, just then I_D also exhibits an exponential-like behavior with V_{GS} as this voltage is increased away from pinch-off.

One explanation for the origin of the term subthreshold is the occurrence of these effects in the region below the formal threshold where simple one-dimensional (1-D) reasoning suggests there should be no current flow. The region is also referred to as the soft pinch-off region.

2.5.3 Consequences of Fitting Well to Some Features of $i_D(v_{GS},v_{DS})$ but Not Others

Most models use analytic equations to fit the measured bias dependence of the drain current $i_D(v_{GS},v_{DS})$ (or $I_D(V_{GS},V_{DS})$) and the extracted equivalent circuit elements. Not all subtleties of the measured forms can be satisfied simultaneously and compromises have to be made. The consequences for the fidelity of the simulation need to be examined. Particular attention should be paid to the temptation to judge a model by the fit it achieves to $I_D(V_{GS},V_{DS})$: the fit may appear acceptable, but, when the derivatives

$$g_{m0} = \frac{\partial I_D}{\partial V_{GS}}\bigg|V_{DS}$$

and

$$g_0 = \frac{\partial I_D}{\partial V_{DS}}\bigg|V_{GS}$$

are calculated, the fit to the experimentally determined variations of these parameters is awful.

Figure 2.3 is a sketch of a measured $i_D(v_{GS})$ characteristic at constant V_{DS} with a fitted model curve. Generally, model equations can be adjusted to fit best in the region of higher curvature near pinch-off (the subthreshold region) or in the more linear region at higher currents. There are merits in both:

1. Given that class A power amplifiers operate up and down a load line from pinch-off to a full open channel with only a small fraction of the cycle spent in the subthreshold region, the best results are obtained when the fit is a good compromise over the entire range of current.

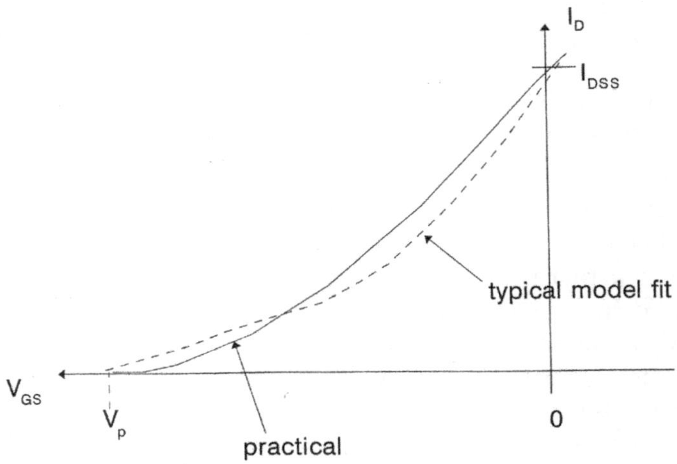

Figure 2.3 Illustrating the general nature of a model fit of practical $I_D(V_{GS}, V_{DS} = \text{CONSTANT})$.

2. For the class B operation, however, the subthreshold region is much more important because that is where the FETs are biased: the application of a signal drives the FET out of its cutoff as-biased state into conduction to amplify the signal. A good fit to the curvature of the $i_D(v_{GS})$ curve near pinch-off is important if the transition from no gain (at cutoff) to finite gain is to be correct. The derivative $\partial i_D/\partial v_{GS}$ is the transconductance, to which the gain of the amplifier is proportional (i.e., the region governs how much output power is obtained for the input signal power).
3. To be able to simulate phenomena in which third harmonic and third-order intermodulation products are important, the equations used to fit the drain characteristics must represent well not just the practical $I(V)$ data itself, but also its first, second, and third-order derivatives with respect to V_{GS}. The observation carries with it the implication that the practical data must be correct down to the third derivative too: what is measured as part of the FET characterization procedure must represent faithfully the third derivative (with respect to V_{GS}) of the $I(V)$ characteristic set along which the FET will operate in its intended circuit application. Pulsed measurement of the $I(V)$ characteristics about the intended operating bias point may be unavoidable if this condition is to be met.
4. Although it appears to be infrequently done, considerations (2) and (3) suggest it would be better to use the basic equations offered by any given model to fit to the transconductance $g_{m0}(V_{GS},V_{DS})$ rather than the current itself.
5. For mixers, which rely for their function on the nonlinearities in the FET at low signal levels, the subtheshold region is also important.
6. For RF FET switches, the entire range is important. The state of the switch FET with no signal applied is $V_{DS} = 0V$. Here, the commonplace circuit model of Figure 2.1(a) is a poor representation, lacking symmetry with respect to the source and drain terminals (which is why a separate equivalent circuit for the switch FET is often used). When the signal is applied, the FET is

driven well away from V_{DS} = 0V and all the nonlinearities come into play, generally requiring a good fit to all the FET's electrical parameters.

2.5.4 Thermal Considerations

Inevitably, when a device has dc electrical power supplied to it to perform a useful microwave function, some of that dc power is dissipated by the device resulting in a rise in temperature. Perhaps because this is an easily understood physical concept, it is often assumed that a FET's characteristics will change due to self-heating if a different bias point is chosen and that, perforce, heating is an important mechanism that must be accounted for. The idea becomes extended through often being invoked as the mechanism responsible for practically observed differences between dc and pulsed characteristics of FETs.

One of the questions unanswered by the foregoing view is what, exactly, is self-heating supposed to change at a physical level which leads to changes at the electrical level? Is it predominantly the Schottky barrier height? Is it the temperature dependence of the ohmic contact resistance? Is it carrier freeze-out into traps that thereby alters the distribution of space-charge (the depletion layers) and consequently the channel current? Is it the scatter-limited electron drift velocity? Is it the low-field mobility? What is the detailed mechanism by which these physical changes lead to the changes in electrical behavior of the FET at the terminals?

Direct measurement of the pulsed characteristics of some FETs about different bias points at which the dc power dissipated is quite different shows that, irrespectively, exactly the same drain characteristics are obtained. An example is given in Figure 3.4. Barring fortuitous canceling of competing effects, such observations refute the simple view of the importance heating. What these measurements imply is that, in some FETs in some applications, heating has no direct importance at all, so the difference between the dc and pulsed characteristics is due to some other mechanism, probably deep levels somewhere in the device. (Heating may then have a secondary importance through changing the relaxation time of the deep levels, that is, the frequency at which the transition from the dc curves to the RF curves occurs as the frequency is increased.)

Where does all this leave us when it comes to large-signal modeling? Should heating explicitly be allowed for? If so, how are the effects of temperature to be characterized by measurement, and how are they to be expressed in the model? And, even if such inclusions are made, how are we to be sure that the magnitude of the electrical changes attributable to temperature in the as-characterized and ready-to-use model are physically justifiable and correct?

Probably due to the inherent difficulty in answering these questions, most of the large-signal models that are reasonably well known make no explicit allowance for the effects of temperature: the Root model attempts it. Consequently, in this review, we are not faced with having to judge the correctness of the models for the way they manage temperature considerations.

A more careful reading of [12] is required before judgment can be passed on the validity of its methods of handling temperature effects in GaAs FETs. Section 11.1.3 of [12] appears to have much in common with D. E. Root's approach and is therefore particularly relevant (although it is not a feature of the other models as we know them formally).

Note: further information on the procedure advocated by Root is given in Section 2.16.4 without comment upon the limits to its validity. Much could be learned by making elevated temperature fast-pulse measurements of the drain characteristics, especially for FETs which exhibit the same pulsed $I(V)$ characteristics even for different starting (bias) points at which the dc power dissipated is very different.

2.5.5 Construction of the Model from Measurements

There is a lack of balance in the way large-signal models are discussed. Much is made of the attributes of the models themselves while leaving largely unquestioned the method of obtaining the values of the elements in them from measurements, a procedure commonly referred to as extraction. Generally, the more complex the model, the more involved is the extraction scheme.

Some models have earned themselves a poor reputation through disagreement of their predictions with experiment yet, although the model itself may not represent the operation of the FET entirely correctly, it is not the prime source of gross inaccuracy: it is the inadequate manner in which the following two steps of the extraction have been implemented:

1. Fitting all elements of the equivalent circuit, which is the basis of the model, to measured microwave data, in whatever form, as functions of frequency;
2. Fitting functions (for later computational use in the model at run time) to the elements obtained from step 1 as functions of bias.

When commercial extractors were first made available in 1989 to 1990, they were unsophisticated and often fell down badly in step 1. Two of the large American software companies began to offer extraction services for customers of their large-signal simulators (EEsof with Xtract and Compact Software with Scout). One motivation was to provide one-off models to customers who had no need for an extractor beyond that one model. Another was probably that, by manual intervention on behalf of the software's originators, a better result to the extracted model could be obtained than by use of the commercial extraction software alone. Even today, it is probably true that the worth of a large-signal model depends much upon the skill and experience of, and the set of tools available to, the person who extracts it.

2.5.6 The Position of Commercial Extractors

The availability of commercial software for the extraction of large-signal models was, in the late 1990s, understood to be:

1. Compact Software still sold its extractor known as Scout.
2. Xtract from EEsof had been withdrawn following the merger with Hewlett-Packard (HP) (and perhaps parts had been merged with HP's High-Frequency Modelling System based on IC-CAP, as below).

3. HP promoted what it referred to as the High-Frequency Modelling System, consisting of microwave parameter extraction and test hardware (HP 85122A) and the HP85190A IC-CAP modeling software.

2.5.7 FET Size Considerations

There is a fundamental difficulty in extracting models for devices with gate widths much over 0.5 mm if, at any stage, the extraction involves reliance on the circuit topology of Figure 2.1(a). The difficulty is best explained by considering what happens to some components of current and resistance in the structure as the FET gate width is increased progressively.

First, the equivalent circuit of Figure 2.1(a) ignores transcapacitance. In common with the usual components of capacitance, the transcapacitances become proportionally greater as the gate width is increased, and so too do the currents they represent—the transreactance currents. Second, for a larger gate width, the series resistances R_s and R_d become smaller.

Now, to correct for the neglect of the transreactance currents (and other topological errors) in the commonplace circuit model, the series resistances R_s and R_d take on bias-dependent values in an unconstrained fit to measured S-parameters. For small gate width FETs (<500 μm), the resistances are comparatively large and the transreactance currents are relatively small; hence the correction by the first for the omission of the second can be accommodated. When the resistances become small and the transreactance currents become large, the circuit defects simply cannot be made good because the resistance values are too small to accommodate the omission by alteration of those values. On these fundamental grounds, models relying on Figure 2.1(a) may be expected to exhibit a deteriorating quality of fit for FETs of a gate width progressively larger than about 500 μm.

Large FETs and HEMTs may consist of perhaps 10 or more cells connected in parallel by busbars. The signal delay down the busbars introduces complexities that cannot be accommodated by the single-circuit model of Figure 2.1(a). A proper description requires the busbars to be modeled also.

To cope with such devices it may be necessary to characterize one cell, chosen to be small enough for all the foregoing objections to be removed, and then to assemble in the circuit simulator a distributed model for the entire device, busbars included. The computer-aided design (CAD) tool of choice is the electromagnetic (EM) simulator.

2.5.8 Model Openness in Construction and Usability

Whether a model gains widespread acceptance seems to have more to do with its openness and usability rather than its faithfulness to practice. There are two aspects to openness: first, whether the model is readily understood in simple physical terms so, for instance, the user can see how to scale it to different gate widths; and second, whether the model is easily constructed without the need for special instrumentation (assuming that general microwave test instrumentation is available, including a vector network analyzer). Other terms used for model construction are "model generation" and "model extraction."

2.5.9 Constraints Placed upon Models by Circuit Simulators

The following restrictions are placed upon a model, be the model provided within the simulator or a user-defined model, by some (but not all) circuit simulators:

1. Negative element values are disallowed;
2. Branch or nodal charges must be conservative.

These two restrictions are bothersome. The first restriction means that, in the circuit of Figure 2.1(a), the parasitic resistances R_s and R_d cannot take on the artificial bias dependence they need to take on to compensate for the omission of transreactance currents. Also, C_{ds} cannot take on the negative values often exhibited by the extracted capacitance function for voltages below the knee in the drain characteristics. The constraint of all-values-positive means that, for some FETs at some bias points, the model will inescapably fit poorly the measured S-parameters from which it was derived.

The second restriction is even more serious because it is unphysical. The principle of charge conservation means that, if the charge Q is a function of two voltages, thus,

$$Q\left(V_{GC}, V_{DG}\right)$$

and if the voltages start off with given values and are then changed arbitrarily but returned to their starting values, Q will return to its starting value also. In real FETs, however, only the gate charge Q_G, equivalently the depletion region charge $Q = -Q_G$, behaves this way [13]. Q cannot be split into two separate charge components, one assigned to the gate-channel branch and the other assigned to the gate-drain branch, such that these components are individually conservative. Equivalently, Q_G cannot be split into two individually conservative nodal charges, Q_S at the source and Q_D at the drain.

In the practical FET, charge is pumped from one electrode (e.g., the source) to the other (the drain), and the property of multiple charge conservation simply does not apply. It is for this reason that circuit simulators which allow the user to specify models incorporating nonlinear capacitances to represent charge storage are much preferred.

Although restricted to their particular equations for charge storage in GaAs FETs, Smith et al. [13] proposed the addition of a current generator that transfers charge between the source and the drain, internally to the device, as a way of entering a physically valid nonconservative branch or nodal charge model into a circuit simulator which insists on all charges being conservative. The principle might be exploitable for other models.

2.6 Rauscher and Willing

The idea of using the circuit of Figure 2.1(a) with nonlinear bias-dependent elements seems to have originated with Willing et al. [14]. They advocated the use of bias-dependent small signal S-parameter measurements to find the incremental (or small-signal) values of the circuit elements and assumed that the circuit model of

Figure 2.1(a) is valid for large-signal operation (except that they had in the circuit a resistive element to represent the Gunn domain that they supposed to exist). In so doing, they gave rise to something that has become frustratingly difficult to query. The most eloquent challenge to the unquestioning use of it to date has come from David E. Root of HP (see Section 2.16).

Note: In the equations of the succeeding sections, it is vital to insert values for the voltages having the correct sign. (For instance, in any expression involving the pinch-off voltage, V_P, it is essential to know whether the author's convention was to regard V_P as a positive number or a negative number.) Reference should be made to the original papers for clarification of the sign convention assumed by any given model.

There is repetition from section to section in the way the models' attributes are described so that each stands complete in its own right.

2.7 The Curtice Quadratic Model

2.7.1 Expression Used for the Modeling Current

$$I_D = \beta\left(V_{GS} + |V_P|\right)^2 \left(1 + \lambda V_{DS}\right)\tanh\left(\alpha V_{DS}\right)$$

2.7.2 Expression Used for the Modeling Capacitance

$$C_{gs,gd} = C_{gs0,gd0}\left[1 - \frac{V}{V_{B0}}\right]^{-1/2}$$

where V_{B0} is the equilibrium Schottky barrier height of the gate-semiconductor junction, and V is the applied gate-source or gate-drain voltage.

2.7.3 Basis

The model is based on the commonplace circuit model of Figure 2.1(a) with the foregoing algebraic equations (having no connection to the physics of FET operation) to fit to measured dependences of the equivalent circuit elements [15].

2.7.4 Underlying Soundness

Because the model is based on the commonplace circuit model of Figure 2.1(a), it has the shortcomings (1) to (5) of Section 2.5.1. Of the criteria (6) to (10), the only one satisfied is (9). Criterion (8) is met in a rather crude way by the contrivance of Figure 2.1(b), if used. Among the practical criteria, (1) and (3) only are satisfied.

The basic quadratic form of the current equation with respect to V_{GS} means that:

1. The fit to $I_D(V_{GS})$ is generally poor.
2. Soft pinch-off, where a progressively higher negative value of V_{GS} is needed to pinch the channel off as V_{DS} is increased, cannot be modeled.

3. The square law means that harmonics only down to the second can be generated.
4. The expression for current can only be used for values of V_{GS} greater than pinch-off. If a value of $V_{GS} < -|V_P|$ is entered, then the model predicts current flow in a well pinched-off device.

The expressions representing the capacitances are rudimentary and do not represent well either the V_{DS} or the V_{GS} dependence of C_{gs} or C_{gd}: there is no functional dependence on V_{DS} to model the increase in C_{gs} with V_{DS} in practice, and the spacing between the C_{gs} curves is modeled as a square law. Similar comments apply to the restrictiveness of the expression for C_{gd}. Simply, the capacitance expressions are the depletion approximation for a 1-D Schottky barrier made to a uniformly doped semiconductor.

2.7.5 Measurements Required

Formally, the Curtice quadratic model draws no distinction between the dc and RF characteristics and so is most frequently extracted from small-signal S-parameter measurements made about dc bias points, together with the dc $I_D(V_{GS},V_{DS})$ characteristics. With the importance of pulsed characterization becoming more widely appreciated, the Curtice quadratic form can be applied equally well to generate an approximate fit to the pulsed $i_D(v_{GS},v_{DS})$ characteristics. If this is done, however, there is a problem in implementing the model in a large-signal circuit simulator which recognizes only one set of drain characteristics. The compromise solution is that shown in Figure 2.1(b).

2.7.6 Openness of Procedure for Extracting the Model from Measurements

All detail of the model is in the public domain, so individual implementations involving steps (1) and (2) of the extraction process of Section 2.5.5 are possible.

2.7.7 Scalability

Being at least loosely related to the physics and construction of the practical FET, the model of Figure 2.1(a) carries with it a confidence that its elements can be scaled with gate width. For instance, the source resistance can be halved in response to a doubling of the gate width, and the gate-source capacitance doubled. If it is acknowledged that the fit of the model to practice is poor in the first place, and the same quality of fit is accepted upon scaling, then the model could be said to possess scalability.

2.7.8 General Comments

Even in the late 1990s, the Curtice quadratic model was more of an historical piece, and it is unreasonable to criticize it for its rudimentary nature. It is fairer to credit it with having initiated activity in large-signal modeling and for having provided impetus to others.

2.8 The Curtice-Ettenberg Model

2.8.1 Expressions Used for Modeling Current

The equation used to fit the dc characteristics is a straightforward extension of the quadratic form [16]. The model is sometimes referred to as the Curtice cubic.

$$I_D = \left(A_0 + A_1V_1 + A_2V_1^2 + A_3V_1^3\right)\left(1 + \lambda V_{DS}\right)\tanh\left(\alpha V_{DS}\right)$$
$$V_1 = V_{GS}\left[1 + \beta\left(V_{DS0} - V_{DS}\right)\right]$$

2.8.2 Expressions Used for Modeling Capacitance

The same expressions are used as in the Curtice quadratic.

2.8.3 Basis

The model is based on the commonplace circuit model of Figure 2.1(a).

2.8.4 Underlying Soundness

The same comments apply as to the Curtice quadratic. The model is based on the commonplace circuit model of Figure 2.1(a), so it has the shortcomings (1) to (5) of Section 2.5.1. Of the criteria (6) to (10), the only one satisfied is (9). Criterion (8) is met in a rather crude way by the contrivance of Figure 2.1(b), if used. Among the practical criteria, (a) and (c) are satisfied, and (b) approximately so. Additionally:

1. The fit to $I_D(V_{GS})$ is generally reasonable to good.
2. The form of the equation for V_1 means that, to some extent, soft pinch-off can be accommodated. The factor β controls $V_P(V_{DS})$.
3. The cubic form provides a reasonable chance of modeling third-order nonlinearities through the term that remains after differentiating the current function three times. The third derivative is thus finite at least, although inflexible: it is constant over the entire range of voltage.
4. Care must be taken to avoid values of V_{GS} that fall outside the fitted range. For example, a very negative value of V_{GS} may lead to current appearing in a pinched-off device or, worse still, negative current being predicted.

For best accuracy, the value of A_3 should be obtained by matching to the third derivative of the current function with respect to V_{GS} or to the second derivative of g_{m0}.

The expressions for capacitance are the same as those used in the Curtice quadratic model, so all the comments made there apply.

2.8.5 Measurements Required

Formally, no distinction is made between the dc and RF characteristics and so the model is most frequently extracted from small-signal S-parameter measurements made about dc bias points, together with the dc $I_D(V_{GS},V_{DS})$ characteristics. With the importance of pulsed characterization becoming more widely appreciated, the cubic form can be applied to fit the pulsed $i_D(v_{GS},v_{DS})$ characteristics. If this is done, however, there is a problem in implementing the model in a large-signal circuit simulator which recognizes only one set of drain characteristics. Once again, the arrangement of Figure 2.1(b) provides a compromise.

2.8.6 Openness of Procedure for Extracting the Model from Measurements

All details of the model are in the public domain, so individual implementations involving steps (1) and (2) of the extraction process of Section 2.5.5 are possible.

2.8.7 Scalability

Being at least loosely related to the physics and construction of the practical FET, the model of Figure 2.1(a) carries with it an expectation that its elements should be scalable with gate width. For instance, the source resistance can be halved in response to a doubling of the gate width, and the gate-drain capacitance can be doubled. If it is acknowledged that the fit of the model to practice is poor in the first place, and the same quality of fit is accepted upon scaling, then the model could be said to possess scalability.

2.9 The Materka-Kacprzak Model

2.9.1 Expressions Used for Modeling Current

$$I_D = I_{DSS}\left[1 - \frac{V_{GS}}{V_P'}\right]^2 \tanh\left[\frac{\alpha V_{DS}}{V_{GS} - V_P'}\right]$$

$$V_P' = \left|V_P\right| + \gamma V_{DS}$$

2.9.2 Expressions Used for Modeling Capacitance

The same expressions are used as for the Curtice quadratic model.

2.9.3 Basis

The model is based on the commonplace circuit model of Figure 2.1(a).

2.9.4 Underlying Soundness

The Materka-Kacprzak model [17] has the shortcomings (1) to (5) of Section 2.5.1. Of the criteria (6) to (10), the only one formally satisfied is (9). Criterion (8) can be met in a rather crude way by the contrivance of Figure 2.1(b). Among the practical criteria, (a) and (c) are satisfied.

The equation for fitting to the current is able to accommodate the soft pinch-off characteristic of practical FETs via the parameter γ. This feature apart, however, at fixed V_{DS} the relationship between I_D and V_{GS} is given by the leading term in brackets (i.e., a square law, which with reference to Figure 2.3 is generally a weak approximation to practical FETs). The equation also requires that V_{GS} is kept within limits to prevent the model predicting erroneous values of current in pinched-off devices.

A more serious limitation is that, for low-threshold devices, the Materka model does not fit near open channel. Obtaining a good current drive from such a device requires the gate to be driven into forward bias, maybe to $V_{GS} = +0.5\text{V}$ or thereabouts. A practical FET so biased may exhibit a positive slope to the $I_D(V_{GS},V_{DS})$ characteristics, whereas the Materka current equation produces a negative slope with respect to V_{DS} for positive gate bias.

The problem arises from the term

$$\left[1 - \frac{V_{GS}}{V'_P}\right]^2 = \left[1 - \frac{V_{GS}}{|V_P| + \gamma V_{DS}}\right]^2$$

When V_{GS} is positive, the magnitude of this expression decreases as V_{DS} increases, thereby giving rise to droop in the $I_D(V_{GS},V_{DS})$ characteristics, which is generally not a feature of low-threshold practical FETs.

2.9.5 Measurements Required

As no distinction was made between the dc and RF characteristics at the time the model was formulated (in the mid-1980s), it was intended to be extracted from small-signal S-parameter measurements made about dc bias points, together with the dc $I_D(V_{GS},V_{DS})$ characteristics. With the importance of pulsed characterization becoming more widely appreciated, however, the expression for the current can be applied to fit the pulsed $i_D(v_{GS},v_{DS})$ characteristics. If this is done, there is once again the problem of implementing the model in a large-signal circuit simulator that recognizes only a single set of drain characteristics. As with the preceding models, the arrangement of Figure 2.1(b) provides a compromise.

2.9.6 Openness of Procedure for Extracting the Model from Measurements

All detail of the model is in the public domain, so individual implementations involving steps (1) and (2) of the extraction process of Section 2.5.5 are possible.

2.9.7 Scalability

Because the circuit model of Figure 2.1(a) is at least loosely related to the physics and construction of the practical FET, it encourages a belief that its elements can be scaled with gate width. For instance, the source resistance can be halved in response to a doubling of the gate width, and the gate-source capacitance can be doubled.

If it is acknowledged that the fit of the model to practice may be poor in the first place, and the same quality of fit is accepted upon scaling, then the model could be said to possess scalability.

2.10 An Illustrated Application

A practical application of models tends to reveal their weaknesses (and maybe the odd strength or two). The following example required a choice to be made, from among the commercially supported models, of a current fitting equation to construct a large-signal model of a low dc power dissipation FET, and the choice similarly of a capacitance fitting equation.

2.10.1 Current Equation: Modified Materka

Further to the shortcoming of the Materka model for low-pinch-off FETs as described in Section 2.9.4, a modification was conceived and implemented by HP in their MDS at the behest of the consortium in [18]. The change consisted of multiplying the as-published Materka model by factor

$$\left(1 + \lambda V_{DS}\right)$$

which is equivalent to adding an output conductance proportional to V_{DS}.

2.10.2 Capacitance Equations: Use of the Statz Expressions

The Statz expressions for capacitance, to be given in Section 2.11.2, at the time of their release were a useful improvement over what had become popular (as distinct from what was known). Bearing in mind that it can take 2 or 3 years for improved models to find their way into commercial software, and a further year for engineers to become proficient in their use, there is a sizeable lag between knowledge and widespread usage. Thus, in 1992, when the practical application from which this example derives was instituted, the Statz capacitance model was the best available in mainstream software.

2.10.3 Results

Figures 2.4 to 2.13 show the measured FET characteristics and the modified Materka-Statz model fit as implemented in HP-MDS [18]. (In connection with Figures 2.11(a), 2.12(a), and 2.13(a), FETLink was the commercial fitting and reverse modeling software application due to GaAsCode Ltd.)

Despite the fact that, all told, probably well over 1 man-year of collective effort went into deciding upon and constructing this model, the correspondence with practical data leaves unanswered the best adjective to describe it. All that can be said in its favor is that with other models it was possible to do worse.

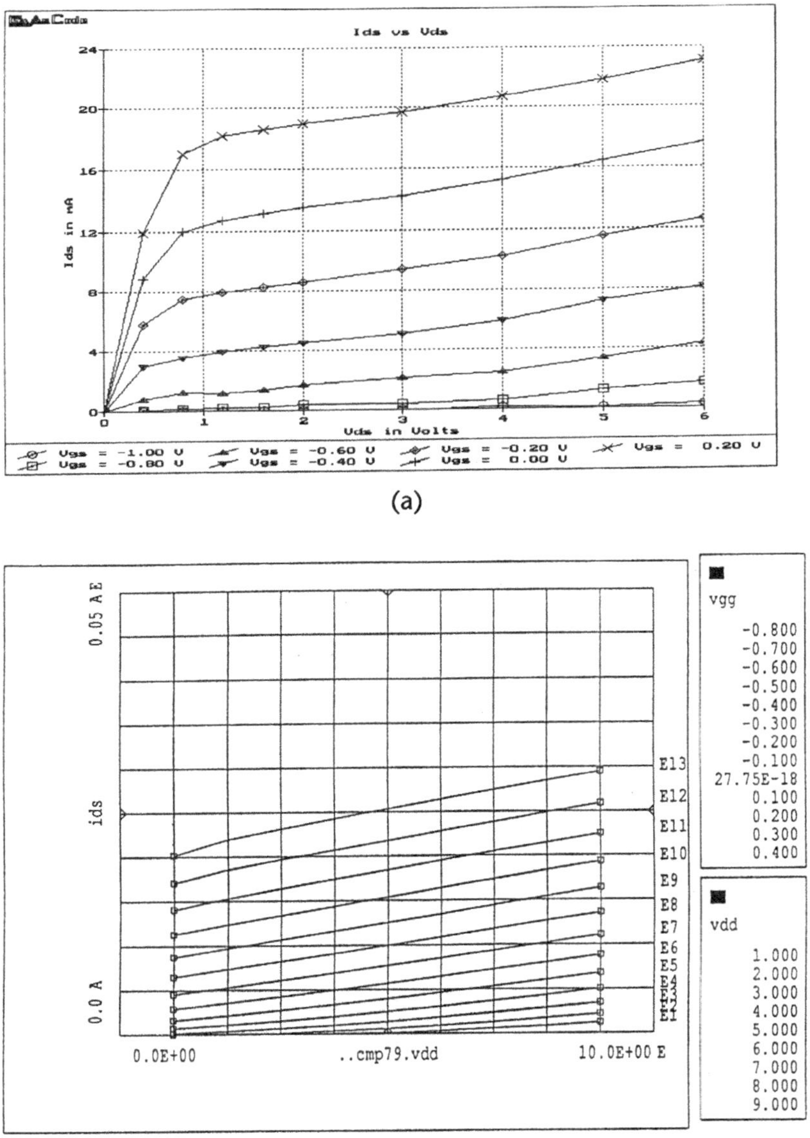

(a)

(b)

Figure 2.4 (a) Measured drain characteristics of a low dc-power consumption FET. (b) Drain characteristics fit using the modified Materka equation.

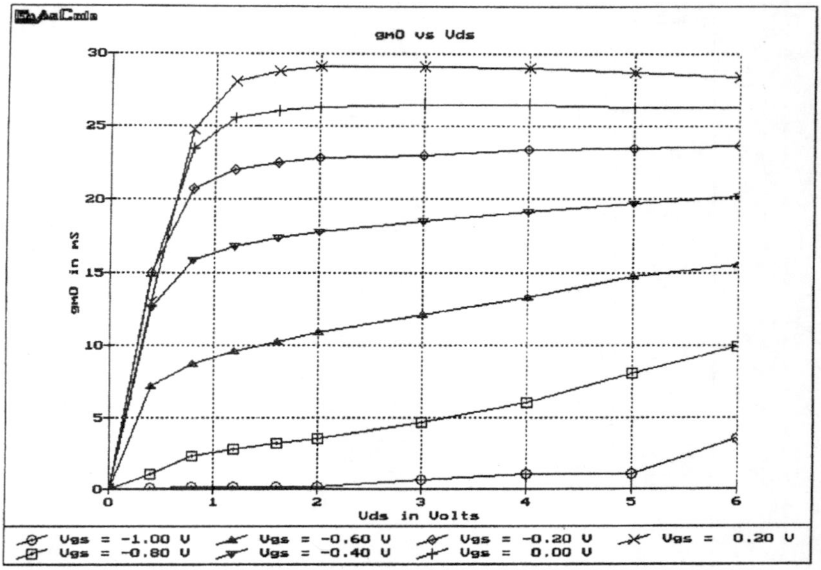

(a)

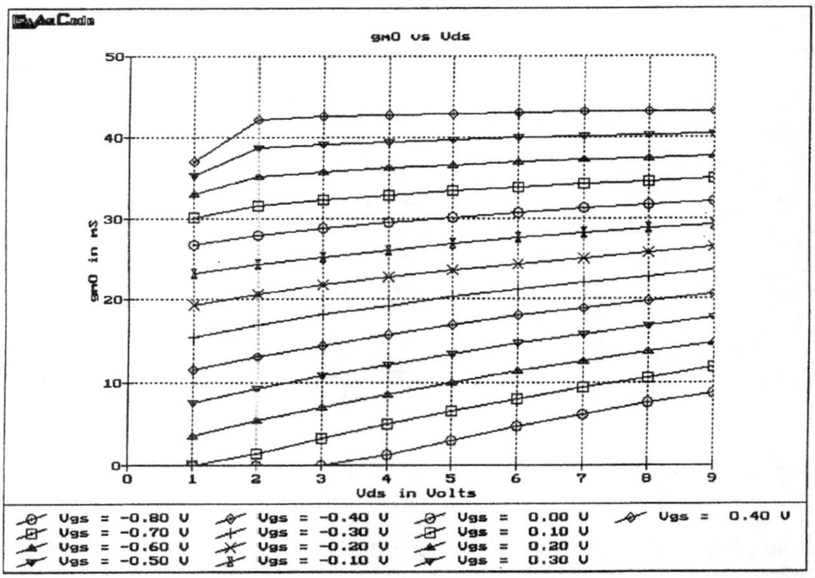

(b)

Figure 2.5 (a) Extracted intrinsic transconductance of a low dc power FET. (b) Fit to the transconductance resulting from the modified Materka equations.

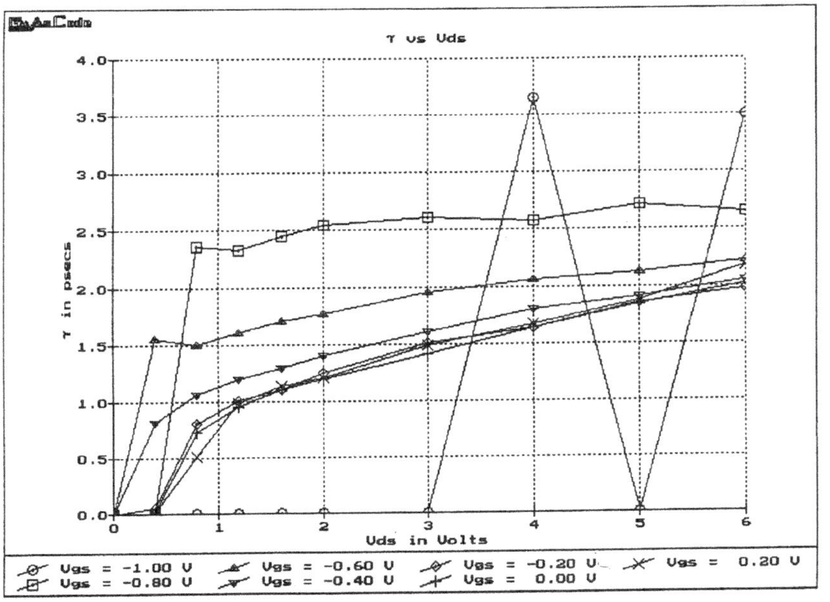

(a)

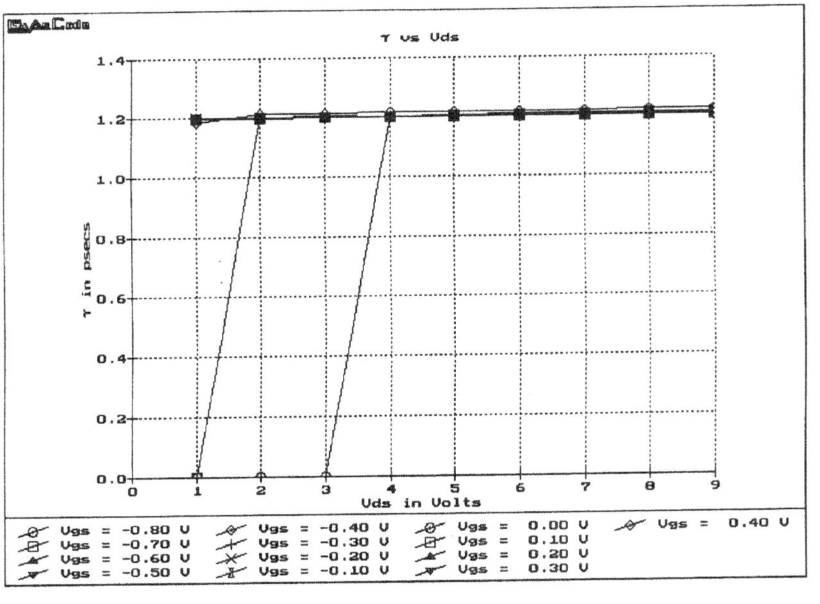

(b)

Figure 2.6 (a) Extracted transconductance delay. (b) Transconductance delay emerging from the fit of the modified Materka-Statz model.

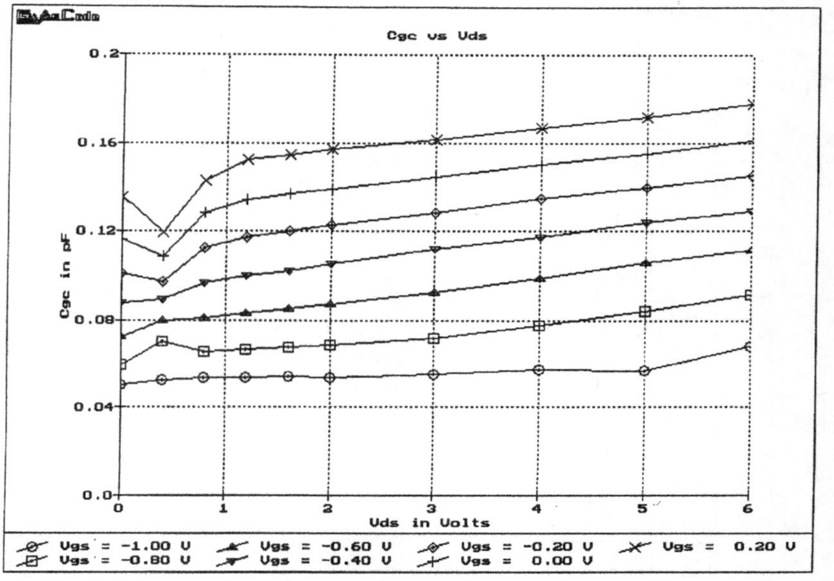

(a)

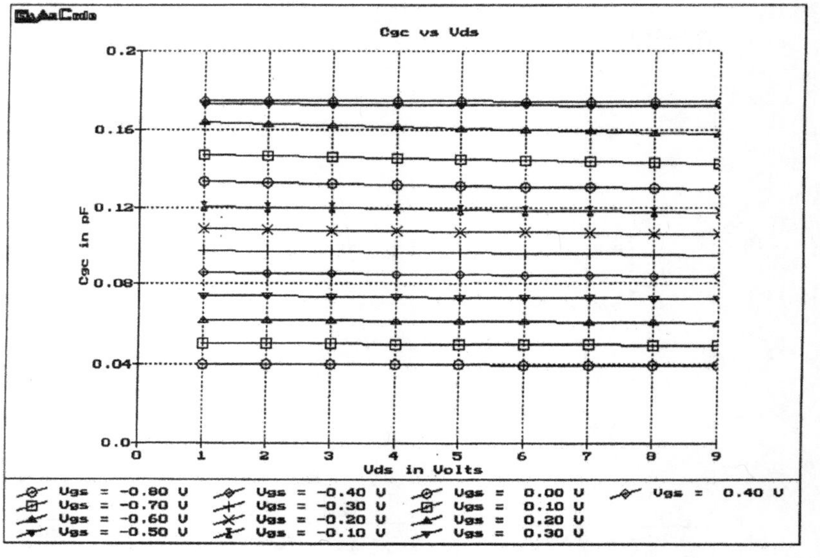

(b)

Figure 2.7 (a) Extracted gate-channel capacitance. (b) Fit of Statz capacitance equations.

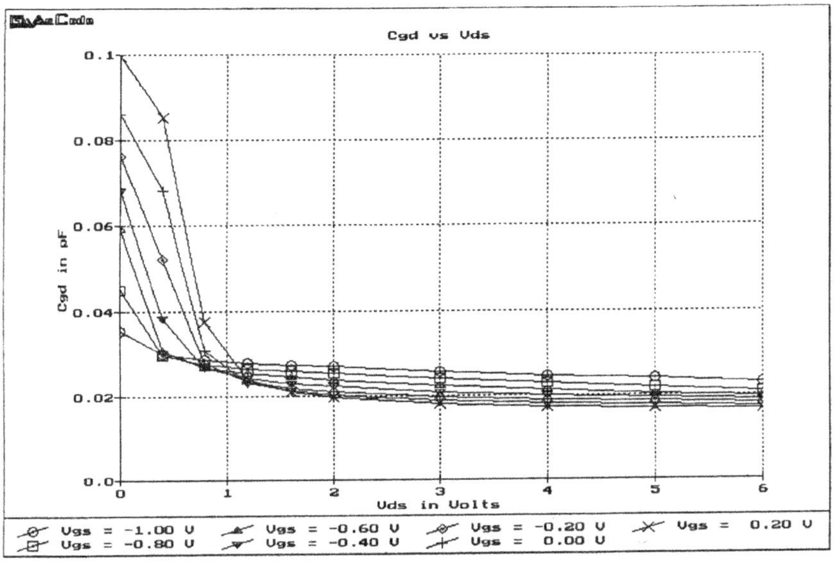

(a)

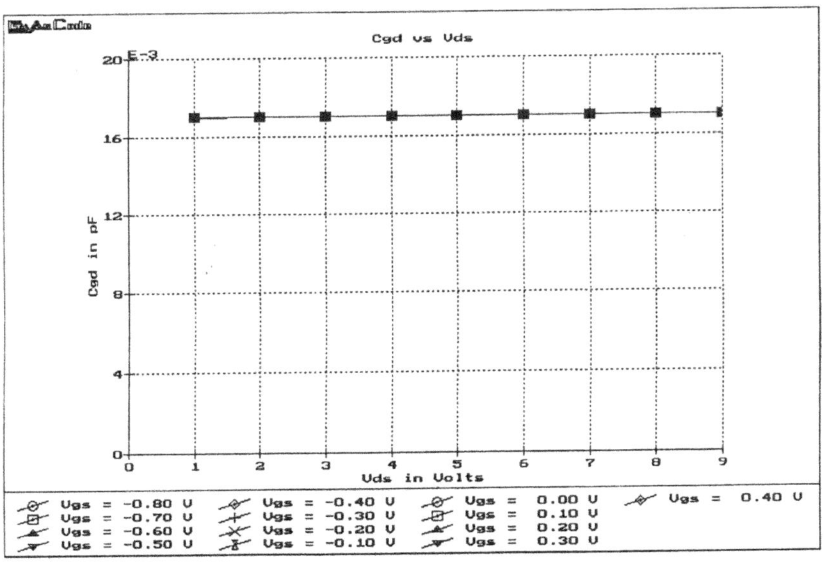

(b)

Figure 2.8 (a) Extracted gate-drain capacitance. (b) Gate-drain capacitance fitted using the Statz capacitance equations.

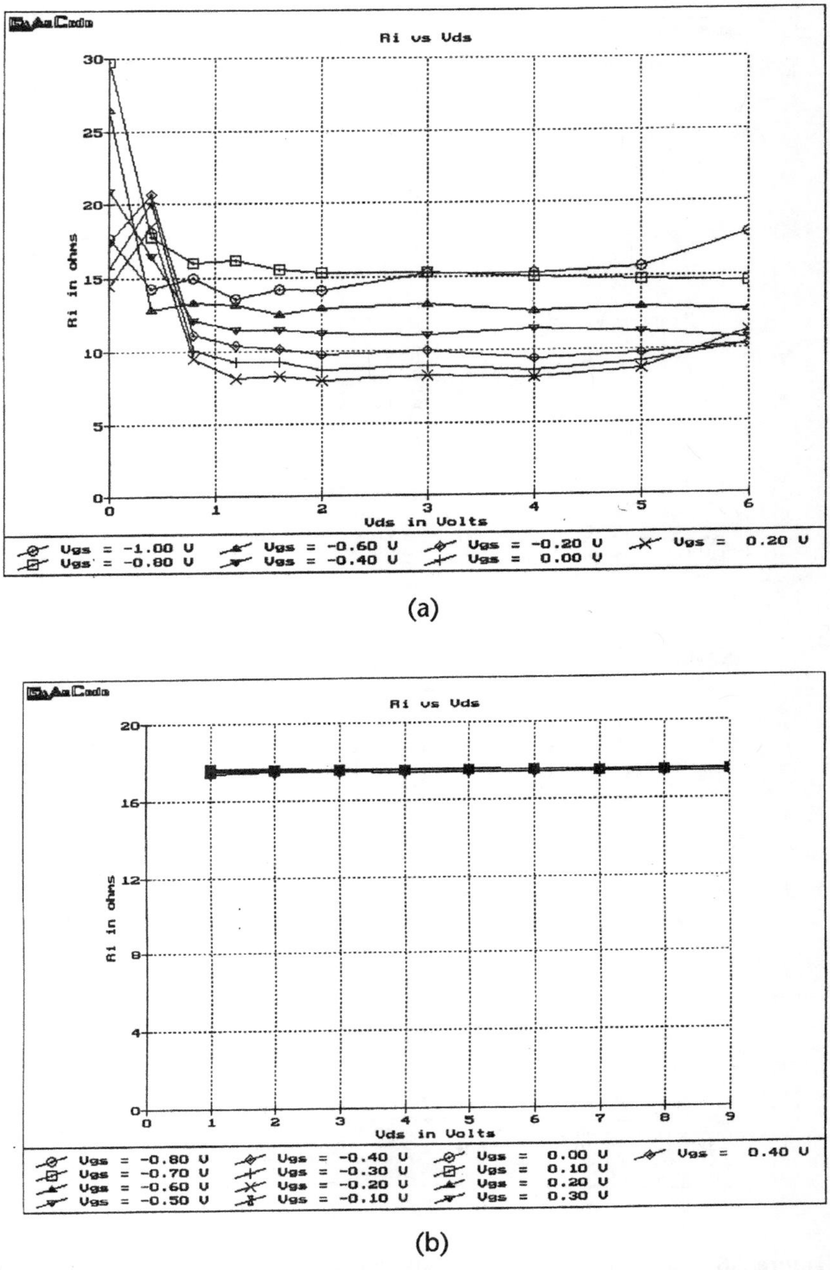

Figure 2.9 (a) Extracted intrinsic channel resistance, and (b) fit of modified Materka-Statz model.

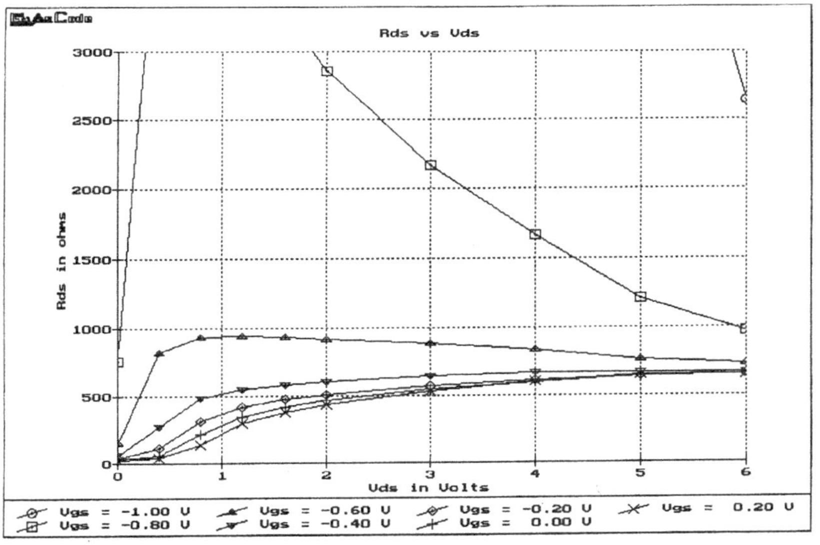

(a)

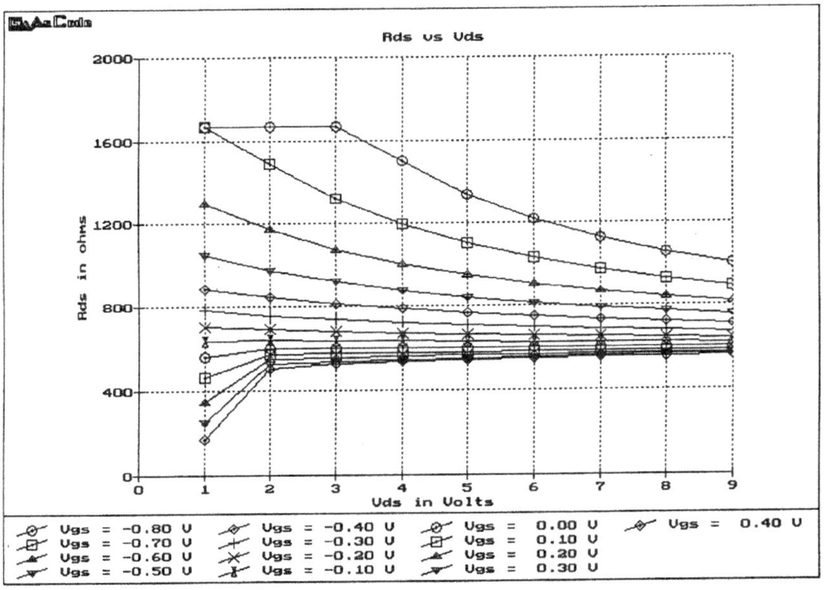

(b)

Figure 2.10 (a) Extracted drain-source resistance, and (b) fit of modified Materka-Statz model.

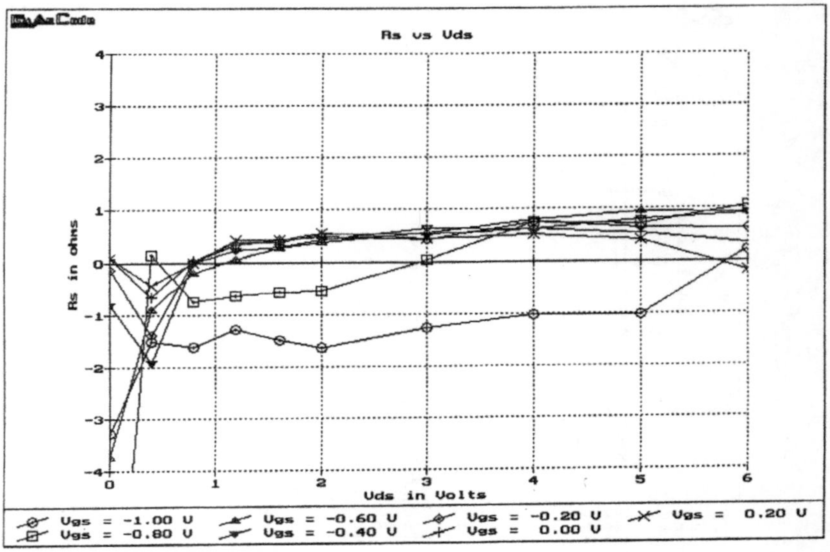

(a)

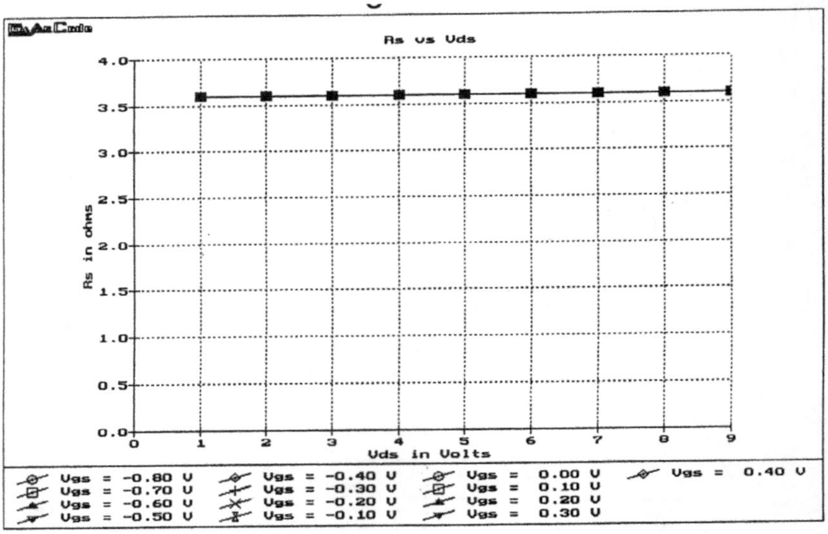

(b)

Figure 2.11 (a) RF source resistance obtained as a free-fit using FETLink, and (b) source resistance used in the modified Materka-Statz model (being the dc value).

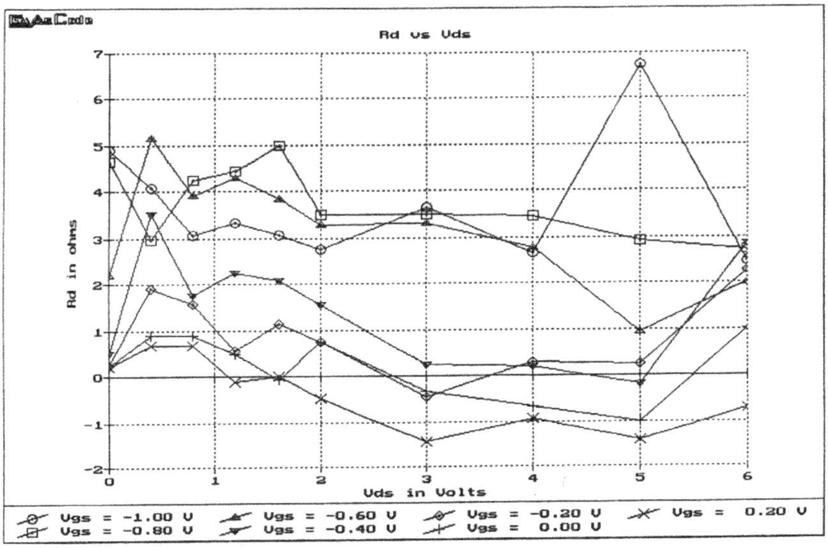

(a)

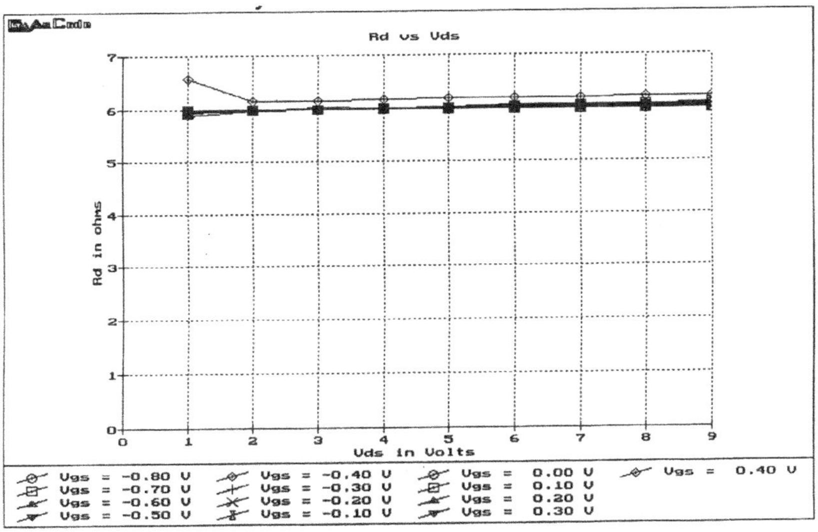

(b)

Figure 2.12 (a) RF value of drain series resistance freely fitted with FETLink, and (b) drain series resistance fitted using the modified Materka-Statz model.

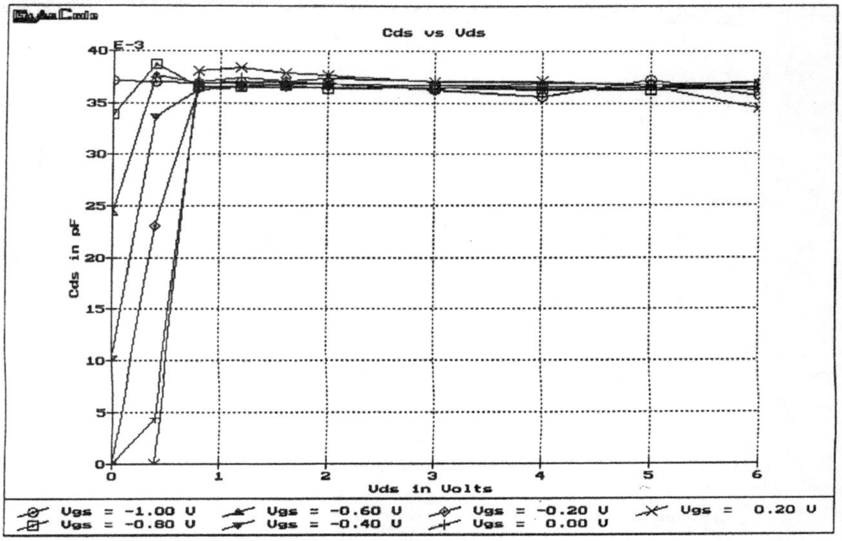

(a)

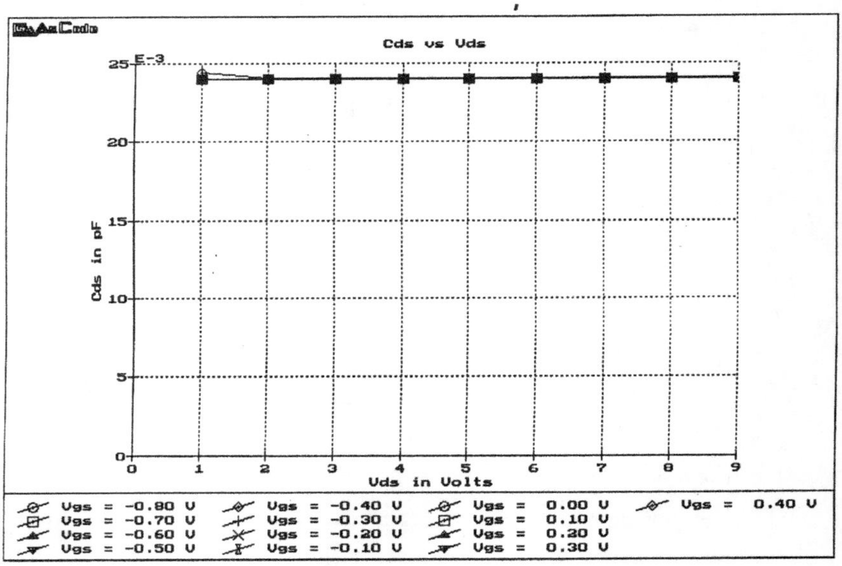

(b)

Figure 2.13 (a) Free-fit obtained using FETLink to the drain-source capacitance. (b) The drain-source capacitance resulting from the fit of the modified Materka-Statz model.

2.11 The Statz Model

2.11.1 Expressions Used for Modeling Current

$$I_D = \left[\frac{\beta\left(V_{GS} + |V_P|\right)^2}{1 + b\left(V_{GS} + |V_P|\right)}\right]\left\{1 - \left(1 - \frac{\alpha V_{DS}}{3}\right)^3\right\}\left(1 + \lambda V_{DS}\right)$$

$$0 < V_{DS} < \frac{3}{\alpha}$$

$$I_D = \left[\frac{\beta\left(V_{GS} + |V_P|\right)^2}{1 + b\left(V_{GS} + |V_P|\right)}\right]\left(1 + \lambda V_{DS}\right)$$

$$V_{DS} \geq \frac{3}{\alpha}$$

2.11.2 Expressions Used for Modeling Capacitance

The simple 1-D depletion capacitance approximation used by the preceding models has the following deficiencies:

1. In the current saturation regime, there is no dependence upon V_{DS} (the model has C_{gs} independent of V_{DS}, whereas practical devices exhibit an upward slope).
2. Below the knee in the $I_D(V_{GS},V_{DS})$ curves, C_{gs} often has a U-shape in practical FETs and C_{gd} rises to values comparable with C_{gs}. The simple depletion capacitance model has both capacitances independent of V_{DS}.
3. If the polarity of V_{DS} reverses, the roles of C_{gs} and C_{gd} reverse: models invoking the simple formula for capacitance have it that the capacitances remain unchanged if the polarity of V_{DS} reverses.

Statz et al. [19] sought to improve the representation of capacitance in GaAs FET models by addressing (2) and (3), but their expressions did nothing about (1). The equations are voluminous. Here we give only summary forms:

$$C_{gs} = \frac{K_1 K_2 C_{gs0}}{\left[1 - \dfrac{V_n}{V_{B0}}\right]^{1/2}} + K_3 C_{gd0}$$

$$C_{gd} = \frac{K_1 K_3 C_{gs0}}{\left[1 - \dfrac{V_n}{V_{B0}}\right]^{1/2}} + K_2 C_{gd0}$$

where the K_i are not constants but are functions of the internal (or intrinsic) device voltages $V_{GS'}$ and $V_{D'G}$.

Figures 2.7(b) and 2.8(b) are examples of capacitance fitted in a real-world application.

2.11.3 Basis

The Statz model persists with the circuit model of Figure 2.1(a) as the basis, but with physics introduced for the first time via the capacitance considerations mentioned in the last section. Formally, the intrinsic resistance R_i in series with C_{gs} is missing as a result of which, like other models where this element is absent, an inability to fit closely to s_{11} can be expected.

2.11.4 Underlying Soundness

The model has the shortcomings (1) to (5) of Section 2.5.1 as it is based on the commonplace circuit model of Figure 2.1(a). Of the criteria (6) to (10), the only one formally satisfied by the model as originally published is (9). Criterion (8) can be met, although unsophisticatedly, by the contrivance of Figure 2.1(b). Among the practical criteria, (a) and (c) are satisfied.

With reference to Figure 2.3, the Statz equation for current is flexible in the type of $I_D(V_{GS})$ behavior that can be fitted. If the factor b is very small, the form of the expression is as a square law. If, however, b is large, the behavior of $I_D(V_{GS})$ is linear. By choice of b therefore, any form between linear and square law can be represented, although it should equally be noted this does not guarantee that any given practical behavior can be fitted (because the curvature that the model provides may be at the other end of the V_{GS} range from that at which the curvature occurs in the practical data). Despite this flexibility of form with respect to V_{GS}, the equation for fitting to the current is unable to accommodate the soft pinch-off characteristic of practical FETs wherein the pinch-off voltage increases in V_{DS}. Even if operation at a fixed V_{DS} could be assured, the model does not provide an independently fixable third derivative of I_D with respect to V_{DS}, so it is not as capable as the Curtice-Ettenberg model of predicting third-order phenomena such as third harmonic generation and third-order intercept point.

As with the previous models, V_{GS} must be restricted in range to prevent the model predicting erroneous results.

The capacitance expressions at first sight are daunting. Qualitatively, they are an improvement over the simple depletion capacitance equations used in earlier models; nevertheless, quantitatively, they represent only very approximately the capacitance functions extracted from practical devices (an example being Figures 2.7(a) and 2.8(a)).

2.11.5 Measurements Required

As no distinction was made between the dc and RF characteristics at the time the model was formulated (in the late 1980s), it was intended to be extracted from small-signal S-parameter measurements made about dc bias points, together with the dc $I_D(V_{GS},V_{DS})$ characteristics. With the importance of pulsed characterization becoming more widely appreciated, however, the current fitting expression can be applied to fit the pulsed $i_D(v_{GS},v_{DS})$ characteristics. If this is done, however, there is once again the problem of implementing the model in a large-signal circuit simulator that recognizes only a single set of drain characteristics. As with the preceding models, Figure 2.1(b) provides a compromise.

2.11.6 Openness of Procedure for Extracting the Model from Measurements

All details of the model are in the public domain, so individual implementations involving steps (1) and (2) of the extraction process of Section 2.5.5 are possible. The prospect of trying to develop an extractor for it, especially the capacitance equations, was at one time seen as rather daunting.

2.11.7 Scalability

Because the circuit model of Figure 2.1(a) is at least loosely related to the physics and construction of the practical FET, it encourages a belief that its elements can be scaled with gate width. For instance, the source resistance can be halved in response to a doubling of the gate width, and the gate-source capacitance doubled. If it is acknowledged that the fit of the model to practice is approximate in the first place, and the same quality of fit is accepted upon scaling, then the model could be said to possess scalability.

2.12 TriQuint Own Model (TOM)

This model was introduced in 1990 [20]. It is one of the oddities of the English language that the "model" inherent in the acronym "TOM" is frequently repeated through reference to the TOM model, namely TriQuint Own Model model.

2.12.1 Expressions Used for Modeling Current

The expression for representing the current (i.e., "the model") embodies some features of the Materka model insofar as it caters for soft pinch-off and was otherwise conceived as a modification of the Statz model. Compared with its predecessors, it fits a broad range of FET types.

The first modification of the Statz equation is designed to improve the capability to fit near the pinch-off by accommodating a pinch-off that is V_{DS} dependent:

$$V_P' = |V_P| + \gamma V_{DS}$$

The second modification allows the slope of the characteristics to be smaller near open channel than near pinch-off via the form

$$I_D = \frac{I_{DS0}}{1 + \delta V_{DS} I_{DS0}}$$

where δ is a model parameter and I_{DS0} is given by:

$$I_{DS0} = \beta \left(V_{GS} - V_P' \right)^Q \tanh\left(\alpha V_{DS} \right)$$

The TOM expression for the drain current is given by these three equations.

2.12.2 Expressions Used for Modeling Capacitance

The Statz capacitance equations are used.

2.12.3 Basis

This model, too, is based upon the circuit configuration of Figure 2.1(a).

2.12.4 Underlying Soundness

As the model is based on the commonplace circuit model of Figure 2.1(a), it has the shortcomings (1) to (5) of Section 2.5.1. Of the criteria (6) to (10), those satisfied are (6), (8), and (9). The transition from dc to RF characteristics is represented approximately by the network of Figure 2.14.

In this scheme, the voltage at the node between the added capacitors and resistors is used to modulate the current generator directly. The authors of the model offered a plausible physical mechanism accounting for the dispersion, but they gave no experimental evidence to support their contention (which briefly was that a redistribution of charge trapped in the substrate gives rise to a backgate voltage and that the channel current is modulated not only by the applied gate voltage but also by this unintended backgate voltage).

Although the model was promoted as a large-signal model, the adjunct of Figure 2.14 is a small-signal fix only and may suffer the type of difficulty illustrated in Figure 2.2.

Among the practical criteria of Section 2.5.1, (a) and (c) are satisfied.

The TOM formally does not include the resistance R_i in the gate-channel branch of the circuit of Figure 2.1(a): this omission makes for a difficulty in fitting to s_{11}.

As with other models using polynomial expressions, the model implementation must restrict the range of V_{GS} to prevent the model predicting nonphysical results.

2.12.5 Measurements Required

Although the original article drew a distinction between the dc and RF characteristics and gave a circuit connection to model the transition between the two (Figure 2.14), no discussion was given to the means of measuring the RF characteristics

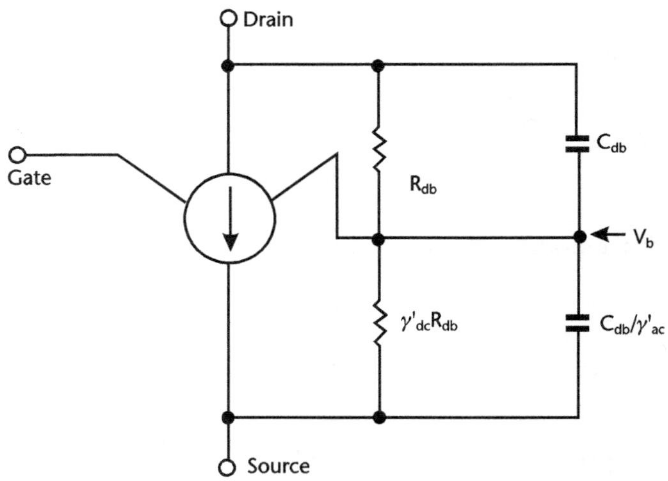

Figure 2.14 Scheme used in the TriQuint Own Model to cater for dispersion.

for insertion into the model. It therefore has to be assumed that what the originators intended was to draw a distinction between the R_{ds} apparent from the dc curves and the R_{ds} extracted from multibias small-signal measurements. Therefore, we assume that the model was to be extracted from small-signal S-parameter measurements made about dc bias points, together with the dc $I_D(V_{GS},V_{DS})$ characteristics. With the importance of pulsed characterization becoming more widely appreciated, however, the TOM expression for the current generator can be applied equally well to generate an approximate fit to the pulsed $i_D(v_{GS},v_{DS})$ characteristics (but if this is done, there is the problem of implementing the model in a large-signal circuit simulator, which does not recognize two sets of drain characteristics; the compromise solution of Figure 2.14 has the type of drawback summarized by Figure 2.2).

2.12.6 Openness of Procedure for Extracting the Model from Measurements

The same comments apply as to the preceding models.

2.12.7 Scalability

The same comments apply as to the preceding models.

2.13 The EEFET3 Model

The model [21] is based on the commonplace circuit model of Figure 2.1(a) and consequently has the shortcomings (1) to (5) of Section 2.5.1. Of the criteria (6) to (10), those satisfied are (6) through (10). Among the practical criteria, (a), (b), and (c) are satisfied.

EEFET3 is the identifier given by EEsof to the last in the line of conventional models using proprietary equation sets to fit the bias dependence of the elements.

2.13.1 Basis

Like many models, claims for its distinctiveness are founded upon the particular heuristic set of closed-form equations used to represent the bias dependence of the equivalent circuit elements fitted to measured S-parameters. Points to note are:

1. The equations of the model take three different forms depending upon the region of V_{GS} being fitted. A claim made for the scheme is that it can simulate the subthreshold current more accurately, an issue of importance in the simulation of digital circuits. It is thought that a function like a cosine taper is used in the subthreshold region to model the gradual decay of current to zero as the gate-source voltage is increased negatively.
2. The drain current is continuous to the second derivative with respect to V_{GS}.
3. Because the model uses the commonplace circuit configuration, currents arising from transcapacitance (i.e., transreactance currents) are not included. Confusingly, however, HP-EEsof referred to the dependence of:

$$C_{gs} \text{ upon } V_{DS}, \text{ and } C_{gd} \text{ upon } V_{GS}$$

as transcapacitance components [21]. Irrespective of the loose use of terminology, EEFET3 claims to be able to represent these voltage dependences.

4. The model is charge-based. Branch charges of the form $Q_{GS}(V_{GS},V_{DS})$ and $Q_{GD}(V_{GS},V_{DS})$ are used. Such a functional form means that both the branch charges will return to their original values if the voltages are returned to their original values, irrespective of the intermediate values that V_{GS} and V_{DS} take. In other words, Q_{GS} and Q_{GD} are conservative functions in the model, whereas in the practical FET it is not possible to identify separately conservative branch charges, only a conservative gate charge balanced by a (conservative) depletion charge. In the drain-source branch, a fixed capacitance is used.
5. The voltages driving the charges are those across the charges themselves, so this means that the charge changes are delayed in their response to the terminal voltages. As a general principle, this is as occurs in real FETs.
6. The model caters for dispersion via the network consisting of C_{bs} and R_{db} in Figure 2.15. At frequencies $\omega \ll 1/C_{bs}R_{db}$, the drain current is simply the dc part, I_D. At frequencies $\omega \gg 1/C_{bs}R_{db}$, the drain current is $i_D = I_D + I_{db}$.
7. The EEFET3 model has only one RF current function: it therefore does not represent the dependence of the RF characteristics upon bias point, which in some devices can be severe.

2.13.2 Underlying Soundness

The model's equations contain a large enough number of fitting parameters to be able to achieve close fits to measured data, a condition that, although a good start, is not sufficient to assure overall fidelity to true life for two reasons:

1. It does not guarantee that the configuration of circuit upon which the model is based is physically valid for a large-signal operation.
2. It does not mean that the measured data is used to construct RF-valid functions in the correct way, even though the element described by that function may be located correctly in a topological sense in the circuit model.

The principal provisos are in these two areas. First, the EEFET3 model uses the commonplace circuit configuration of Figure 2.1(a) and is subject to the shortcomings listed in Section 2.5.1. Second, conservative branch charges are assumed. The consequences of this particular assumption are difficult to assess without comparisons of the results with and without this assumption. Third, there is just one RF drain current function, whereas many devices exhibit a pronounced dependence of the characteristics upon starting bias point.

2.13.3 Openness of Procedure for Extracting the Model from Measurements

HP's extractor for this model is the IC-CAP modeling software, to which must be added part number HP85192B, the EEFET3, EEHEMT1 model generator. Enough information on the model and its needs has been made available, however, for those with more time than money for capital purchases to be able to extract a model conforming to the EEFET3 description without these specialist parts.

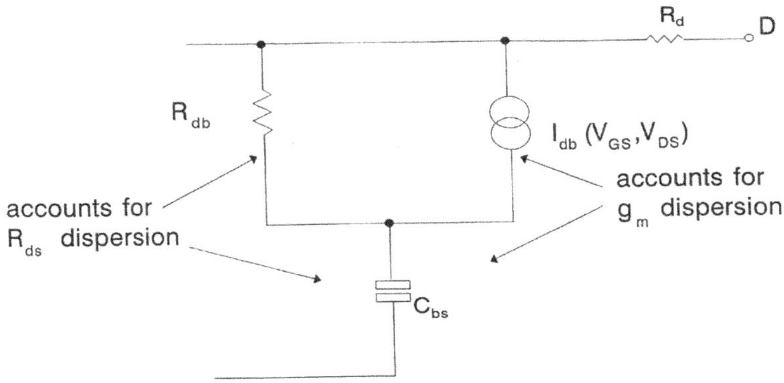

Figure 2.15 Scheme used to model dispersion in the EEFET3 model.

Purchasers of the large-signal circuit simulator LIBRA have available (in one of the manuals) the equation sets which together describe the EEFET3 model implemented in the simulator. There is no instruction given as to how to extract the model from scratch. For an experienced engineer with access to software for fitting the standard equivalent circuit to bias-dependent measured S-parameters and to general optimization tools, the equation set provides just enough information to be able to extract a model using those equations (which may, however, end up with different values for the coefficients from those which the HP extraction system would give).

Although the equations have no basis in physics insofar as they are not related in any way to the FET structure, they have a heuristic logic to them. Regions of strong curvature such as humps and bends in the fitted curves or characteristics are controlled by readily identifiable individual coefficients, essentially a logical extension of the principles surrounding the earlier models of Curtice, Materka, Statz, and colleagues. EEsof's extension of the EEFET3 model for HEMTs, EEHEMT, is reputed to have 48 fitting parameters in its equation set. However, the traceability to geometric features in the curves to be fitted means that the process of determining values for them is manageable.

It is thought that the equations extrapolate reasonably well beyond the data range.

2.14 Other Models Using the Commonplace Equivalent Circuit

There are other works worthy of note. Some are implementations of models considered above but where the practical challenges of characterization have been highlighted. Others are little more than equations for modeling drain current. The following is a short list of less well-known models that had come to attention by the late 1990s; it is far from being exhaustive.

2.14.1 Dortu-Muller Method

This work [22] centered on the use of commercially available software to extract large-signal parameters and the use of load-pull measurements to implement a version of the Materka model.

2.14.2 Rodrigues-Tellez

Taking yet another variation on the conjure-an-equation theme, Rodriguez-Tellez et al. offered the following [23]:

$$I_D = \beta\left(V_{GS} - V_P + KV_{DS}\right)^2\left(1 + \lambda V_{DS}\right)\tanh\left(\alpha V_{DS}\right)$$

$$C_{gs} = C_0\left[\frac{1 + A_2A_3 10^{A_1V_{GS}}}{1 + A_3^2 10^{A_1V_{GS}}}\right]\left(1 + A_4V_{DS}\right)$$

$$C_{gd} = C_0\left[\frac{1 + A_2A_3 10^{A_1V_{DS}}}{1 + A_3^2 10^{A_1V_{DS}}}\right]\left(1 + A_4V_{GS}\right)$$

While these equations have a pleasing symmetry, no examples are to hand of how well they fit typical empirical data.

2.14.3 Tajima

Tajima deserves credit for being one of the early contributors [24]. His expression for the drain current consists of a nested set of equations which, compared with some, is extensive.

For capacitance, Tajima et al. used the simple 1-D junction capacitance expression as used by Curtice.

2.14.4 University of Cantabria Model

Emphasis was given in [25] to measurements that allow a model to be constructed possessing the same broad attributes as observed in practice, notably rate effects governing the transition from the dc $I_D(V_{GS},V_{DS})$ characteristics to the RF $i_D(v_{GS},v_{DS})$ characteristics. Importantly, pulsed measurements are made as part of the characterization and model extraction procedure, including the bias point dependence of the pulsed characteristics that constitute the large-signal RF characteristics.

2.14.5 University College Dublin Model

This model [26] offers yet further variation on the closed-form expressions for drain current and the capacitances:

$$I_D = \beta V_{\text{eff}}^{\varphi}\tanh\left(\alpha V_{DS}\right)$$

$$\varphi = \frac{1}{1 + \mu V_{\text{eff}}V_{DS}}$$

$$V_{\text{eff}} = 0.5\left[V_{GS} - V_P + \left(V_{GS} - V_P\right)^2 + \delta^2\right]$$

$$V_P = V_{P0} - \gamma V_{DS}$$

Here, α, β, γ, δ, and μ are fitting factors, and V_{P0} is the pinch-off voltage at $V_{DS} \rightarrow 0$V.

The parameter V_{eff} provides a smooth transition of the drain current to pinch-off as V_{GS} is increased negatively. It also caters for a negative slope of I_D with respect to V_{DS} at open channel, as occurs in some FETs.

$$C_{gs,d} = \frac{C_1}{\left[1 - \frac{V_{GS,D}}{V_{B0}}\right]^m} + C_2\left[1 + \tanh a\left(V_{GS,D} - V_{P0}\right)\right]$$

V_{B0} is the built-in Schottky barrier potential. C_1, C_2, and m are fitting factors.

Particular attention is also paid to representing the bias variation of the intrinsic channel resistance with the equation:

$$R_i = R_{i0} + K_g\left[1 - \tanh a_1\left(V_{GS} - V_{P0}\right)\right] + K_d\left[1 - \tanh a_2\left(V_{DS} - V_{DS0}\right)\right]$$

where V_{DS0} is the drain voltage at which velocity saturation occurs, and R_{i0}, K_g, K_d, a_1,a_2 are yet more fitting factors.

Apart from these features, which do not confer upon the model a firm distinction from the others, the originators claim distinctiveness in the extraction, which is systematic and carefully done. Although no detail of the extraction procedure is known outside of the University of Dublin, it appears to involve no new principle or technique, and it is not based on any advanced physics, just simple physical principles. The method for all that results in very smooth functions for the extracted elements [26], even though in minor ways they are sometimes unphysical (nonzero transconductance at V_{DS} = 0V, for instance). The particular interest in this model, therefore, is in the attention paid to methods for its extraction, not just in the model itself.

2.15 The Parker-Skellern Model

Among the models of the mid-1990s, two stand out as applying thought and fresh principles beyond inventing variations of preexisting algebraic equations to improve the fit to the current-voltage characteristics of sometimes very specific FET types. One is the Parker-Skellern model [27], and the other is the Root model. Both, to a degree, started again and examined from the first principles where existing models were failing, with Parker addressing the lack of smooth continuity in the derivatives of the functions used to represent the current with respect to the terminal voltages, and Root identifying some of the potentially important physical processes occurring in real FETs that are omitted from the much-used circuit model of Figure 2.1(a).

2.15.1 Shortcomings in Previous Practice

Note that, in this section, $I_D(V_{GS},V_{DS})$ and $i_D(v_{GS},v_{DS})$ are used interchangeably as the arguments apply equally to both the dc and RF characteristics.

Parker began by observing that, for modeling third-order intermodulation and other third harmonic phenomena, the third derivative of the equation(s) used to model the $i_D(v_{GS},v_{DS})$ characteristics must not only be continuous but must also match the experimental third derivative extremely well. Further, Parker observed that past models that had attempted to use a single equation to represent $I_D(V_{GS},V_{DS})$, $i_D(v_{GS},v_{DS})$ failed in this regard. Worse, models that use different forms of equations for each of the three distinct zones of FET electrical behavior, as outlined next, have first derivatives that are discontinuous at the boundaries of the zones, leaving absolutely no hope of getting the second and third derivatives right. According to Parker, the three zones can be summarized as:

1. The controlled resistance region (below the knee) where $I_D(V_{DS})$ obeys a power law;
2. The controlled current region above the knee where $I_D(V_{DS})$ obeys a linear law;
3. The subthreshold region, where I_D exhibits an exponential dependence upon V_{GS}.

Following from these observations, Parker concluded that:

1. The likelihood is that no single algebraic expression will fit $I_D(V_{GS},V_{DS})$, $i_D(v_{GS},v_{DS})$ in all the respects required of it and its derivatives.
2. The best way to achieve different functionality appropriate to the three different regions of operation defined above is via a series of transformations of the voltage variables. Each transformation may be succeeded by another, or nested within another, to ensure continuity in the derivatives of $I_D(V_{GS},V_{DS})$, $i_D(v_{GS},v_{DS})$ with respect to both voltages to as high an order as desired.

2.15.2 Parker's Scheme: Nested Transformations

Gaining a good appreciation of Parker's approach requires several readings of his published work or, equivalently, an equally long transcription into terms with which the individual reader may be more comfortable. However, the principle may be illustrated by considering just the transformation of the actual or applied gate control voltage, here called V_{GS}, to a model control voltage V_C:

$$V_C = \sigma \ln\left[1 + \exp\left(\frac{V_{GS}}{\sigma}\right)\right]$$

The parameter σ adjusts the rate at which the model control voltage V_C reduces exponentially as pinch-off is approached: as V_{GS} is reduced (i.e., becomes more negative), the exponential becomes very small and $\ln(1 + y)$ becomes approximately equal to y. So, in the expression for V_C above, V_C becomes related to V_{GS} approximately by an exponential. When $V_{GS} \gg \sigma$, however, V_C approximately equals V_{GS}.

Now, to take the simplest possible form, if I_D is made a linear function of V_C (rather than of V_{GS}), then, as the real, practical, variable V_{GS} varies from pinch-off to open channel, the drain current will increase exponentially with V_{GS} at first and then change smoothly to a linear dependence, thereby re-creating the general form found in practice.

In the final implementation of the model, a power law dependence of I_D upon V_C is employed.

Similar considerations are applied by Parker to transforming the drain-source voltage.

2.15.3 Expressions Used for Modeling Capacitance

A modified Statz expression is used for the gate-source capacitance. C_{gd} is set to a constant. The gate-source and gate-drain capacitances interchange smoothly when the drain-source voltage changes polarity.

2.15.4 Basis and Underlying Soundness

The model is based on the commonplace circuit model of Figure 2.1(a) and consequently has the shortcomings (1) to (5) of Section 2.5.1. The intrinsic resistance R_i in series with C_{gs} is missing, as a result of which, like other models where this element is absent, an inability to fit closely to s_{11} can be expected. Of the criteria (6) through (10) of Section 2.5.1, those satisfied are (6) through (9).

Regarding points (6), (7), and (8), Parker implements rate effects not by introducing a subnetwork like those of Figures 2.1(b), 2.14, and 2.15, but by introducing an additional state variable, what he refers to as a body-bias voltage (the terminology derives from MOSFET engineering). The body-bias voltage, V_b, is a component of the gate-control voltage attributable to the drain voltage. There is assumed to be a delay in V_b responding to any change in V_{DG} via the differential equation

$$V_b(t) = V_{DG}(t) - \tau \frac{dV_b}{dt}$$

Solving this equation requires the previous state of the body-bias voltage to be stored; hence, the body bias has attributed to it the character of a state variable.

Temperature rises due to self-heating are similarly modeled by storing past values of the effective average power dissipation.

Concerning the practical criteria of Section 2.5.1, there are no simple answers. Criterion (a), whether the equations extrapolate well to regions not covered by measurement, should be satisfied because much attention is paid to getting high-order derivatives of I_D correct over the measurement range. No proof or example is available, however. As far as (b) is concerned, whether the model caters well for FETs used in all operational modes, the model would appear to be of mixed merit. Although C_{gs} and C_{gd} swap over as V_{DS} passes from positive to negative, C_{gd} does not have a dependence upon V_{DS}. Moreover, there are no resistive elements in series with the capacitances as would be necessary to model a switch FET. However, the basic nonlinearity upon which a mixer depends is well modeled by the Parker-Skellern scheme. With regard to point (c) of Section 2.5.1, measurement equipment required to put values to the model, Parker's written accounts do not make clear how the RF characteristics are to be obtained and utilized. Consultation with Parker reveals that pulsed measurements are made (but how to incorporate them into the model remains unclear).

2.15.5 Measurements Required

For the majority of FETs where there are differences between the dc and RF characteristics, pulsed measurements must be made, as discussed in the preceding section, in addition to the usual small-signal S-parameter measurements at multiple dc bias points.

2.15.6 Openness of Procedure for Extracting the Model from Measurements

Although the principles of the nested transformation of variables method have been made openly available, extraction of the model seems not to have been published as a logical, easily implemented, step-by-step procedure. Such a criticism is not reserved for the Parker-Skellern model; it applies to all the others too. However, the problem for the would-be user of the Parker-Skellern model as distinct from the others is that there are so many fitting coefficients attendant to a mathematically involved model that there is nowhere near the same visibility of how the model might be extracted as there is for the TOM, for example.

2.15.7 Scalability

The model is said to be fully scalable.

2.15.8 General Comments

In discussion with Parker in November 1994, there was an expectation that his model would be implemented in HP-MDS. Whether it is available in circuit simulation software from other U.S. vendors is not known. Note, however, that the early application was in SPICE.

2.16 The Root Model

2.16.1 Basis

D. E. Root's model is based on a fixed equivalent circuit topology with very general elements. Figure 2.16 replicates Root's own diagrammatic summary.

As far as can be told from Root's published works [28, 29], the term "general elements" appears to relate to charge and current functions treated as elements having connection to three terminals or nodes. Announcement of the model followed publication by Root and Hughes of some generic physical principles by which FETs work [28]. A reasonable assumption is that the model builds upon that physics, physics that was the first to describe with sufficient clarity to be widely understood the principle of transcapacitance and with the current flows associated with it.

The bias-dependent values of the charge and current functions are stored in tabular form, and a method of interpolating between the values based on splines is employed [29]. Thus, the model undeservedly earned itself a reputation as being a tabular model having no physical basis.

The Root model deals with transitions from dc to RF and can simulate these transitions in the time domain.

2.16.2 Underlying Soundness

A most important feature of the Root model is that it evidently synthesizes a set of RF characteristics that are dependent upon the bias point, a property of many practical FETs. As far as we are aware, at the time there were no examples on release of how well these synthesized characteristics compared with directly measured (fast-pulsed) characteristics, so no comment on their efficacy could be made. Among the commercially available models, Root's model stands alone in recognizing the bias-point dependence of the large-signal RF characteristics and in attempting to model them in other than a simplistic way (whereby "simplistic" is meant the scheme of Figure 2.1(b)).

Root began by drawing attention to the omission of transcapacitance from all previous large-signal models. His papers dealt properly with the issue of dispersion in the drain characteristics wherein, in simple terms, the slope of I_D with respect to V_{DS} increases several fold when the frequency is raised from near dc to microwave, with a transition centered on typically a few kilohertz. Although the implementation is described as a measurement-based tabular model, the tabular label probably refers only to the manner in which the values of the elements in a fundamentally based equivalent circuit are stored (rather than them being represented via simple analytic equations like all the other models). We may guess, therefore, that the model is, in fact, equivalent-circuit based, and that the circuit does not propagate some of the most serious omissions and shortfalls of the commonplace configuration of Figure 2.1(a).

One possible omission is that, according to Root in 1994, there was no built-in relationship between the conservative current and charge functions within the model.

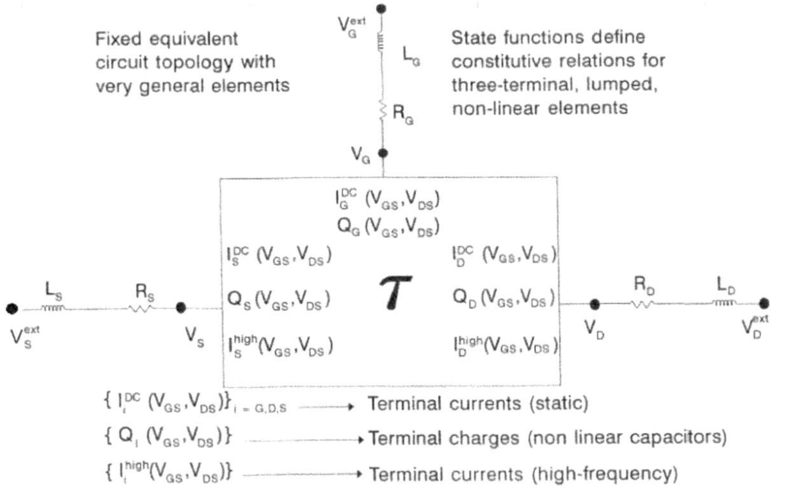

Figure 2.16 Diagram of the Root model.

Detailed analysis of the physics of FET operation suggests that there should be such a relationship. The consequences of the omission cannot be told without intimate knowledge of how the conservative charge and current functions are recovered from measurements, with examples drawn from specific devices. To take a guess at the answer, however, if the conservative charge and current functions are individually well recovered from the measurements, any physically implied relationship between them is probably obeyed closely enough for practical purposes.

A topical theme around the time these notes were originally written was whether the Root model separates the conservative depletion charge function $Q_G(V_{GS},V_{DS})$ into two nodal charges $Q_S(V_{GS},V_{DS})$ and $Q_D(V_{GS},V_{DS})$, each a stand-alone conservative function. If so, it raised the question of how such a division of the total charge could be made without violating the principle of HFET operation wherein charge is pumped from one electrode or node (e.g., the source) to another (the drain) with only the gate charge being conservative.

2.16.3 Measurements Required

The Root model uses nothing more than the dc $I_D(V_{GS},V_{DS})$ characteristics and commonplace multibias S-parameter measurement (i.e., small-signal S-parameters measured at a dc bias point and not pulsed S-parameter measurements). The series parasitic elements in the equivalent circuit of Figure 2.16 are obtained by analysis of S-parameters measured with the gate forward biased with a dc current of 200 mA/mm flowing [29].

To make the measurements required a modular dc source (HP 4142B) together with a microwave vector network analyzer (HP8510). Other equipment included an HP 700 series computer to control the instrumentation and special wideband bias tees with Kelvin probe voltage sensing. In addition, the IC-CAP software was required (HP describes this as the HP85190A Modelling Suite, which includes software for driving the instruments) and the HP85191A HP Root FET Model Generator. For someone starting from scratch, the whole amounted to a hefty capital outlay, notably in some countries outside the United States.

2.16.4 Thermal Effects

In May 1994, the following method of allowing for temperature had been implemented in HP-MDS by Root as a symbolically defined device (SDD) model [29]. It is not known whether any subsequent release of HP-EEsof's software offers a fully integrated version.

The scheme is a phenomenological approach, advocated in the absence of any detailed device physics. A simple linear temperature dependence of I_D is assumed. Recognizing that the lattice temperature of the FET in a real-world application (what Root calls the ambient "chuck" temperature T_C) is different from the conditions under which the device was measured for the purposes of extracting the FET model, T_M:

$$I_D(V_{GS},V_{DS},T_M) = I_D(V_{GS},V_{DS},T_C) + K_{ID}(V_{GS},V_{DS})(T_M - T_C)$$

The relationship between T_C and T_M is

$$T_M = T_C + P_{\text{TOT}}\left(V_{GS}, V_{DS}, T_M\right) R_M$$

where R_M is the thermal resistance of the FET in the mounting in which it was measured (or characterized). The various unknowns in these equations can be found by making measurements on a hot stage (or chuck) at two or more elevated temperatures. Similar assumptions of linear dependence upon temperature must be made for the gate current, impact ionization current, charge storage in the depletion region, and so forth, all yielding their own values for the coefficients expressing the linear dependence of each electrical quantity upon temperature.

Once all the coefficients are determined, corrections to the as-measured nonisothermal characteristics can be made to create the isothermal characteristics according to which the device is assumed to operate in the intended circuit application.

2.16.5 Openness of Procedure for Extracting the Model from Measurements

The Root model was (and maybe still is) a closed model that required the special software to construct it from measurements. The model extractor, most of what is implemented therein, was unpublished. All that is known for certain is that, if the measured S-parameters are denoted by

$$S^{\text{meas}}\left(V_{GS}, V_{DS}, \omega\right)$$

where ω represents the angular frequency and if the dc drain characteristics are denoted by

$$I_D^{\text{meas}}\left(V_{GS}, V_{DS}\right)$$

then the conservative gate charge function is calculated, or reconstructed, from both the measured functions, thus:

$$Q_G\left(V_{GS}, V_{DS}\right) = f\left(S^{\text{meas}}\left(V_{GS}, V_{DS}, \omega\right)\omega, I_D{}^{\text{meas}}\left(V_{GS}, V_{DS}\right)\right)$$

and similarly for the conservative drain current function the model requires. Here, $f(\ldots)$ signifies an entire procedure, not simply an algebraic function.

There can be little doubt that extraction of the model relies heavily on the physical analysis published by Root et al. [28, 29] but, even with this background, it is not possible to deduce how the model extraction is done. The HP8519A "HP Root FET Model Generator" is required to extract the model from measurements.

2.16.6 General Comments

The facts that the internal workings of the Root model remain hidden from would-be users, and that extra expenditure on instrumentation and extraction software is necessary to acquire the model for practical application, have meant that the uptake of the model probably has not been what it deserves to be.

There were (and maybe still are) three aspects to Root's model that engendered debate, gossip, or conjecture depending upon who was doing the talking:

1. The model is based on innovative and sound physics.
2. No pulsed measurements need to be made to ascertain what the large-signal RF characteristics of a device will be.
3. The model is tabular or based on tables of data.

And so to sources of bafflement:

a. Between (1) and (3), why was the model promoted as table-based when there is a proper basis in physics? Does a physically justifiable circuit model emerge from the background analysis and, if so, why is it not made obvious and open?
b. Between (1) and (2), why was special measurement equipment apparently not needed when, at the time, the need for pulsed $I(V)$ measurements was known to many?
c. Would greater openness have served the cause better?

2.17 The Angelov Model

This section, an addition to the foregoing notes from the late 1990s, deals with the Angelov (or Chalmers) model, which has since proved deservedly popular. It not only fits a wide variety of dc $I(V)$ and pulsed $I(V)$ characteristics, but it also has equations that nicely fit the empirical C_{gs} and C_{gd} functions [30] upon which the standard equivalent circuit model relies.

An appealing feature is the general form

$$I_D = f_G\left(V_{GS}\right) \cdot f_D\left(V_{DS}\right)$$

that is, the drain current $I_D(V_{GS},V_{DS})$ is represented by the product of two separate coordinate functions: $f_G(V_{GS})$, which is a function of V_{GS} only, and $f_D(V_{DS})$, which is a function of V_{DS} only. Although Angelov et al. were not the first to adopt such a formalism, there is a symmetry to their use of it that was logical and well explained. The same formalism is applied to modeling capacitance, a fair proposition insofar as it reflects what happens in the real device where current and charge storage are interlinked.

2.17.1 Expression Used for Modeling Current

$$I_D = I_{pk}\left(1 + \tanh(\psi)\right)\left(1 + \lambda V_{DS}\right)\tanh\left(\alpha V_{DS}\right)$$

$$\psi = \sum P_n\left(V_{GS} - V_{pk}\right)^n \quad n = 1,2,3\ldots$$

The power series in V_{GS} means that the drain current is continuously differentiable with respect to gate voltage for modeling third-order intermodulation and other harmonically related measures.

2.17.2 Expression Used for Modeling Capacitance

$$
\begin{aligned}
C_{gs} &= C_{gs0}\left[1 + \tanh(\psi_1)\right]\left[1 + \tanh(\psi_2)\right] \\
C_{gd} &= C_{gd0}\left[1 + \tanh(\psi_3)\right]\left[1 - \tanh(\psi_4)\right] \\
\psi_1 &= \sum P_i (V_{GS})^i \qquad i = 0,1,2,3\ldots \\
\psi_3 &= \sum P_j (V_{GS})^j \qquad j = 0,1,2,3\ldots \\
\psi_2 &= \sum P_k (V_{DS})^k \qquad k = 0,1,2,3\ldots \\
\psi_4 &= P_{1CC} V_{GS} V_{DS} + \sum P_l (V_{DS})^l \qquad l = 0,1,2,3\ldots
\end{aligned}
$$

where, to quote the original text, "The term $P_{1CC} V_{GS} V_{DS}$ reflects the cross-coupling of V_{GS} and V_{GD} on C_{gd}."

2.17.3 Basis

The model is based on the commonplace circuit model of Figure 2.1(a) with the foregoing equations to fit to measured dependences of the equivalent circuit elements.

2.17.4 Underlying Soundness

Because the model is based on the commonplace circuit model of Figure 2.1(a) it has some of the shortcomings of Section 2.5.1. Among the practical criteria, (a) and (c) are satisfied.

The expressions for capacitance represent well the V_{DS} dependence and the V_{GS} dependence of C_{gs} and C_{gd}. There is functional dependence on V_{DS} to model the increase in C_{gs} with V_{DS} and the crossover in the C_{gd} curves, which occurs in the vicinity of the knee, is catered for.

2.17.5 Measurements Required

The original paper [30] drew no distinction between the dc and RF $I(V)$ characteristics. In common with the majority, the model requires small-signal S-parameter measurements made about multiple dc bias points, together with the dc $I_D(V_{GS},V_{DS})$ characteristics. With the importance of pulsed characterization now widely understood, the equations of Section 2.17.1 can be applied equally well to fit the pulsed $i_D(v_{GS},v_{DS})$ set. If this is done, however, there is a problem in implementing the model in a large-signal circuit simulator that recognizes only one set of drain characteristics. One compromise solution is that shown in Figure 2.1(b).

2.17.6 Openness of Procedure for Extracting the Model from Measurements

Beyond a simple statement to the effect that the parasitic elements were extracted from the S-parameters measured at V_{DS} = 0V, [30] contained no detail about the extraction procedure, saying only that "The different parameters can as a first approximation be easily obtained by inspection ... ." Whereas such a facility is clearly an asset, it does not convey how well the model can do when effort is put into extraction. Section 2.17.8 gives a couple of examples.

2.17.7 Scalability

Being at least loosely related to the physics and construction of the practical FET, the model of Figure 2.1(a) carries with it a confidence that its elements can be scaled with gate width. For instance, the source resistance can be halved in response to a doubling of the gate width, and the gate-source capacitance can be doubled. If the limited validity of the standard model in practice is acknowledged and the same limited applicability is accepted upon scaling, then the model could be said to be scalable.

2.17.8 General Comments

As a mark of the flexibility of the Angelov model, Figure 2.17 shows a fit to dc characteristics exhibiting negative slope, a not uncommon occurrence among some device types. Importantly, the use of such a set of characteristics would be inappropriate for RF circuit simulation as the negative differential conductivity is generally not broadband; it is strictly a slow-rate phenomenon. It follows that the measurements and the fitted set have to serve only the bias calculation. That being so, the divergence

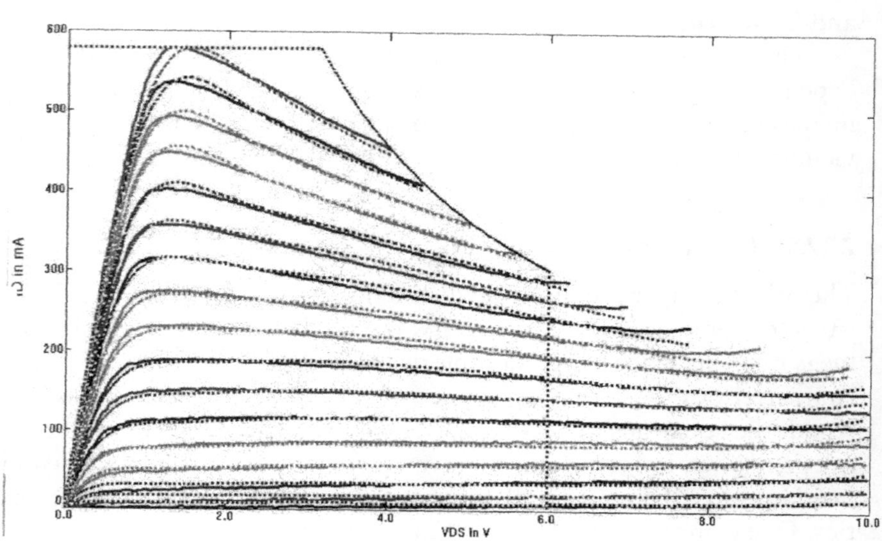

Figure 2.17 Fit of the Angelow model to measured dc characteristics (solid lines).

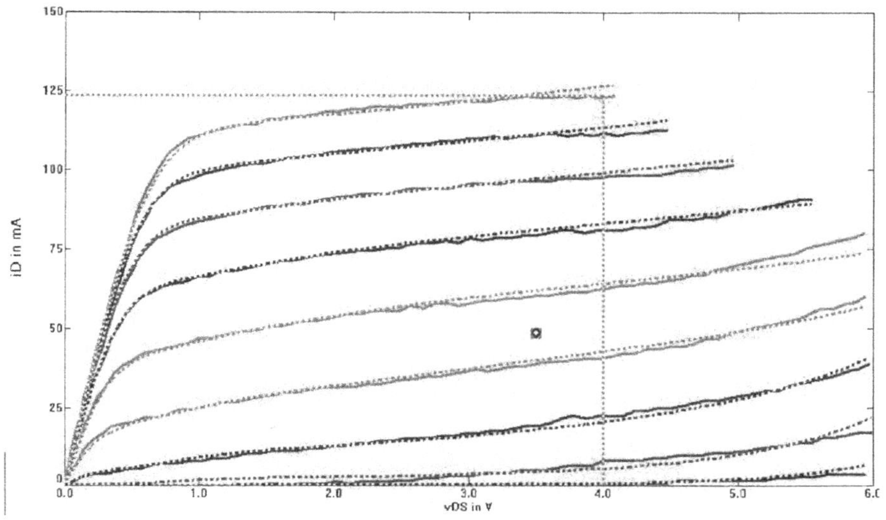

Figure 2.18 Angelov model fit to pulse-measured characteristics.

of the fit, which is minor anyway, is of no consequence; the derivatives $\partial^n I_D/\partial V_{GS}^n$ (or rather, $\partial^n i_D/\partial v_{GS}^n$ to be strictly correct) only come into play in the RF solution.

In both Figures 2.17 and 2.18, the fitted model curves are the broken lines.

An example fit to a set of fast-pulse measured characteristics is shown in Figure 2.18. The result speaks for itself.

2.18 Conclusion

From the foregoing review, the following points stand out:

1. The use of large-signal models is probably governed more by usability and simplicity than by fidelity to practice and principle. Simplicity is attractive; complexity is daunting.
2. Fidelity does not come cheaply: the models of Parker-Skellern and Root are high on principle, but they are much more difficult to understand.
3. The ease with which all the parameter values needed for a model can be recovered from measurements, the process known as measurement and extraction, can make or break the popularity of a model. Complex models have to appear open with a detailed and well-documented extraction process to compete in popularity.
4. Acceptance of a model seems to depend heavily upon it being based on an equivalent circuit. The importance of having a circuit model is that it offers the user understandable rules by which to scale the FET to different gate widths simply by multiplying or dividing the elements representing charge storage, charge flow, and energy dissipation. For the majority of users, the importance of a circuit model is unlikely to be an all-consuming interest in the detailed physical origin of its elements.

5. Although it was shown in 1987 [13] that only the gate charge is Q_G conservative, later models persisted with the idea that conservative source and drain charges Q_S,Q_D could be identified as well. Among the reasons might be: circuit simulators insist upon a charge-based formulation; simulators fail to converge given a capacitance-only model; insistence on a single conservative nodal charge is an unreasonable imposition in an already difficult circuit design field.

References

[1] Singh, R., and C. M. Snowden, "A Quasi-Two-Dimensional HEMT Model for DC and Microwave Simulation," *IEEE Transactions on Electron Devices*, Vol. ED-45, No. 6, June 1998, pp. 1165–1169.

[2] Ladbrooke, P. H., *MMIC Design: GaAs FETs and HEMTs*, Norwood, MA: Artech House, 1989.

[3] Cripps, S. C., *RF Power Amplifiers for Wireless Communications*, 2nd ed., Chapter 9, Norwood, MA: Artech House, 2006.

[4] www.modelithics.com.

[5] Ladbrooke, P. H., A. J. Hill, and J. P. Bridge, "Negative Rs and Rd in GaAs FET and HEMT Equivalent Circuits," *Electron. Lett.*, Vol. 26, No. 10, May 1, 1990, pp. 680–681.

[6] Qi, H., J. Benedikt, and P. J. Tasker, "Nonlinear Data Utilization from Direct Data Lookup to Behavioural Modelling," *IEEE Transactions on Microwave Theory and Techniques*, Vol. MTT-57, No. 6, June 2009, pp. 1425–1432.

[7] Pedro, J. C., et al., *Nonlinear Circuit Simulation and Modeling*, Cambridge, U.K.: Cambridge University Press, 2018.

[8] *Modelithics Advanced Simulation Models—Overview and Tutorial*, www.modelithics.com.

[9] www.awr.com.

[10] www.keysight.com.

[11] Aaen, P. H., J. A. Plá, and J. Wood, *Modeling and Characterization of RF and Microwave Power FETs*, Chapter 7, Cambridge, U.K.: Cambridge University Press, 2007.

[12] Anholt, R., *Electrical and Thermal Characterization of MESFETs, HEMTs and HBTs*, Norwood, MA: Artech House, 1995.

[13] Smith, I. W., et al., "On Charge Nonconservation in FETs," *IEEE Transactions on Electron Devices*, Vol. ED-34, No. 12, December 1987, pp. 2565–2568.

[14] Willing, H. A., C. Rauscher, and P. de Santis, "A Technique for Predicting Large-Signal Performance of a GaAs MESFET," *IEEE Transactions on Microwave Theory and Techniques*, Vol. MTT-26, No. 12, December 1978, pp. 1017–1923.

[15] Curtice, W. R., "A MESFET Model for Use in the Design of GaAs Integrated Circuits," *IEEE Transactions on Microwave Theory and Techniques*, Vol. MTT-28, No. 5, May 1980, pp. 448–456.

[16] Curtice, W. R., and M. Ettenberg, "A Nonlinear GaAs FET Model for Use in the Design of Output Circuits for Power Amplifiers," *IEEE Transactions on Microwave Theory and Techniques*, Vol. MTT-33, No. 12, December 1985, pp. 1383–1394.

[17] Materka, A., and T. Kacprzak, "Computer Calculation of Large-Signal GaAs FET Amplifier Characteristics," *IEEE Transactions on Microwave Theory and Techniques*, Vol. MTT-33, No. 2, February 1985, pp. 129–135.

[18] DTI LINK Project PC012 *Evaluation of Semiconductor Technologies for the Radio Functions of Personal Communications Network Applications*, U.K Department of Trade and Industry, 1992–1995.

[19] Statz, H., et al., "GaAs FET Device and Circuit Simulation in SPICE," *IEEE Transactions on Electron Devices*, Vol. ED-34, No. 2, February 1987, pp. 160–169.

[20] McCamant, A. J., G. D. McCormack, and D. H. Smith, "An Improved GaAs MESFET Model for SPICE," *IEEE Transactions on Microwave Theory and Techniques*, Vol. MTT-38, No. 6, June 1990, pp. 822–824.

[21] Roberts, G. I., "Understanding the EEFET3 Model," *The HP-EEsof High Frequency Modeling and Simulation Seminar, Digest of Technical Papers, Hewlett-Packard*, 1994, pp. 21–31.

[22] Dortu, J. M., and J. E. Muller, "Accurate Large-Signal GaAs MESFET Modeling for a Power MMIC Amplifier," *Microwave Journal*, April 1993, pp. 74–84.

[23] Rodriguez-Tellez, J., M. Al-Daas, and K. A. Mezher, "Assessment of MESFET Models for Nonlinear Microwave Circuit Design," *IEEE Circuits and Systems Conference Proceedings*, Chicago, IL, 1993, pp. 1278–1281.

[24] Tajima, Y., B. Wrona, and K. Mishima, "GaAs FET Large-Signal Model and Its Application to Circuit Designs," *IEEE Transactions on Electron Devices*, Vol. ED-28, No. 2, February 1981, pp. 171–175.

[25] Mediavilla, A., A. Tazon, and J. L. Garcia, "Measurement and Non-Linear Modelling of Power MESFETs," *Microwave Eng. Europe*, August-September 1993, pp. 45–54.

[26] Cojocaru, V. I., and T. J. Brazil, "A Large-Signal Model for GaAs MESFETs and HEMTs Valid at Multiple dc Bias Points," *European Gallium Arsenide and Related III-V Compounds Applications Symposium Proceedings*, Torino, Italy, April 28–30, 1994, pp. 419–422.

[27] Parker, A. E., and D. J. Skellern, "A Realistic Large-Signal MESFET Model," *Asia Pacific Microwave Conference*, Tokyo, December 6–9, 1994.

[28] Root, D. E., and B. Hughes, "Principles of Nonlinear Active Device Modeling for Circuit Simulation," *32nd IEEE MTT ARFTG Conference Proceedings*, Tempe, AZ, December 1988, pp. 3–24.

[29] Root, D. E., "Measurement-Based Active Device Modelling for Circuit Simulation," *The HP-EEsof High-Frequency Modelling and Simulation Seminar, Digest of Technical Papers, Hewlett-Packard*, 1994, pp. 33–41.

[30] Angelov, I., H. Zirath, and N. Rorsman, "A New Empirical Nonlinear Model for HEMT and MESFET Devices," *IEEE Transactions on Microwave Theory and Techniques*, Vol. MTT-40, No. 12, December 1992, pp. 2258–2266.

CHAPTER 3

Practical Behavior of FETs

Compound semiconductor FETs and HEMTs can be so complex in their behavior that they confound as semiconductor devices. In this chapter, we review why.

To begin with, what might reasonably be expected to be the simplest of properties, the current-voltage or $I(V)$ characteristics, turn out to be far from simple. Aside from being nonlinear, they appear different when measured slowly (i.e., at dc) from when they are measured quickly, a phenomenon known as frequency dispersion. When measured at microwave frequencies, HFETs are found to possess charge storage and power-dissipative properties, or capacitance and resistance in circuit terms, which also are highly nonlinear.

To one degree or another, all transistors exhibit dispersion. Dispersion refers to the difference that exists between the large-signal characteristics measured at dc and those measured in a much shorter time, typically submicroseconds (for the time being, we use the term "large-signal characteristics" in this strictly limited sense). To distinguish one from the other, at dc the relationship is written as $I(V)$, whereas the dynamic relationship is written as $i(v)$. As will described in Chapter 6, for many years, instrumentation has been available for measuring not only the dynamic $i_D(v_{GS}, v_{DS})$ properties of field-effect transistors but also the transients governing their decay to the dc $I(V)$ set, loosely termed extended or long-term memory effects, with the qualifiers often omitted to just memory effects.

When it comes to microwave properties, the main problem is how to put what is measured into a form useful for circuit design and analysis. One way is to use the measurements themselves, often scattering or S-parameters. That can work well enough for uncomplicated linear circuits, but less well for nonlinear circuits. Instead, what would be nice to have is a model of some sort, particularly a network of recognizable circuit elements: capacitors, resistors, inductors, voltage generators, current generators, and so on. Only then does the realization dawn that such a quest brings an entirely new set of challenges: how accurate must the measurements be, how we figure out a network of elements that is truly equivalent, and what functional forms these circuit elements need to take.

3.1 dc $I(V)$, Dynamic $I(V)$, and RF Properties

In this section, we review evidence that FETs and HEMTs generally possess current-voltage characteristics that are different depending upon the speed with which they are measured.

3.1.1 Example Differences Between dc $I(V)$ and Dynamic $i(v)$

Figure 3.1 is a typical result, in this instance, for a GaN FET. The set of characteristics plotted as solid lines are the dc $I(V)$, while the dotted lines are the dynamic set measured with 200 ns pulses about the point marked by a diamond symbol (except that only half the diamond symbol is visible at 25V on the horizontal axis). The dynamic or instantaneous set of characteristics is not unique but depends upon the starting or quiescent point about which the characteristics are measured. Consequently, a nonlinear model has to incorporate such dependence for analysis and design in a circuit simulator.

For both the dc and pulse-measured set the voltage sweep in Figure 3.1 has been limited to avoid degradation of the device by being overstressed, notably by self-heating at dc.

The 200 ns pulsed and dc characteristics in Figure 3.2 are for an early SiC FET, where again the heavy diamond symbol marks the quiescent point about which the dynamic set of measurements was made. To prevent overheating, a cutoff power limit was set in the instrument for the dc voltage sweep. Compared with the GaN device, there is pronounced dispersion. General conclusions as to the relative merits of the two can only tentatively be drawn as the differences are technology-dependent, and they vary from manufacturer to manufacturer (and often within a single manufacturer).

Following the foregoing brief introduction to the large-signal current-voltage characteristics of compound semiconductor HFETs, there are nuances worth pursuing.

First, with reference to Figure 3.3, the dc and RF curves should coincide at the quiescent point, the starting point from which the pulses emerge (see Section 6.2), as marked by the + symbol. If they do not, it means there is something amiss with the measurement setup [1].

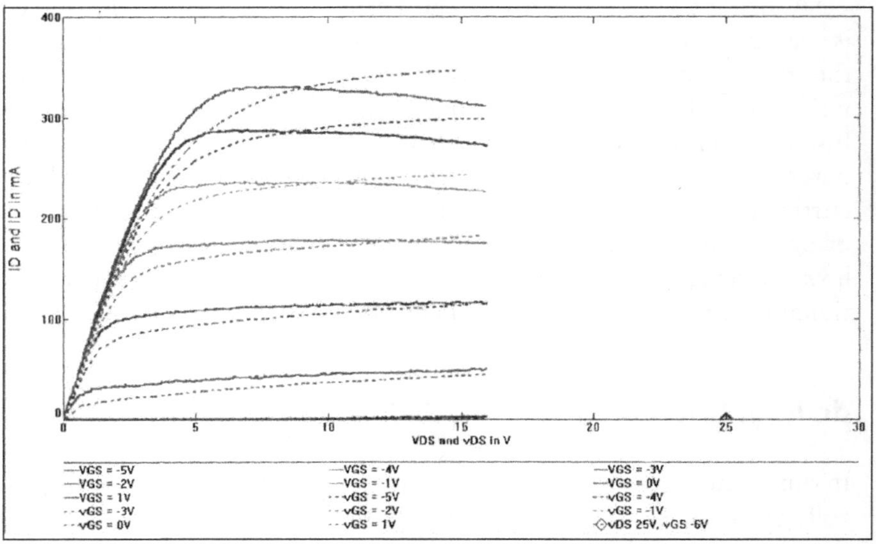

Figure 3.1 GaN HFET dc characteristics (solid lines) compared with pulsed characteristics measured about the point V_{DS} = 25V, I_D ~ 0 mA. Pairs of curves are for V_{GS} = +1V, 0V, −1V, −2V, −3V, −4V, and −5V.

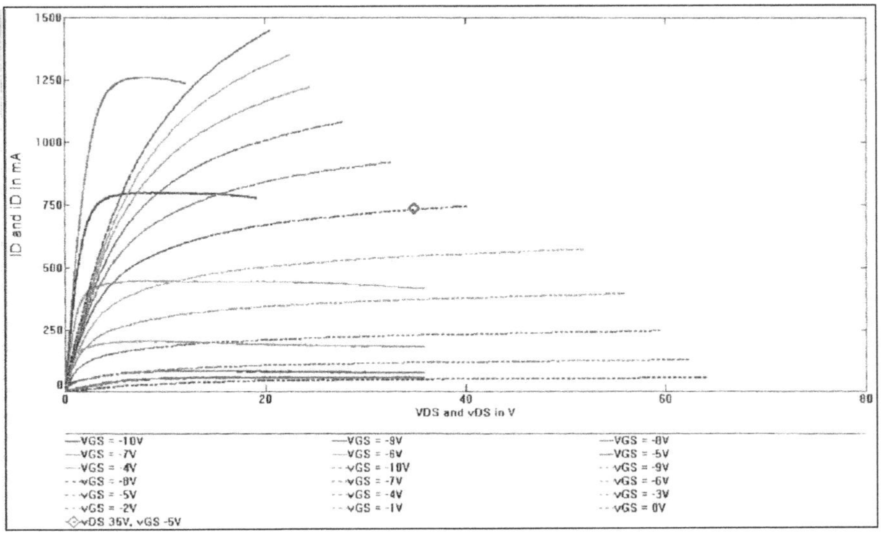

Figure 3.2 Pulsed and dc characteristics of an early SiC FET. From top to bottom, the dc curves run from −4V to −10V in 1 V steps, and the pulsed set run from 0V to −10V in 1 V steps.

Second, clues as to the physical source of dispersion can be found using the test instrument. In Figure 3.4, for instance, there are two sets of fast pulse curves belonging to an ion-implanted GaAs FET cell, each measured from an entirely different quiescent point (both marked with + symbols). Despite the power dissipation being different in the two cases, and therefore the degree of self-heating being different similarly, the two sets of dynamic characteristics are the same.

The clue thus provided is that charge exchange with deep levels located somewhere within the device is the likely cause of the dispersion. Looking again at Figure 3.3, the second fact to note is that there is a near-vertical line where the dc and pulsed characteristics intersect from which the two quiescent points in Figure 3.4 have been chosen. By implication, whatever it is that is causing the dispersion therefore appears to be set or determined by the drain-source voltage. More critical inspection of Figure 3.3 reveals the crossing trajectory to follow a line of constant V_{DG}, strongly suggesting a dispersive mechanism operating in the gate-drain space. Charge moving in and out of surface states, driven by the intense electric field at the drain-edge of the gate, is the obvious contender.

The existence of a line of intersection where the dc and instantaneous curves cross is an eventuality which we revisit in Section 10.2.

Dispersion is present in the overwhelming majority of devices investigated, although the degree varies even among devices of the same nominal type from different fabrication houses.

Outliers are always fascinating. Figure 3.5 is a rarity. It shows the dc characteristics of an example from a PHEMT process together with the device's pulsed characteristics measured about the point +. It is impressive for being dispersionless, a result that applies for a wide range of starting bias points.

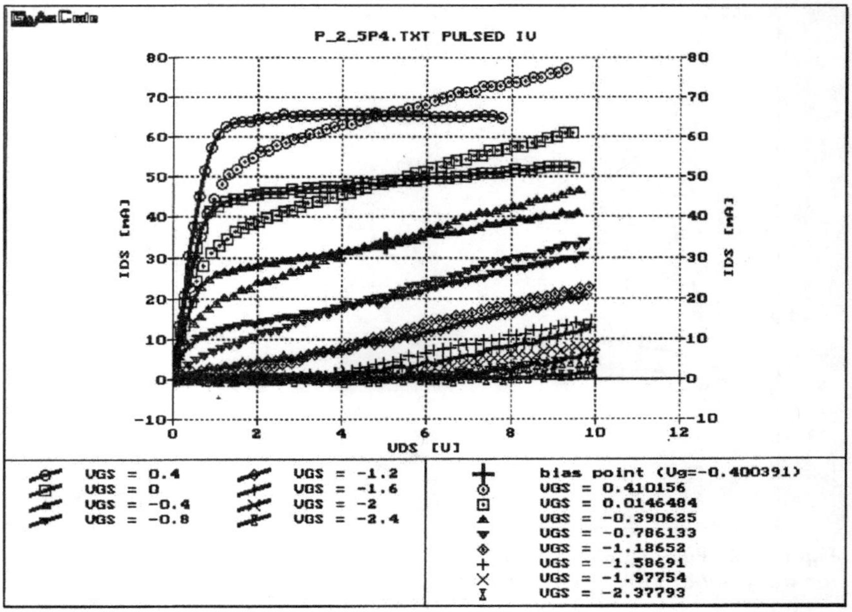

Figure 3.3 Illustrating the points of intersection between the dc and RF curves both above and below the quiescent point +.

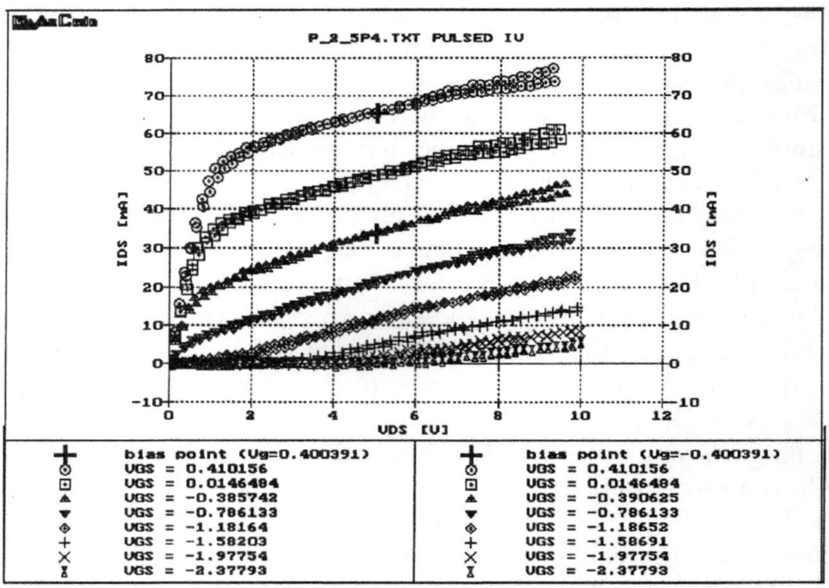

Figure 3.4 Demonstrating coincidence of two pulsed *i*(*v*) sets of characteristics despite the difference in bias points.

At the other extreme is the outlier of Figure 3.6, a FET cell intended as the basis of a larger device for a power amplifier operating in Class B. The bias point, which is indistinctly marked, is at $V_{DS} = 7V$, $I_D \sim 0$ mA. Technologically, the fault is believed to have arisen from an inappropriate gate recess structure and surface treatment.

In summary, the consequences of a FET being dispersive are twofold: there are implications for circuit design, and there is a diagnostic need to work backwards to try and identify the material or technological origin of dispersive behavior. Both are served by a pulsed-measurement instrument or a dynamic $i(v)$ analyzer. For circuit analysis and design, the dc $I(V)$ are required by a circuit simulator to establish the bias condition, followed by the dynamic $i(v)$ characteristics along which the device will work in an RF circuit starting from that bias or quiescent point. Many sets of $i(v)$ characteristics may have to be measured, one set for each bias point at which the device could settle.

Finally, Figure 3.7 shows the current measured for a FET gate driven into forward conduction. Unsurprisingly, perhaps, it is a function of both V_{GS} and V_{DS}.

In the lower part of Figure 3.7, on the vertical axis where V_{DS} is zero, the gate current has a diode-like dependence upon V_{GS}. Focusing upon the topmost curve, I_G decreases with increases in V_{DS}. Increasing V_{DS} has the effect of reducing the forward bias on the Schottky barrier toward the drain end of the gate, so the conduction current there decreases correspondingly. Current is now distributed nonuniformly along the gate, being smaller at the drain end and larger at the source end where the current density (per unit length) near enough retains its original value. Overall, the result is a drop in the average conduction current. Exactly the same applies to the remaining curves. The upper part of Figure 3.7 puts these findings into a perhaps more familiar form.

Either way, it is clear that gate conduction current cannot be accounted for in a model by a simple diode across the gate-source branch or even by two diodes as is sometimes proposed (the second across the gate-drain branch). In their place, two current generators are required, both being functions of (V_{GS}, V_{DS}) to give the correct division of I_G between the source and the drain. Happily, in many devices, I_G is only weakly dispersive.

3.1.2 Breakdown Different at RF from dc

An early manifestation of dispersion was that the maximum RF power measured for GaAs FETs was less, and in some cases much less, than the power estimated from the product $I_{D\max}$ and $V_{DS\max}$ evident from the dc $I(V)$ characteristics. It took much effort by semiconductor technologists and device physicists to find out why and to render the problem manageable.

The ideas in this book grew out of a program of research, begun in 1978, to find out why the RF power available from GaAs FETs often fell short of the estimate available from the dc $I(V)$ characteristics, namely,

$$P_0 = \frac{I_F V_{DSB}}{8}$$

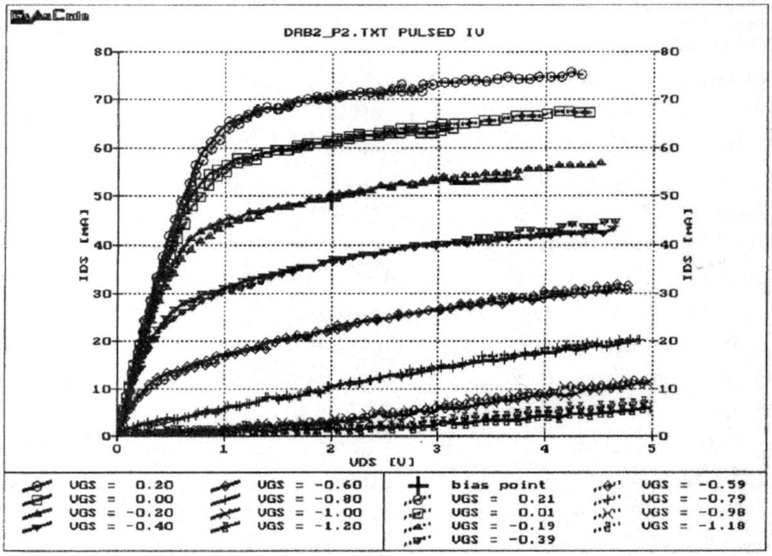

Figure 3.5 A dispersionless PHEMT.

where I_F is the maximum drain current that the FET can pass when the gate is driven positively, and V_{DSB} is the drain-source voltage at which breakdown occurs.

The factor of 8 in the denominator is to convert peak-to-peak electrical quantities, assumed sinusoidal, to rms.

Much of the problem was found to lie with V_{DSB}. In the first place, breakdown turned out not to be sudden and catastrophic but soft, where instead of the remaining cutoff for gate-source voltages at or beyond pinch-off, the device would start

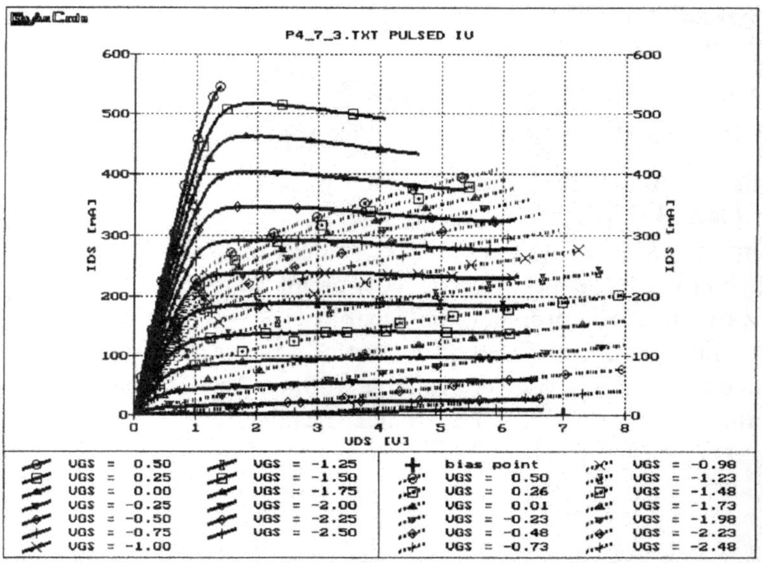

Figure 3.6 RF current collapse in a GaAs FET cell when biased for Class B operation.

to pass current that increased with V_{DS}, thereby compromising the current swing available. It soon became clear that soft breakdown was (and is) a property dependent on the rate at which it is measured, the mechanism being self-regulating impact ionization or self-regulating avalanche. Once it was found that under dynamic conditions the self-regulating avalanche sets in at lower voltages than at dc, the power output discrepancy became better understood. What was causing it took man-years of effort to determine.[1]

[1] Apart from work conducted by others in the United States, Japan, and Europe, personal involvement entailed work done at Plessey Research (Caswell) Ltd., University of New South Wales, Sydney, Australia, Telecom Research Laboratories, Melbourne, Australia, Cornell University, Ithaca, United States, and GEC Research, Wembley, England.

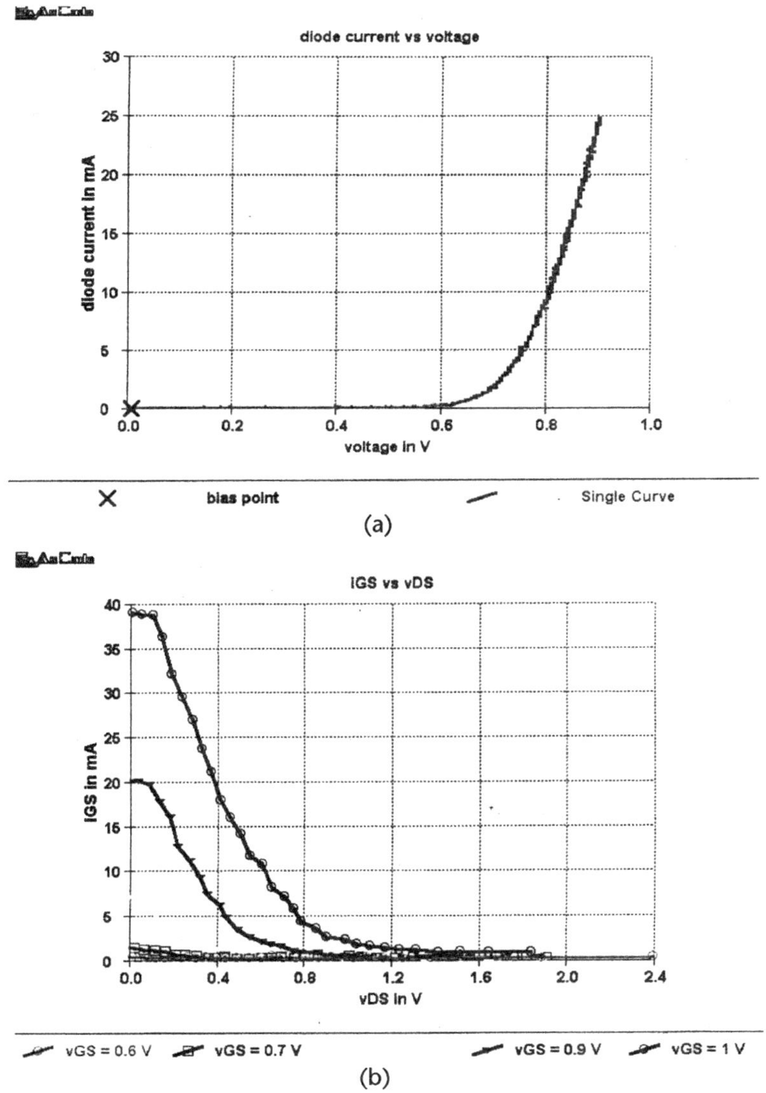

Figure 3.7 Gate conduction characteristic of a GaAs FET cell.

Quite by accident, an errant device provided an early clue. A 0.8 μm gate length FET was able to sustain 82 V dc applied to the drain without ill-effect, well over four times the expected breakdown voltage (see Figure 3.8, reproduced from Figure 6.61 of [2]).

When the manually applied voltage was increased quickly, the drain current would start to appear at voltages a fraction of 82V. Oddly, if further increase in the drain-source voltage was halted as soon as the low-level avalanche current started to appear, the current would gradually relax and disappear, a procedure that could be repeated by increasing the voltage in increments of a few volts all the way up to 82V. Small steps are just discernible in the trace of Figure 3.8 where the sweep was temporarily suspended while the current decayed.

A very high density of surface states in the gate-drain region was found to be responsible for this rather odd behavior. Such surface states capture and hold electrons emitted from the gate metal when avalanche is initiated, thereby increasing the negative charge in the gate-drain region. A negative charge held in surface states offsets the positive ionic charge in the depletion layer extension, leading to a weakening of the intense electric field at the gate corner. At each stopping point, the surface states slowly charged, thus suppressing the very mechanism giving rise to the excess electron current in the first place. Thus, each step of a few volts progressively deferred the onset of a soft breakdown. At RF the device was useless because the surface states could not respond in the same way.

Follow-on experiments examined the influence of secondary electrodes on breakdown as in Figure 3.9. Just as a negative charge held in surface states provides an alternative to the gate on which electric field can terminate, so too does the secondary electrode. Hence, a dual-gate FET, with the drain-side second electrode held at a negative potential, is a logical means of increasing the gate-drain breakdown voltage for FET power.

If deep levels of any kind, either surface or bulk, are responsible for such odd effects in the breakdown properties, then it is reasonable to expect the self-regulating avalanche characteristics to be different when measured with different length pulses. Early proof is due to Barton, whose results are reproduced in Figure 3.10 [3].

In these experiments, the gate pulse was applied just before the drain pulse, being sufficiently negative to ensure the channel was well pinched off. When current began to flow in response to increases in the amplitude of the drain pulse, simultaneous monitoring of the drain and gate currents showed them to be identical, confirming the phenomenon is confined to the gate-drain region as predicted by computer simulation.

For these experiments, Barton used a Mitsubishi small-signal FET of 1 μm gate length mounted in a microwave test fixture. The whole assembly could be cooled in a cryostat to investigate the temperature dependence of the transient implied by the pulse-length dependence evident in Figure 3.10. Lowering the temperature lengthened the transient, thereby demonstrating that breakdown is associated with a thermally activated process of some sort.

Few cases are so extreme as that in Figure 3.8. Irrespective, historical findings like these remain relevant to diagnosing faults in contemporary high-power HFETs in the newer material systems, GaN and SiC. Uncontrolled material defects, like

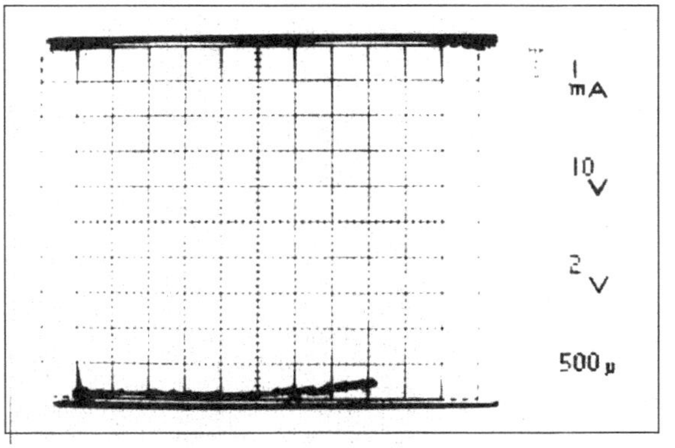

Figure 3.8 Slowly applied drain sweep showing 82V applied between drain and source of a 0.8 μm gate length GaAs MESFET. (*From:*[2].©1989 Artech House. Used with permission.)

surface states specifically and interface states generally, can be unstable and so become reliability and outright failure issues.

3.1.3 Memory Effects: Surface States, Deep Levels, and Self-Heating

"Memory" refers to a condition whereby the state of a device is dictated not just by the immediate present but also by what has gone on in the past. A simple example is when, after operating at high power, the driving voltage is reduced to small-signal while the FET is still hot. The small-signal behavior of the FET will be different from what it would have been if the chip temperature had been ambient. What the device is doing now depends upon what was happening to it before.

In power FETs fabricated in silicon, memory arises from the self-heating that a device experiences in operation. In compound semiconductor FETs of all types, more complex physical processes occur as well, generally described as charge-exchange with deep levels, or traps. Crystal lattice imperfections, discontinuities at the interfaces between substrates and active layers, between the active and passivating layers, and at free surfaces, create states that can capture then release electrons on timescales that are slow, meaning with characteristic times typically ranging from seconds to submicroseconds.

To grasp the effect on terminal characteristics, imagine a device to be switched instantaneously from one operating state to another by applying different voltages in the form of flat-topped pulses. The drain current typically will respond immediately and then relax to a different value. Precisely this kind of behavior is evident in Figure 3.1, where for the topmost curve (V_{GS} = +1V) at V_{DS} = 15V the current relaxes from around 350 mA to about 310 mA. We could hope, falsely as it turns out, that the decay would be something like that in Figure 3.11, a simple exponential (the figure is from a different device).

Looking again at the topmost curve in Figure 3.1 but this time at V_{DS} = 5V, the instantaneous current is lower than the elapsed time value, so the current transient will be the other way up. Apart from having to think about how to handle these effects in a circuit simulator, we need also to unravel how deep levels affect the

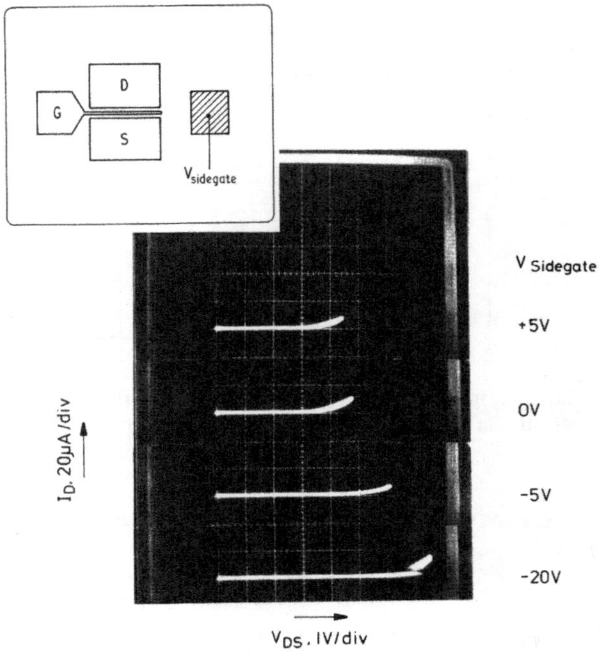

Figure 3.9 Effect of side gate potential on gate-drain breakdown. (*From:* [2], Figure 6.58].)

physical processes going on in the FET to give rise to different currents in the first place (broadly as in Section 10.2).

A transient like that in Figure 3.11 can often be followed by extended time changes as different long-term mechanisms come into play. To add to the puzzle, different mechanisms interact giving rise to transients that are, from beginning to end, far from simple.

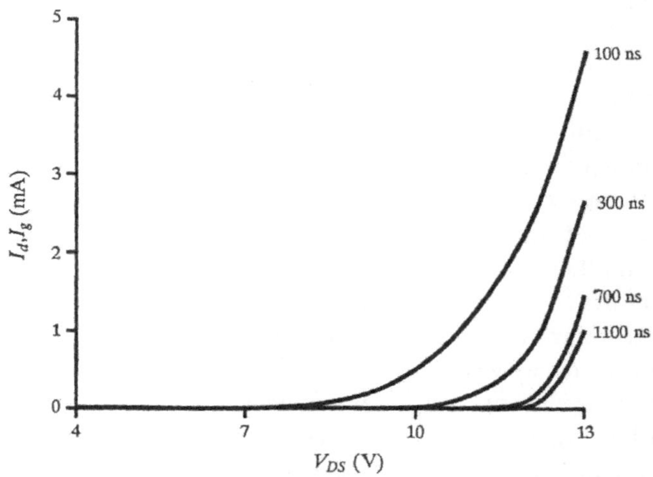

Figure 3.10 Illustrating the dependence of soft breakdown on pulse length. (*From:* [3]. Reprinted with permission.)

At the present introductory level, it is sufficient to note that long-term transient effects exist (a more detailed account is given in Section 6.3). Their importance lies in the consequences for some circuit performance attributes including amplitude modulation-to-amplitude modulation (AM-to-AM), amplitude modulation-to-phase modulation (AM-to-PM), adjacent channel power ratio (ACPR), and third-order intermodulation (IM3). In the present work, the effect of long-term transients on circuit behavior has been investigated at a preparatory level only, with simple (exponential) time constants to gain a first impression of how such effects can be dealt with in a circuit simulator. Finding a comprehensive and durable solution was, for the time, a challenge for the future and maybe it remains so.

3.1.4 S-Parameters: dc Bias and Pulsed Bias

Scattering parameters, often shortened to S-parameters and occasionally to simply S2P (signifying S-parameters, 2 port), are four parameters that characterize the RF or microwave properties of a linear device or network terminated in a known impedance, usually 50Ω. Each parameter is complex, consisting of a real part and an imaginary part, or magnitude and phase, whichever is preferred. For linear characterization, an active device must be held at fixed bias, while the small-signal microwave excitation is applied. Nonlinear FET models centered on an equivalent circuit rely heavily on S-parameters acquired at multiple bias points right cross the $I(V)$ plane even though they are, to emphasize the point, linear measures. We shall defer until Chapter 4 the question of whether multibias S-parameters are in principle valid as a basis for a nonlinear equivalent circuit and concentrate instead on the quality desired.

Although this is not a book about RF measurements per se, the omission in no way diminishes their importance: they are crucial to what we need to carry out. A readable account of the instrumentation and techniques involved in S2P measurements has been given by Aaen et al. [4].

To ally the microwave properties of FETs with the dispersive behavior described in Section 3.1.1, it follows that the S2P measured for any given values of gate-source voltage and gate-drain voltage stand to be different at dc from those measured under dynamic conditions. For any voltages away from the bias point (V_{GSB},V_{DSB}), we have already seen that the instantaneous value of the drain current, i_D, generally differs from the dc value, I_D. It implies that the depletion layer and associated electric fields are different in shape and magnitude in the two cases, so it is reasonable to expect the RF properties to be different too.

Presuming (still) that S-parameters have a role in constructing a nonlinear model, the question arises as to which should be used, the multibias dc S2P or the dynamic variants and, whatever the answer, is it practicable?

Assuming the FET to be operating at frequencies well above those at which the long-term memory mechanisms can respond, the state of the device is set by the dc condition (V_{GSB},V_{DSB}): the bias or quiescent point. From there, large signals with instantaneous values (v_{GS},v_{DS}) swing the operating point at any instant well away from (V_{GSB},V_{DSB}), meaning the device's behavior is now determined by four voltages, (V_{GSB},V_{DSB},v_{GS},v_{DS}). At each instant, the local S-parameters dictate what the

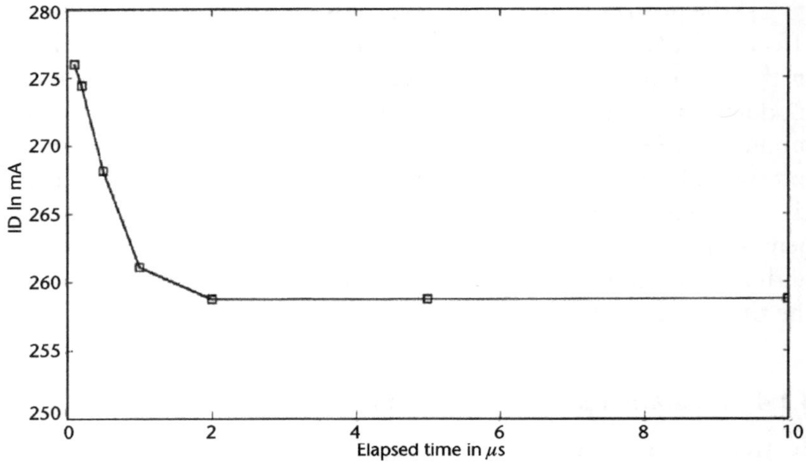

Figure 3.11 Transient in drain current due to deep levels.

device will do. In other words, ideally, it is the pulse-biased S-parameters that we should be using, what might be called a dynamic characterization.

Herein lies a difficulty: for different bias points, there is a different set of pulsed S-parameters, which means that we should have to develop an equivalent circuit model specific to each conceivable dc bias point, a prospect that is undesirable if not verging on the unmanageable. We are thus faced with having to rely on the multibias dc S-parameters, of which there is just one set. In so doing, we have to accept two shortcomings of principle:

1. It is a static, not dynamic, characterization.
2. It is a multibias linear characterization, whereas what we really need is a full nonlinear scheme.

How much these shortcomings of principle matter in practice can only be told by trial. The evidence of Chapter 11 suggests that multibias dc S2P are, contrary to worst fears, an adequate basis for a nonlinear model.

3.1.5 Device-to-Device Variations

Reference [5] was much concerned with how production variations affect the yield of linear circuits. That issue remains for nonlinear circuits, but the problem is at a higher level of difficulty. For linear circuits, it is only necessary to construct a model at a predetermined bias point of how the FET's RF characteristics vary due to variations in its material and technological makeup. The task is made comparatively easy because in most, if not all, linear circuits, the devices are operating in the velocity-saturated regime where the physics is relatively straightforward. For nonlinear circuits, however, any model must cover the whole $I(V)$ plane, which requires an all-encompassing device physics and a far more sophisticated device simulator. As will be argued in Chapter 16, such fast-executing simulators have shown promise but not attained the maturity required to rely on them.

In the absence of a yield-prediction scheme akin to that for linear circuits, the only dependable way is to extract a model for a denumerable number of nominally identical devices and to simulate the behavior of the complete circuit for each in turn. That will only give good answers, though, if the model for every device is extracted with complete precision. Were that not so, some of the spread in the final circuit performance would be due to model inaccuracies and not to genuine device-to-device variations.

Not all large-signal design is concerned with production spreads. Regardless, little is lost and much is gained by achieving the best accuracy possible at each stage of building a model. Our focus, therefore, is on accuracy and precision as opposed to calculated compromise as may be justified, for instance, to ensure ready workability in a simulator. That said, starting from a principled base, compromise is always possible in the interests of ease of model extraction, simulator run time, or other foreseeable benefit.

3.2 Bias Dependence of the Elements

From its beginning in 1978, the practice of basing analysis and design of circuits on an equivalent circuit model of compound semiconductor FETs has persisted. It has relied on the fundamental elements of resistance, conductance, and capacitance in the equivalent circuit being dependent upon the voltages that exist at the terminals of the device, voltages that naturally vary by virtue of bias condition and applied signals. When the amplitude of the signals becomes large, the circuit elements vary nonlinearly, and that raises a problem: how to make measurements that encapsulate the nonlinearities to befit the model for any application.

Until comparatively recently, the answer was that it could not be done. Arguably, even now true nonlinear characterization of universal applicability proves elusive; it tends to be restricted to circuit applications akin to those conditions under which the measurements were made. For good reason, then, overwhelmingly the nonlinear effort has been based on bias-dependent linear characterization. With prolonged usage has come the push to ever more demanding applications, necessitating improvements to the basic method with varying degrees of success.

Not only because of its expansive history but also because the approach underlies the developments now under discussion, the remainder of this chapter considers what to expect from bias-dependent linear characterization.

3.2.1 Common Practice: The Beginning with Rauscher and Willing

Credit for inaugurating the equivalent circuit approach to nonlinear modeling is normally given to Willing, Rauscher, and de Santis [6]. It may also be that the ever-present claim to accuracy began with them. Their article described a scheme based on dc measurements and small-signal multibias S-parameters to impart to the elements in an equivalent circuit a voltage dependence that would then account for the nonlinear behavior of GaAs MESFETs in circuits. Importantly, deriving the elements and their voltage dependence from small-signal measurements amounts to a bias-dependent linear characterization that assumes terms like $i\delta R$ and $v\delta C$ are small and can be ignored. Rauscher and Willing referred to their model as quasi-static.

Various plausibility arguments of how the basic equivalent circuit arises have surfaced over the years. Perhaps the most useful ab initio [7] is that shown in Figure 3.12.

What is commonly called the standard or commonplace equivalent circuit, then, is simply obtained by abstracting a flat network from Figure 3.12. It appears in Section 3.2.3 with elements relabeled with the symbols catalogued in the List of Symbols.

The central objective of [2] was to connect each element with a definable physical process occurring within the device, thereby allowing the RF behavior of an HFET to be related by understandable and practically workable device physics to its material and technological makeup. If that could be done, it would allow whole circuits to be designed from the ground up for manufacturability, at least for small-signal circuits.

Looking ahead, however, that work revealed conceptual difficulties with some of the elements that could compromise the model's use for nonlinear applications. Irrespective, it retained and proved its usefulness for yield-driven design of small-signal circuits, including the examples given in Chapter 1.

Reservations aside as to authenticity for nonlinear applications, given the enduring popularity of the model, it behooves us to do the best possible job of recovering the voltage dependence of the elements, starting with the bias-dependent, small-signal S-parameters.

3.2.2 Fitting to S-Parameters: Examples

Almost by definition, an equivalent circuit attempts to represent in the form of an electrical network the processes that occur physically in an actual device, for instance, charge storage, power dissipation, and conductance. If it is well configured, how these circuit elements vary with bias can tell us a great deal about the nature of those physical processes.

The bias dependence of the elements can be found by matching the S-parameters of the equivalent circuit to measured multibias S-parameters, a procedure we call fitting. (All the fits in this book were done with FETLink, a commercial product of GaAsCode Ltd. The workings of the fitter are described in Section 9.2.1.) By virtue of its ubiquitousness, the commonplace or standard equivalent circuit serves the need well. As will be discussed in the next section, although there are deficiencies in this standard model topology that are bothersome for nonlinear applications, they are of comparatively minor consequence for linear use. As we are concerned with fitting to multibias S-parameters, which are linear measures, the commonplace equivalent circuit can hardly be bettered.

We need the best possible rendition of how the elements in the equivalent circuit behave. To achieve it, we should accept compromise at no stage, so:

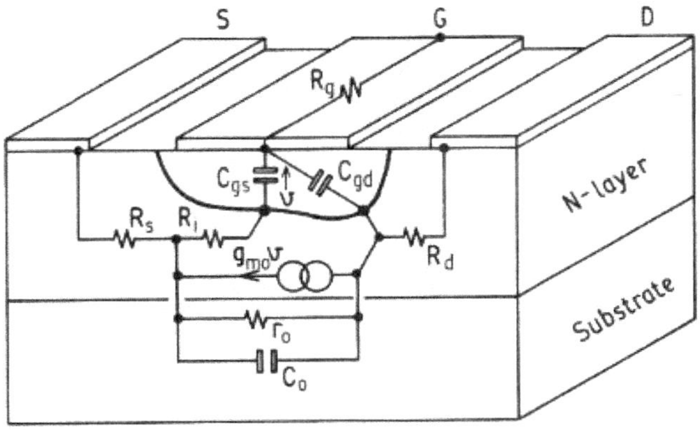

Figure 3.12 Illustrating qualitative origin of equivalent circuit elements for a GaAs MESFET. (*From:* [2].©1989 Artech House. Used with permission.) Note that the symbols for the elements in some cases differ from those used in the text.

- The network analyzer used for measuring the S-parameters must be well calibrated and the measurements themselves must be properly de-embedded to leave the S2P of the FET without parasitics.
- Fitting the equivalent circuit to the device's S-parameters must be exact.

In judging closeness of fit, plots on Cartesian coordinates are preferable to polar coordinates despite requiring eight separate frames to display the amplitudes and phases of the four S-parameters as opposed to the four frames of polar plots. In the latter, errors in magnitude and phase may not be obvious because they can compensate one another over much of their range, the only clue being that the ends of the measured and fitted plot lines do not coincide.

To illustrate the closeness of fit desired, we make use of results from a packaged device, not just a chip, the purpose being to illustrate that the methods in this book are not restricted to chips. Subject to the device size limits outlined in Section 1.3.3, they can equally well be applied to developing nonlinear models of complete packaged HFETs.

Figure 3.13 shows the fit ultimately achieved on s_{11}, both magnitude and phase. It must be stressed that, for a packaged device, the quality of fit does not begin thus but is the aim of several to-and-from iterations as outlined below. Fits of this quality are the yardstick for when a satisfactory solution has been reached. The other three S-parameters must conform to the same standard, as in Figures 3.14 through 3.16.

By an iterative procedure of fit and strip, an equivalent circuit for the package is developed concurrently with the equivalent circuit for the chip within [8]. A satisfactory overall solution also requires the S-parameters at the ports of the chip to match those of the chip equivalent circuit with agreement comparable to that in the figures.

Reference [8] is a detailed account of how the fit-and-strip scheme works. Briefly, it relies on the equivalent circuit fitter with a reverse modeling tool described in Section 9.2.1 for the chip. Working progressively from the outside in, a model for the package is evolved in stages. At each step, a best estimate but temporary equivalent circuit is fitted to the chip after stripping off the last iteration for the package elements. In the early stages, the fit at chip level is poor, meaning that the reduced set of S-parameters do not (yet) belong to a FET. The procedure of fit and strip is repeated until the reverse modeling tool finds a reduced set of S2P consistent with the S-parameters of a physical HFET.

An essential feature of the method is that the network for the package is of fixed topology, so the foregoing procedure only decides the values of the elements within it and not the topology (there would be too many degrees of freedom). Therefore, the network configuration has to be carefully thought out in the first place.

Figure 3.17, with values now shown for the present example, is typical of a package model topology. Once values have been assigned to its elements, the package can be stripped from multibias S-parameter measurements to leave a reduced set valid at the ports of the FET chip, from which a nonlinear model can eventually be developed.

For the present, at the single bias point for which Figures 3.13 to 3.16 apply, the equivalent circuit for the HEMT chip is as in Figure 3.18.

As a final check, the reverse-modeled gate length has come out at 0.256 μm, which is in agreement with the device manufacturer's nominal value.

Lastly, it must be stressed that the iterative procedure followed for a packaged device is unnecessary for cases where the measured S2P are already referenced to the two ports of the chip itself. Fitting to a single set of S-parameters then takes typically a few seconds only; the fitter used throughout was designed specifically to fit the standard model of Figure 3.18 at the chip level.

3.2.3 The Commonplace Model

Coverage of the standard or commonplace model at this juncture is mainly to help to understand the results in the following section.

When inductances and geometric capacitances are omitted, the standard or commonplace model looks like Figure 3.19. Some of the elements in the figure have a clear origin in physical terms, while others are convoluted to the point where it is difficult to unravel them, as considered in Section 4.1. As will be repeated in many places in this book, the problem is that the model does not allocate one circuit element to each separately identifiable physical process occurring in the FET. Consequently, some elements have to adjust to make good the deficiencies, meaning that they take on apparently nonphysical values. Reference [9], for example, described a mechanism showing how negative values of R_s are a perfectly reasonable expectation. Providing such nonphysical values are accepted for the corrective entities they are, for small-signal applications, the standard model works well (actually, extremely well) because it then reproduces the measured S-parameters faithfully, as in the preceding section.

The elements most obviously affected are R_s and R_d across the whole range of bias, and C_{ds} principally below the knee. Less obviously, C_{gd} is affected too but in a manner that is difficult to untangle (a feat attempted in Section 15.4.2). These, then, are the elements to watch for in the typical bias-dependent results of the next section.

To place the utility of the model fully in context, although it has proven worth as a small-signal characterization, it is too compromised for nonlinear use. Even in the small-signal sense, it has a limitation: it is not scaleable, as discussed in Chapter 2. Neither of these two limitations detract from its usefulness in bias-dependent, small-signal applications of a single-sized device; put simply, the trouble begins when attempts are made to use it out of bounds.

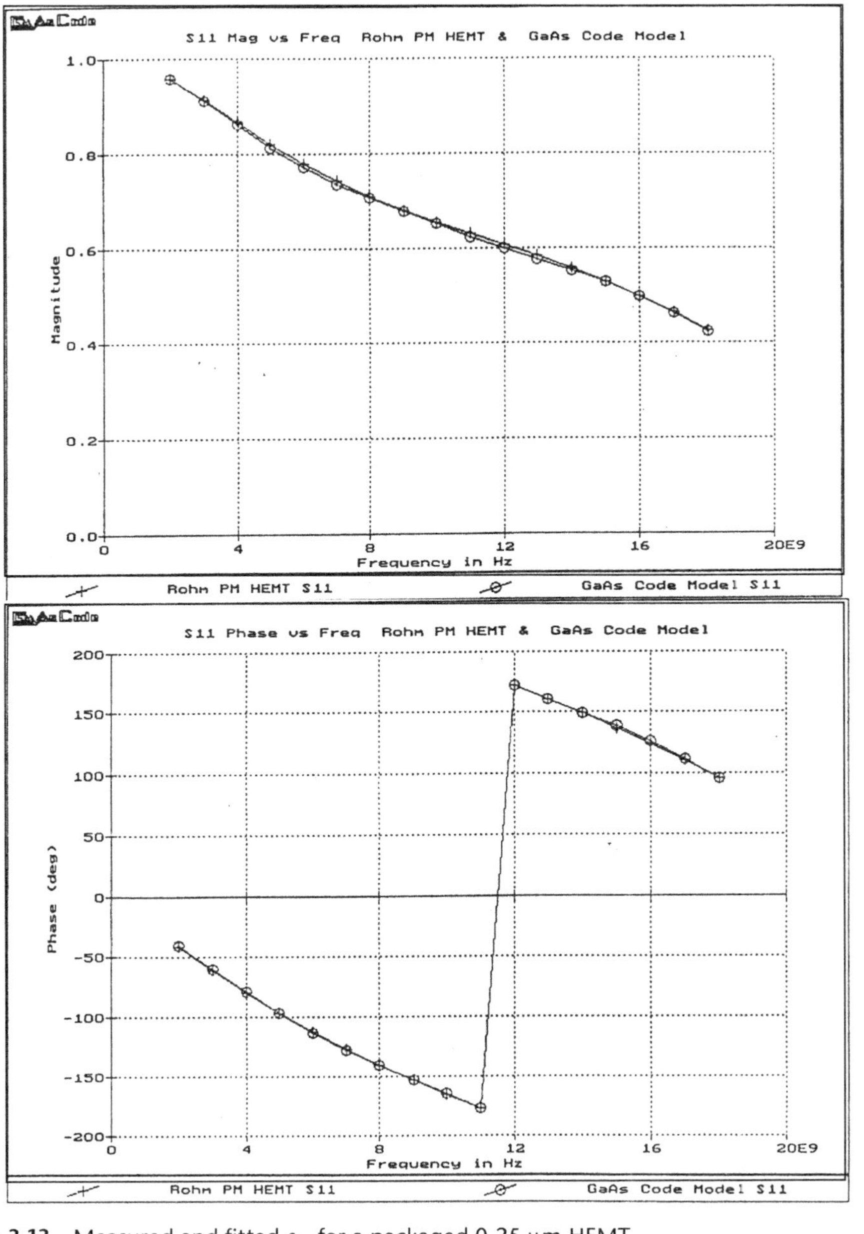

Figure 3.13 Measured and fitted s_{11} for a packaged 0.25 μm HEMT.

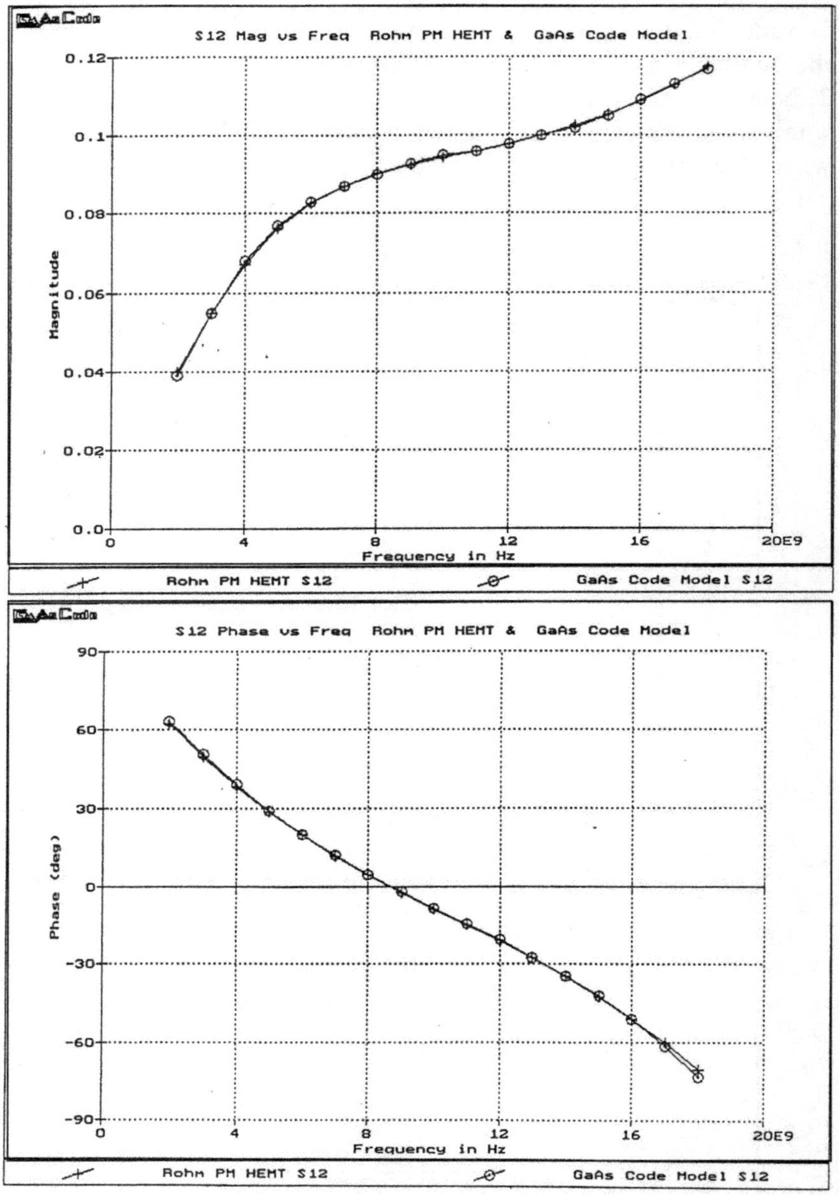

Figure 3.14 Measured and fitted s_{12} for a packaged 0.25 μm HEMT.

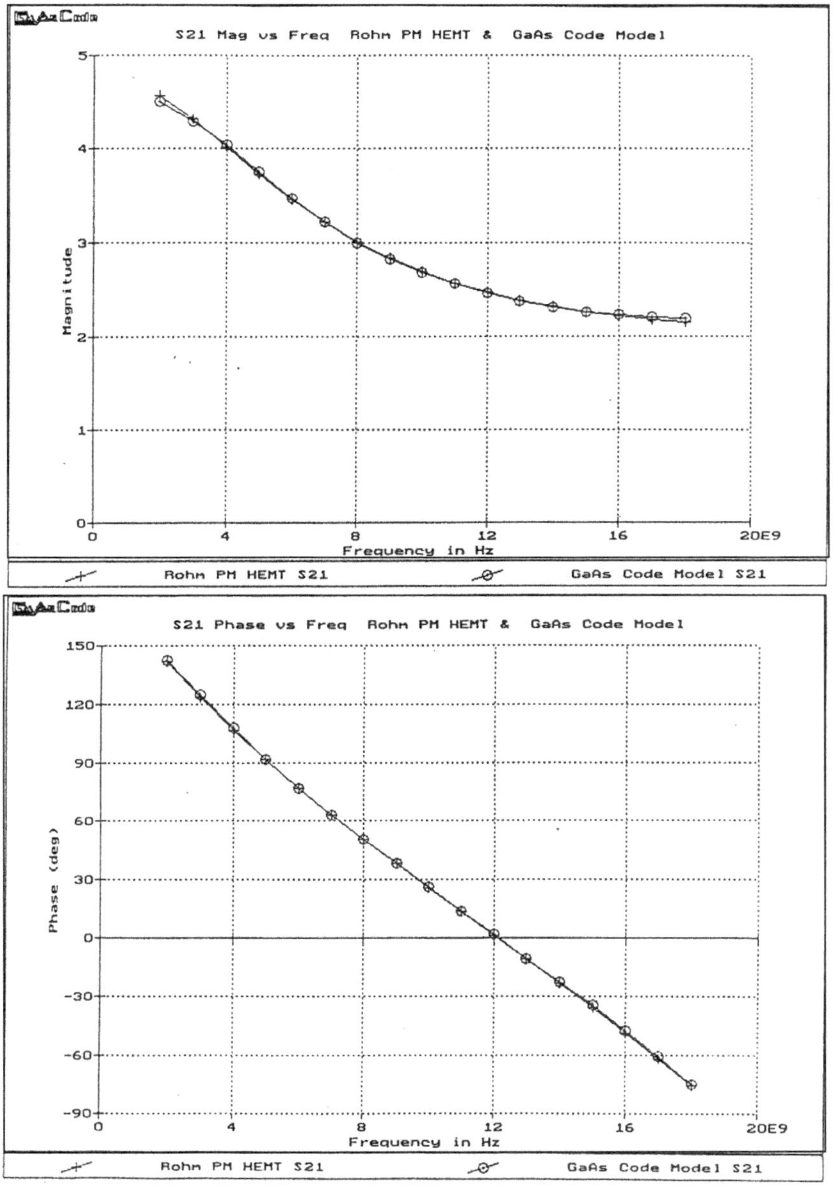

Figure 3.15 Measured and fitted s_{21} for a packaged 0.25 μm HEMT.

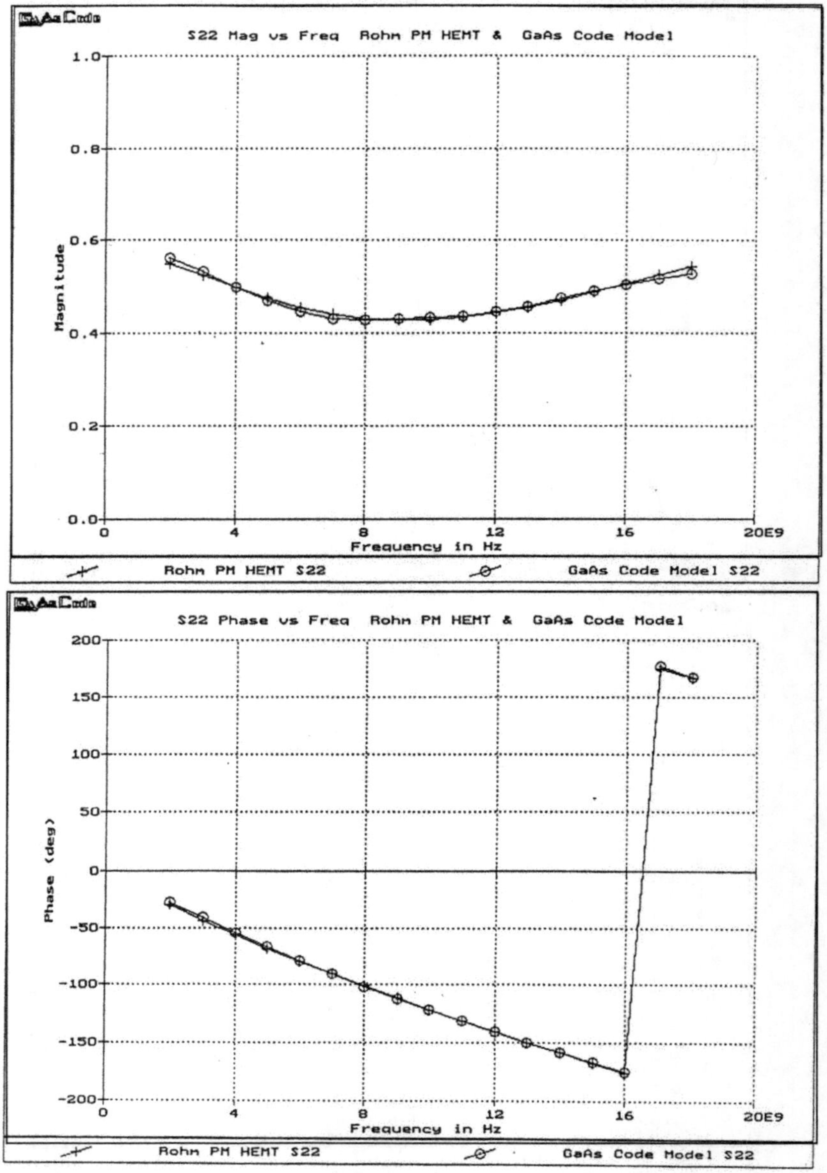

Figure 3.16 Measured and fitted s_{22} for a packaged 0.25 μm HEMT.

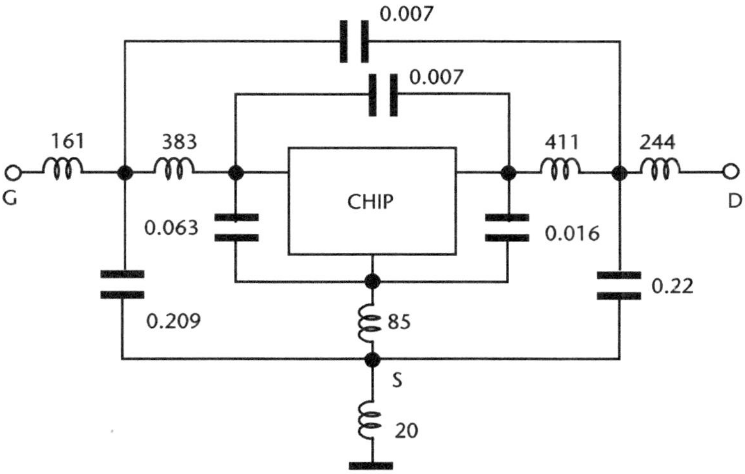

Figure 3.17 Package model for a 0.25 μm HEMT. Units are pH and pF.

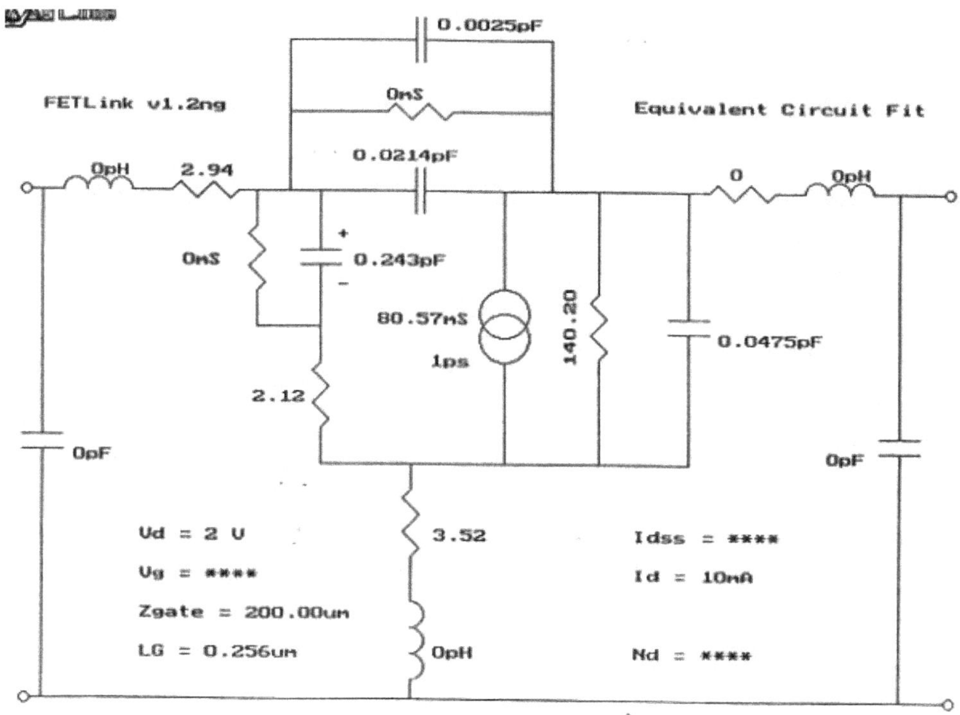

Figure 3.18 Equivalent circuit of a 0.25 μm HEMT chip retrieved by iterative fit and strip from S-parameters of the packaged device.

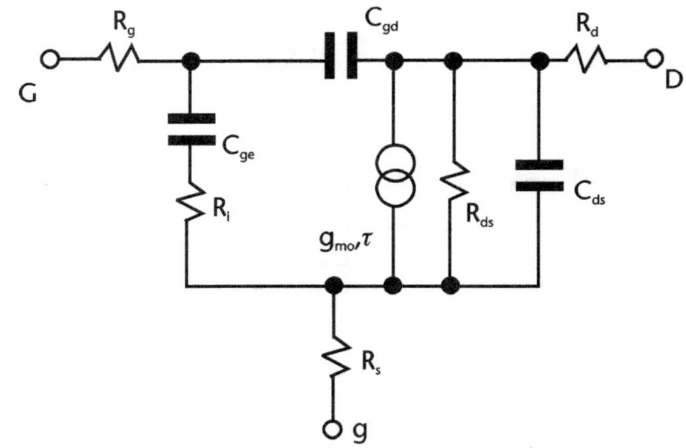

Figure 3.19 The standard or commonplace model. The series inductances L_g, L_s, and L_d are omitted for clarity (they are minor elements).

3.2.4 Bias Dependence of the Elements: Examples

The following results are reproduced exactly as they first appeared by fitting to multibias S-parameters; they have not been doctored in any way (see Figures 3.20 to 3.28). They exhibit smoothness where it can be expected and, to illustrate where fitting may be difficult or uncertain, erratic excursions just as they occurred. Nothing has been hidden. All elements except R_d come from the same device and were fitted unconstrained. The gate resistance R_g (not shown) was the only fixture.

To begin with the major elements g_{m0} and C_{gc} should consist of smooth curves, separated from one another for most devices (but not in all; see Section 4.2.2).

Below $V_{DS} = 1V$ in Figure 3.22, there is mild erratic behavior which occurs commonly (another example is given in Section 4.1). Further, in this particular device the curves for different V_{GS} are reasonably well separated, again not always the case. The ordering with respect to V_{GS} is consistent with the channel being gradually pinched as the gate is made more negative.

When extracted from multibias S2P, C_{gd} almost always has the shape of Figure 3.23. The extent to which the upsweep is representative of a true physical process is debateable. Irrespective, once again the curves should be smooth and distinct.

An interesting property of the output resistance of Figure 3.24 is that it exhibits a close correlation with the delay element τ, suggesting they have a common origin (see Figure 13.33).

We now come to the awkward elements, beginning with R_s in Figure 3.25.

In this particular example R_s is positive over much of the range but with a value smaller than that expected from the geometry of the device. This, together with the negative values, is a result of the defects in the standard model topology outlined in Section 3.2.3.

Typically, the space-charge component of C_{ds} is small, so the capacitance can, although not necessarily, be dominated by the geometrical component contributed by the drain pads and other metallic part of the structure. When that happens, C_{ds}

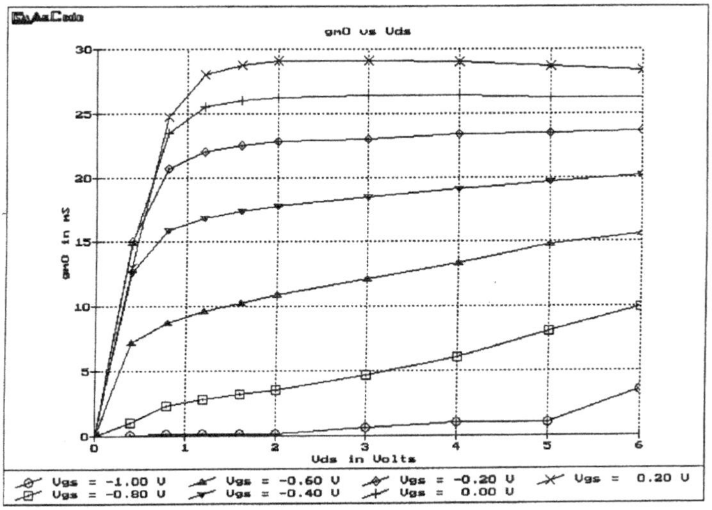

Figure 3.20 Bias dependence of transconductance g_{m0}.

should be only weakly dependent on voltage, which is what emerges in Figure 3.26 above $V_{DS} \sim$ 1V. Such constancy can be misleading as to the origin and nature of C_{ds}, however. As will be discussed in Section 15.4.2, there are grounds for believing that C_{ds} is a manifestation of fitting a strictly nonphysical model to data having a definite physical origin.

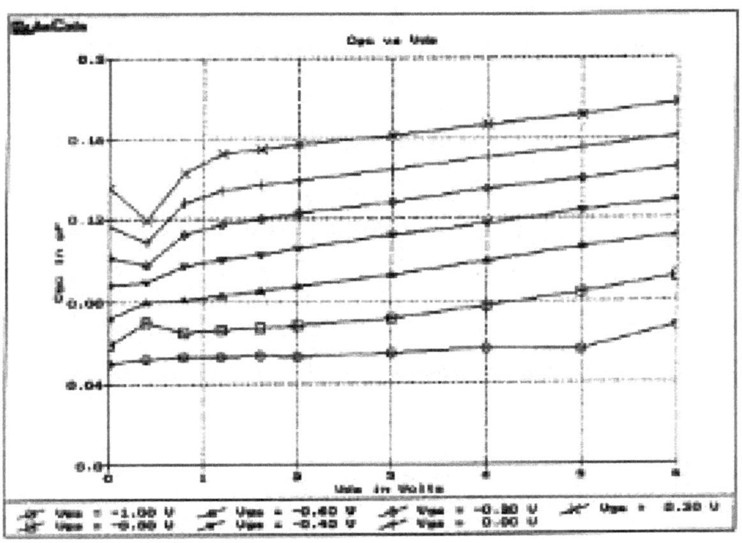

Figure 3.21 Bias dependence of the gate-channel capacitance C_{gc}.

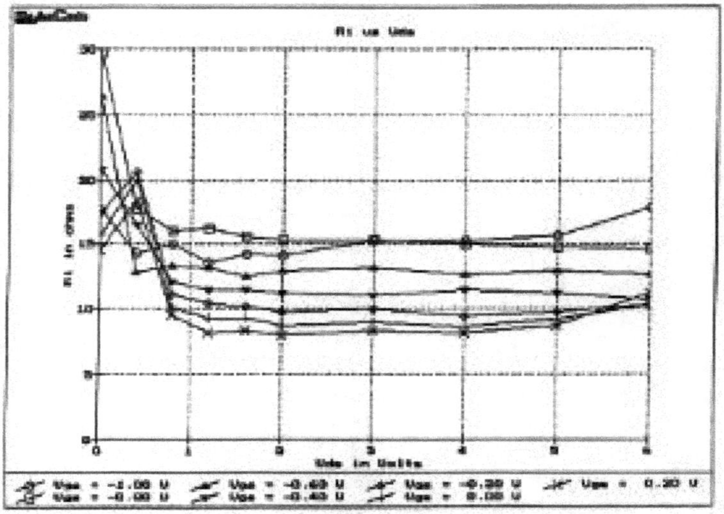

Figure 3.22 The intrinsic or channel resistance R_i.

Inexplicably, no record was kept of R_d for the foregoing device. To illustrate what can happen when R_d is freely fitted, we have to resort to Figure 3.27, which is from a different device. Negative values of R_d are quite common and, as for R_s, they arise as a consequence of the commonplace circuit's topology not correctly modeling each individual physical process occurring within the device so, by omission, it mixes them [9]. Thus R_d, like R_s, is an unfathomable composite, which, in fitting to measured S-parameters, de facto makes good the deficiencies.

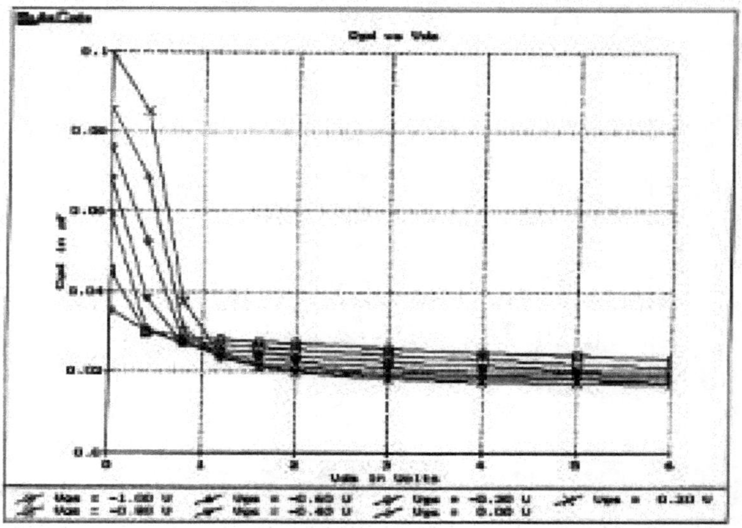

Figure 3.23 Bias dependence of the gate-drain capacitance C_{gd}.

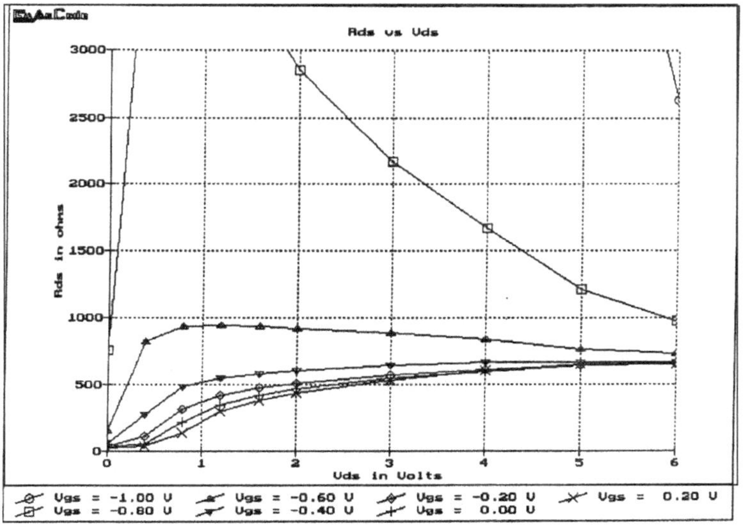

Figure 3.24 The output resistance R_{ds}.

Figure 3.27 also typifies the excursions or noise frequently observed in fitting the minor elements.

Finally, Figure 3.28 is a plot of τ, often treated as a troublesome inconvenience affecting the transconductance and sometimes disregarded altogether. It is an open question how many tricks have been missed through that view. Demonstrably, τ is a key factor governing the detail, much of it first order, of how HFETs as a genus work.

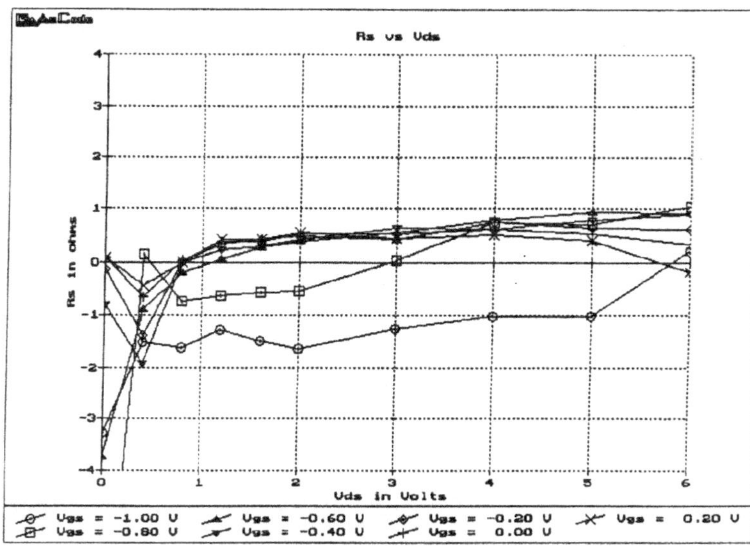

Figure 3.25 Source resistance showing apparently unphysical negative values.

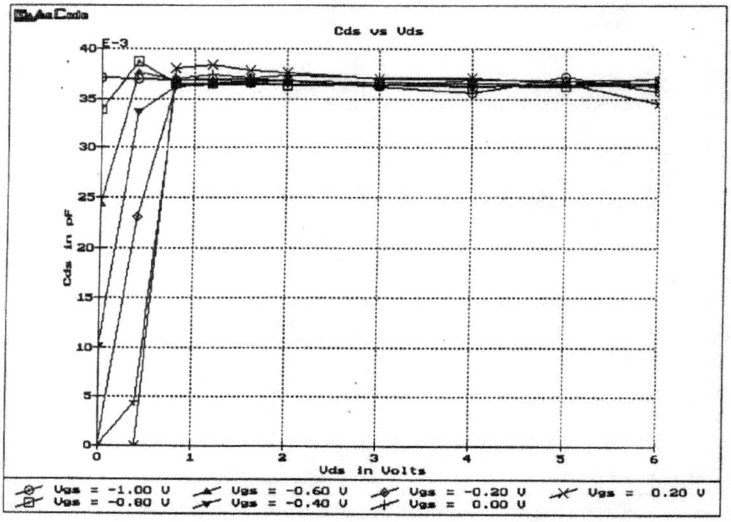

Figure 3.26 Drain-source capacitance C_{ds}.

The outstanding exception to the indifference rule was probably the Root model, although, being a closed model, it is difficult to know exactly what his τ stood for and how it was used [10].

3.3 τ: A Vital But Overlooked Physical Variable

τ and X are synonymous, differing only by a constant, $2v_{sat}$. Given that working out how FETs function is an iterative endeavor alternating between experiment with reverse modeling then forward modeling, it is helpful to regard each of X and

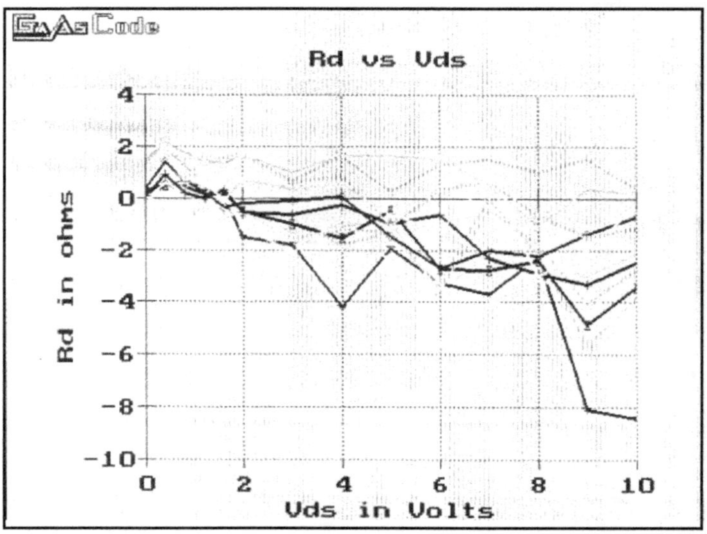

Figure 3.27 Bias dependence of drain resistance for a different device.

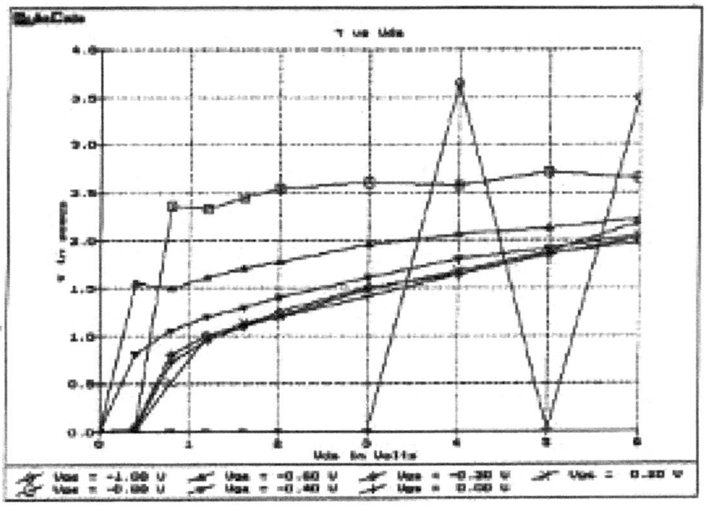

Figure 3.28 The delay element τ.

τ as a function of the other, so we may write $\tau(X)$ and $X(\tau)$. The first is truer in the sense of physical dependence, whereas the second has great value in the reverse sense. By distilling $\tau\ (V_{GS},V_{DS})$ from S-parameter measurements, we automatically know $X(V_{GS},V_{DS})$. Plainly speaking, we hereby have a direct method of finding the depletion extension into the gate-drain region and seeing how it varies with the voltages applied at the gate and drain. By turns, the 2-D electric field originating on the positive ionic charge within X governs how the channel opening $(W - d)$ varies with the voltages and that, in turn, determines R_{ds} and affects g_{m0} and C_{gc}. Looked at this way, τ is a major source of intelligence on the physics of FET operation.

Section 13.8 is devoted to this most important variable.

References

[1] *Pulsed I(V) Measurement of Semiconductor Devices with Applications*, Accent Optical Technologies, 2004.

[2] Ladbrooke, P. H., *MMIC Design: GaAs FETs and HEMTs*, Norwood, MA: Artech House, 1989.

[3] Barton, T. M., "Characterisation of the Physical Behaviour of GaAs MESFETS," Ph.D. Thesis, University of Leeds, May 1988.

[4] Aaen, P. H., J. A. Plá, and J. Wood, *Modelling and Characterization of RF and Microwave Power FETs*, Cambridge, U.K.: Cambridge University Press, Chapter 3, 2007.

[5] Cameron, N. I., et al., "Selectively Dry Gate Recessed GaAs Metal-Semiconductor Field-Effect Transistors, High Electron Mobility Transistors, and Monolithic Microwave Integrated Circuits," *J. Vac. Sci. Technol. B*, Vol. 11, No. 6, November/December 1993, pp. 2244–2248.

[6] Willing, H. A., C. Rauscher, and P. de Santis, "A Technique for Predicting Large Signal Performance of a GaAs MESFET," *IEEE-MMT-S International Symposium Digest*, Ottawa, Canada, June 27–29, 1978.

[7] Liechti, C., E. Gowen, and J. Cohen, "GaAs Microwave Schottky-Gate Field-Effect Transistor," *ISSCC Digest of Technical Papers*, February 1972, pp. 158–159.

[8] Bridge, J. P., P. H. Ladbrooke, and A. J. Hill, "Characterisation of GaAs FET and HEMT Chips and Packages for Accurate Hybrid Circuit Design," *IEE Proceedings-H*, Vol. 139, No. 4, August 1992, pp. 330–336.

[9] Ladbrooke, P. H., A. J. Hill, and J.P. Bridge, "Negative Rx and Rd in GaAs FET and HEMT Equivalent Circuits," *Electron. Lett.*, Vol. 26, No. 10, May 1990, pp. 680—681.

[10] Root, D. E., and B. Hughes, "Principles of Nonlinear Active Device Modeling for Circuit Simulation," *32nd IEEE MTT ARFTG Conference Proceedings*, December 1988, Tempe, AZ, pp. 3–24.

CHAPTER 4

The Standard Model: Deriving the Elements

It was once said that, even after many years of practice and refinement, large-signal models of FETs and HEMTs meet with varying degrees of success. A commonly held feeling was that, when there is disagreement between the simulated and practical responses on nonlinear RF and microwave circuits, it is the device model and not the simulator itself that is mostly responsible. Restricting the discussion to schemes based on the commonplace equivalent circuit, it is therefore appropriate to see if there are reasons why it may still be so.

First, a positive note. When allied with pulsed $I(V)$ measurements, the standard model performs adequately in the following circumstances:

1. When the frequencies of interest, including any harmonics, are low enough for the model's function in a circuit to be dominated by its conductance properties with its reactive properties playing little part. "Low enough" means a fraction of the cutoff frequency of the device, typically less than $f_T/3$ but in some cases up to $f_T/2$.
2. When the device is not overdriven so it self-biases.
3. When the circuit performance specifications are simple.

As each of these conditions is relaxed, the standard model becomes progressively unreliable. Nevertheless, the relative simplicity of its circuit configuration and its ubiquitousness augur well for its continued popularity.

4.1 Element Functions Obtained by Fitting: True or Askew?

We begin by looking at the extent to which the individual elements in the commonplace equivalent circuit tell us something about the physical processes, individual or otherwise, occurring within the actual device.

It turns out that the elements most closely representing identifiable physical processes are τ, g_{m0}, g_0, and C_{gc}. Those least closely associated are R_s and R_d, leaving the intermediate cases as C_{gd}, C_{ds}, R_g, and R_i. It is now necessary to state briefly why.

Of the leading four, τ is the time taken by electrons traveling at saturated drift velocity to charge (or discharge) the depletion edge in region II–III. From it can be found the x dimension of region III, which we call X, the extension of the space-charge layer into the gate-drain region. Although often treated lightly if at all in equivalent circuit models, τ plays a key role in determining how HFETs work.

Overlooking the fact that the transconductance g_{m0} is dispersive, to first order it is a measure of how the depth of depletion d changes when V_{GS} changes (to second order it responds to changes in $V_{D'G}$, or V_{DS} as well). These comments mainly apply in the region above the knee in the current-voltage characteristics. Below the knee, more complex processes are active.

g_0, although also dispersive, which we here overlook similarly, results from the lifting of the deletion edge by the 2-D electric field, which originates in the depletion layer extension X, and which increases with increase in $V_{D'G}$ or V_{DS}.

C_{gc} is a composite of three components: C_{IB} (a branch capacitance), $g_{m0}\tau$, and C_N (which is an orthogonal branch- or transcapacitance).

Although C_{gc} does in one sense represent one individual physical process, namely, charge storage, that charge is stored in three separately identifiable locations within the depletion layer. Usefully, the fitted C_{gc} can be deconstructed from the practical data to yield information on each and to serve as a check that the results so far are physically sound.

Concerning the intermediate cases, although C_{gd} and C_{ds} appear to reflect charge changes in the space-charge layer above the knee, below the knee their form may be as much dictated by the equivalent circuit itself as by physical process. Expressed another way, over part of the voltage range, the form of these two capacitances is an admixture of other elements in the equivalent circuit, a result that comes about by having to determine them, in the process of fitting to measured S2P, by looking through the other parts of the equivalent circuit.

Of the two resistances R_g and R_i, R_g might be expected to correspond to the resistance of the gate metallization, modified to account for distributed effects across the width of the gate strip [1]. Sometimes, however, the freely fitted value can be affected in a way similar to R_s and R_d, as below. R_i is a difficult case because, below the knee where g_{m0} is small, there is too little to separate R_i from R_s in the process of fitting the equivalent circuit to measured S-parameters, so the division of resistance between R_i and R_s can become troublesome.

An expression for R_i based on first-order physical considerations is given in Section 6.3.2 of [2]. A more elaborate determination of the form of R_i, as in Figures 4.1 and 4.2, is obtained from a macro-cell solver, a fast-executing form of physics-based simulator described in Chapter 16.

For gate biases near or beyond pinch-off, R_i varies inversely with V_{DS}. Such dependence is consistent with the idea that d-lift will open up the channel in region I beneath the gate as X increases with increases in V_{DS}. This behavior is completely missing in the practical fit of Figure 3.22 for example. Why is hard to tell. Hints of the shape expected are present in Figure 4.3, albeit not for the near-pinch-off curves. Nothing consistent emerges from studying devices of different subtypes from different manufacturers.

As will be touched upon in Section 14.5, there is in any case uncertainty as to what R_i represents in the standard equivalent circuit. Some of the confusion surrounding it is removed by the reformed topology version.

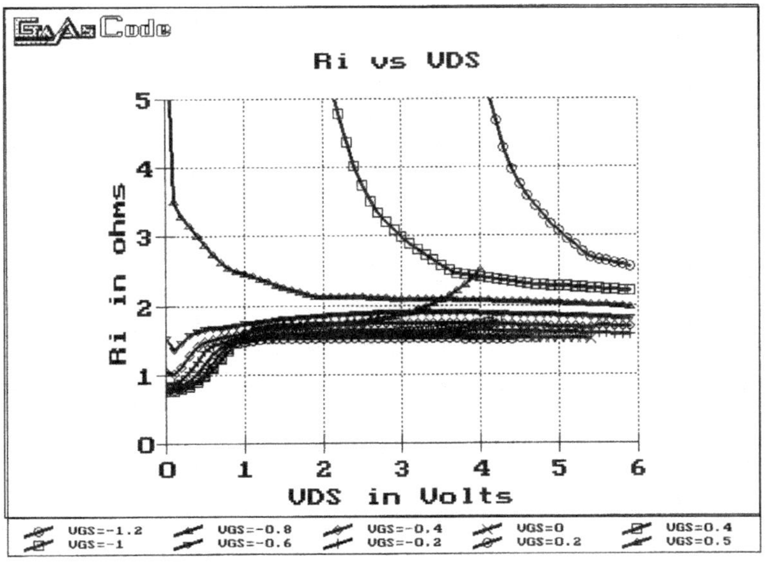

Figure 4.1 Simulated R_i as might be fitted to measured S-parameters (PHEMT example).

Lastly, both R_s and R_d often, if not always, take on nonphysical values. De facto, it is their lot to compensate for topological deficiencies in the standard model to the point where all physical association is lost. Notably, they may take on negative values. This too can be expressed inversely: if physical values of R_s and R_d are set as unchangeable when fitting to measured S-parameters, a good fit may be impossible.

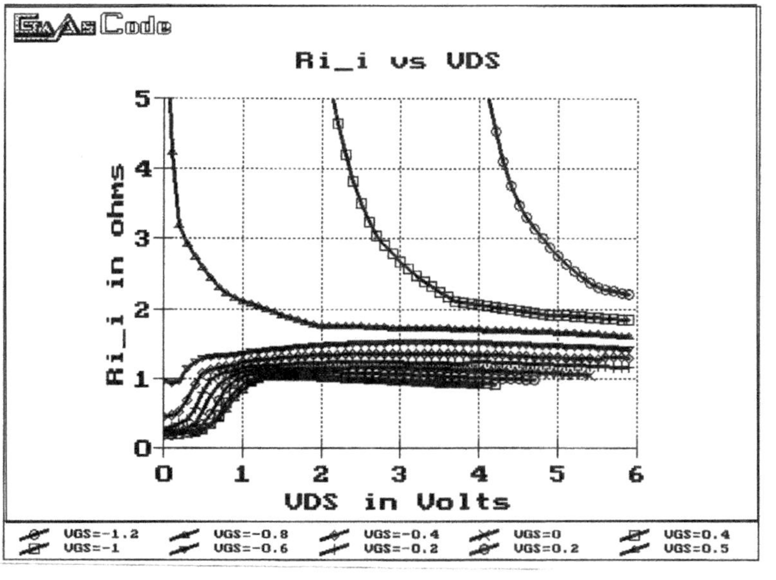

Figure 4.2 Likely form for the intrinsic (or true) R_i resulting from a physics-based simulator (PHEMT example).

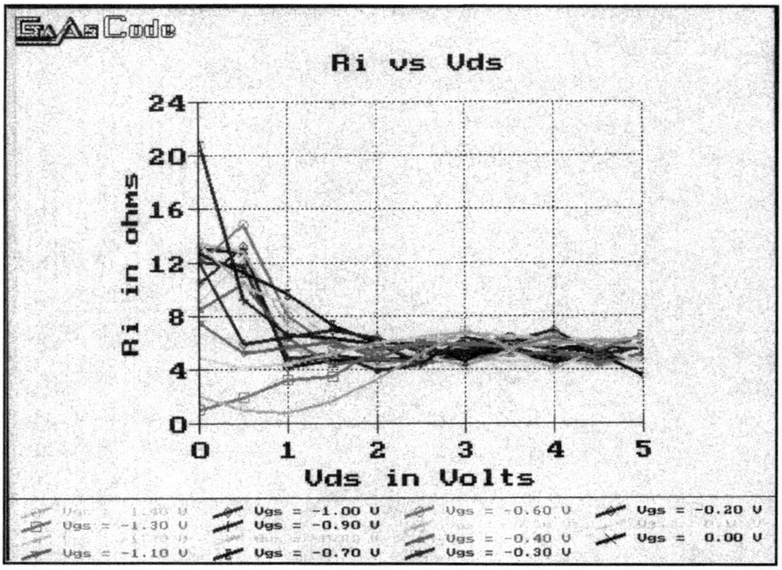

MATCOM 0.3μm FET

Figure 4.3 Fitted R_i for a 0.3 μm gate HFET.

4.2 Neglect of Nonlinear Terms

4.2.1 The Problem of Nonlinear Extraction

There are at least four aspects of large-signal modeling where errors of principle can be identified. All apply to equivalent circuit-based models, but not all, for obvious reasons, apply to behavioral models. In no particular order of importance, they are:

1. When characterization measurements are made for the purpose of developing or extracting a model, an operational state is forced upon the device, which may be quite different from the state it will be in when operating in its intended application. Here, the word "state" refers to the totality of all the physical variables which together determine how much current the device will pass, including the unobvious or hidden variables of trapped charge held in bulk deep levels or surface states and the crystal lattice temperature, as well as the obvious variables of depletion region size and shape, electron drift velocity in the various layers of the channel, and so on. The solution to this problem is to make the basic characterization measurement under pulsed conditions. The key is to be able to make such measurements about a predefined bias point, which sets the hidden variables.
2. The functional forms used to fit the measured current-voltage and capacitance-voltage characteristics may be incapable of fitting with sufficient accuracy. Although it has long been appreciated that the derivatives of the current with respect to the gate-source voltage control harmonic generation [3], and although functional fitting schemes to address this truth have been conceived [4], the use of equations that are simple and quick to implement endures from a workability point of view.

3. The topology of the widely used standard model does not represent correctly the physical processes that actually occur in FETs and HEMTs. Several of the elements in the equivalent circuit are composites of mechanisms active in different parts of the device, with the result that the model gives rise to the wrong currents flowing in the wrong branches of the equivalent circuit.
4. Due to the difficulty of an alternative, the practice of extracting a bias-dependent linear model and pressing it into use as a nonlinear model is widespread. It is fundamentally in error. A bias-dependent linear model is not at all the same as a true nonlinear model. The issue is independent of the dispersion problem alluded to in item 1.

Our principal concern for the moment is with the fourth of these defects.

Referring to the commonplace equivalent circuit of Figure 3.19 in Chapter 3, the widespread procedure by which the elements are found as functions of the device voltages (V_{GS},V_{DS}) is as follows:

1. Small-signal S-parameters are measured either at multiple dc bias points, in which event we denote their functionality as (V_{GS},V_{DS}), or at multiple operating points (v_{GS},v_{DS}) achieved by fast-pulsing to those point from a fixed, known, bias point (V_{GS},V_{DS}), in which case the S-parameter functionality is written (v_{GS},v_{DS},V_{GS},V_{DS}). The distinction between the two is that, provided that fast enough pulses are employed, pulsed S-parameter measurements eliminate the error at item 1.
2. By a process of iteration, the S-parameters at any given bias point (or operating point, as the case may be) are fitted by adjusting the value of each element in Figure 3.19 in turn. When adjusting the value of any one element, all the others are treated as constant.
3. The process is repeated for all the measured points, by which time each of the elements is available as a function of the voltages, for example,

$$C_{gc}\left(V_{GS},V_{DS}\right) \text{ or } C_{gc}\left(v_{GS},v_{DS},V_{GS},V_{DS}\right)$$
$$g_{m0}\left(V_{GS},V_{DS}\right) \text{ or } g_{m0}\left(v_{GS},v_{DS},V_{GS},V_{DS}\right)$$

   and so on.
4. The element functions so obtained are lodged into a nonlinear circuit simulator as one or another large-signal model (Angelov, Curtice, Cobra, Parker, and Skellern, as in Chapter 2).

The error of principle that concerns us here is the constraint from point 2 that when adjusting the value of any one element, all the others are treated as constant. Such a constraint ignores terms of the type

$$i\delta R$$

which not only may be non-negligible but also may be dominant in cases where elements are highly nonlinear. The central point can be made by considering the simpler functional case of

$$C_{gc}(V_{GS},V_{DS}),\ g_{m0}(V_{GS},V_{DS})\ldots$$

Using self-evident electrical quantities, from Figure 3.19:

$$\delta v_{GS} = \delta i_G R_g + \delta v_C + \delta v_{Ri} + \delta v_{Rs}$$

Recognizing that R_i and R_s are nonlinear elements, δv_{Ri} and δv_{Rs}, composed of two terms each:

$$\delta v_{Ri} = \delta i_{GC} R_i + i_{GC}\delta R_i$$
$$\delta v_{Rs} = \delta i_S R_s + i_S \delta R_s$$

In each case, the last term derives from the nonlinear nature of the element itself. These terms are ignored in the commonplace model extraction procedure.

There are also terms like

$$\delta Q = C_{gc}\delta v_C + v_C \delta C_{gc}$$

where, similarly, the second of the two terms is ignored in the widely used extraction method.

The neglect of the terms originating from the nonlinearity of the elements themselves has repercussions that are difficult to assess generally. Even detecting it presupposes the existence of some independent means or method, which in itself is problematic. The impetus to find qualitative if not quantitative answers came from a nondispersive PHEMT representative of its type. Fitting the standard model to measured S-parameters yielded transconductance and gate-channel capacitance functions that decreased with increasing forward gate bias. Not having encountered characteristics of the kind before, which to add to the puzzle seemed inconsistent with basic FET operating principles, a way had to be found of deciding whether we were looking at a genuine property of the PHEMT structure or an artefact arising from the whole measurement and fitting method.

Enlightenment came from a physics-based HEMT simulator by which can be calculated:

- The true nonlinear element functions as they exist internal to the device;
- The element functions as would be obtained at the terminals by fitting the commonplace equivalent circuit model to measured S-parameters.

Section 4.2.2 is the account of how, piece by piece, the simulator enabled a picture to be built of the contrast between terminal and internal device behavior.

4.2.2 Extracted Versus True Nonlinear Element Functions

Figure 4.4 shows the gate-channel capacitance that resulted from fitting the standard model to S-parameters measured at multiple bias points for a PHEMT that is almost entirely dispersionless. The shape or form of this extracted element function is intriguing: as the gate-source voltage is made increasingly positive, the gate-channel capacitance eventually starts to decrease. Whereas such behavior may seem implausible, the question is can C_{gc} actually decrease with increasing forward bias?

In the limit the answer is clearly yes, because I_D itself has a definite limit. Approaching the limit, the change in I_D per unit change in V_{GS} must decrease consistent with a smooth transition between well-spaced curves of $I_D(V_{GS})$ and the end condition where the limit is reached. There is progressively less charge change at the depletion edge per unit change in V_{GS}, so C_{gc} should decrease. However, if it can be shown that the true, nonlinear, space-charge capacitance is unlikely to have such a form, that form might therefore be an artefact of fitting the commonplace model to measured S-parameters (as in Section 3.2).

One way to shed light on the problem is to compare the fitted $C_{gc}(V_{GS},V_{DS})$ function with predictions available from a device simulator. (Given that the layer composition and plan-view structure of the practical device in Figure 4.4 were not known, a guess had to be made for the simulator. Also, in the results to follow, sharp transitions and occasional wild excursions arise from forming differentials $\delta Q/\delta V$ and $\delta i/\delta V$, which are sensitive to coarse-mesh and allied approximations in the simulator.) Predictions can take two forms: first, calculations for the equivalent circuit elements as they would appear at the device's terminals to compare with practically derived data; and second, elements calculated locally within the device itself. We refer to the first as external elements and the second as internal elements. A method of using the simulator to predict external elements to compare directly with fitted practical elements is to synthesize bias-dependent RF behavior at the device terminals and reuse the fitting routine to fit the equivalent circuit to the synthesized S-parameters.

Figure 4.5 shows the result from the simulator for the external capacitance. Much the same characteristic is evident as in the practical plot: $C_{gc}(V_{GS},V_{DS})$ eventually starts to decrease for continued increase in V_{GS}.

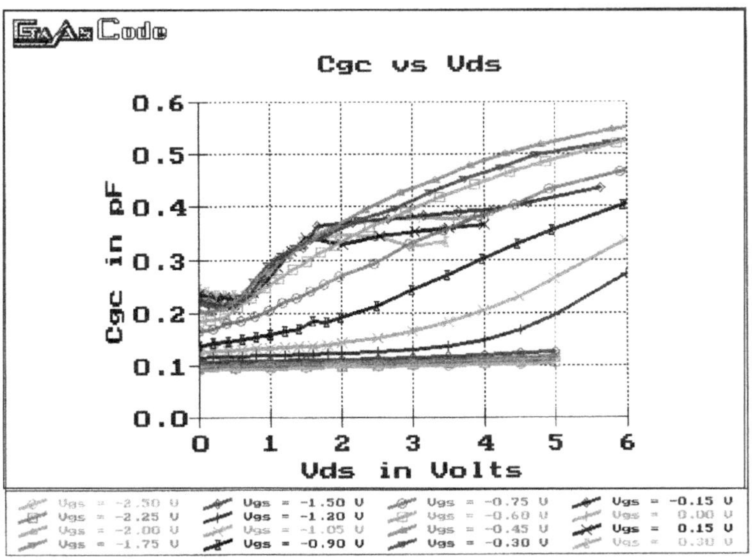

Figure 4.4 Fitted C_{gc} for a PHEMT.

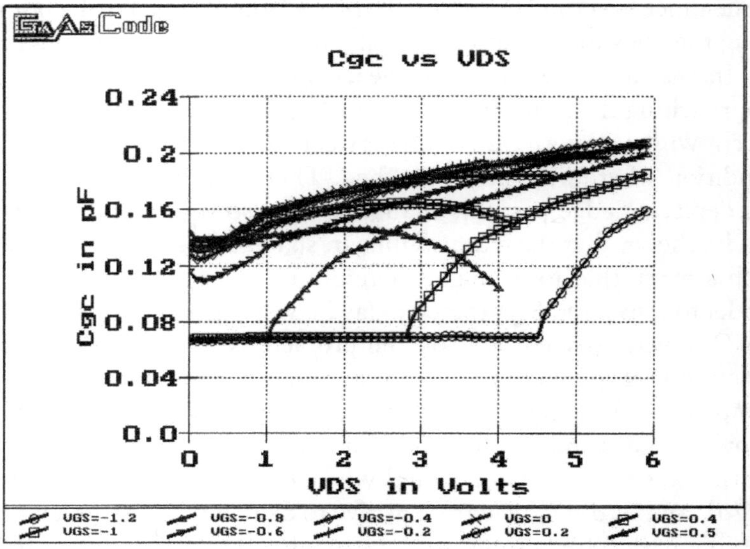

Figure 4.5 Simulated C_{gc} for a PHEMT structure like that in Figure 4.4.

In sharp contrast, Figure 4.6 shows the simulator's result for the true nonlinear gate-channel capacitance calculated as

$$C_{gc} = \frac{\partial Q}{\partial V_{GC}}$$

where Q is the total space charge, in all layers of the device, deriving from the presence of the Schottky gate electrode. Although the capacitance tends to saturate

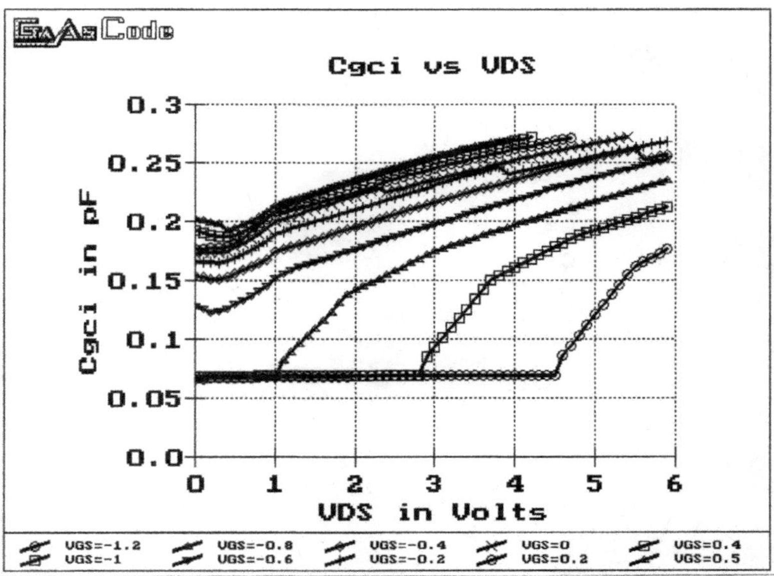

Figure 4.6 Intrinsic gate-channel capacitance: the true nonlinear element.

as V_{GS} is increased positively, there is no crossover and C_{gc} does not diminish in value anywhere.

Thus, the practical feature of C_{gc} that we set out to investigate is not the way the true element behaves. Without question the contorted form of gate-channel capacitance appears at the terminals of the device, but why? Is it a real effect or a distortion brought about by forcing distributed charge storage internal to the device into the simple topology of the standard equivalent circuit model of discrete elements? Either way, it is not the capacitive element that we need for a valid nonlinear model.

Similar functional curiosities are found for the intrinsic transconductance for the PHEMT example. Figure 4.7 shows the simulated version of what would result from routine fitting of the standard equivalent circuit to multibias S-parameters. The transconductance starts to decrease as V_{GS} is increased positively.

Figure 4.8 shows the true (or intrinsic) nonlinear element function calculated internally to the device structure directly in the region where current modulation occurs. In a manner similar to the gate-channel capacitance, the true nonlinear transconductance tends to saturate but it does not decrease as V_{GS} is made progressively more positive.

Looking ahead to the results for R_s (Figure 4.16), the intrinsic (or true) transconductance holds the key for the C_{gc} crossover that we are trying to explain. From elementary circuit theory, the extrinsic transconductance at the terminals is related to the intrinsic or true transconductance by the relationship

$$g_{m0}^{E} = \frac{g_{m0}^{T}}{\left(1 + g_{m0}^{T} R_S\right)}$$

In Figure 4.16 the source resistance increases nonlinearly near open channel. It is this fact which, via the foregoing simple equation, connects Figures 4.7 and 4.8 and, following Figure 4.7, Figure 4.5. Thus, the crossover effect is real.

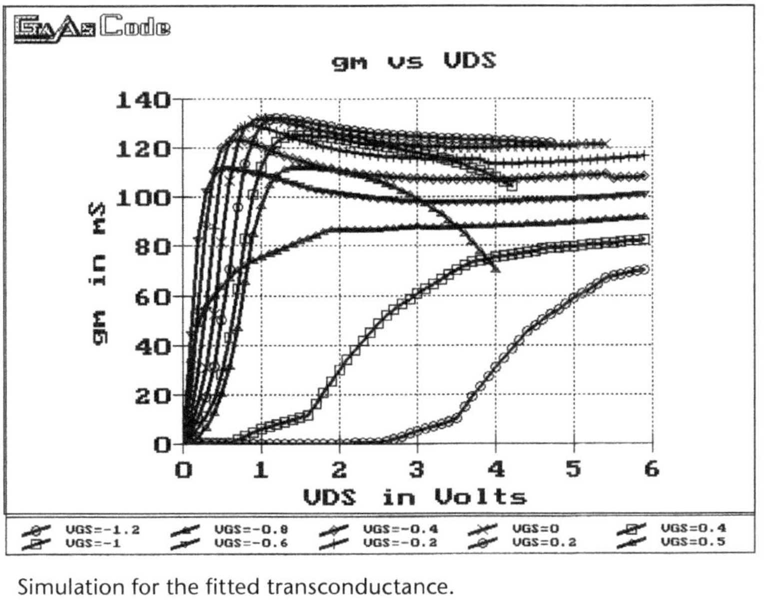

Figure 4.7 Simulation for the fitted transconductance.

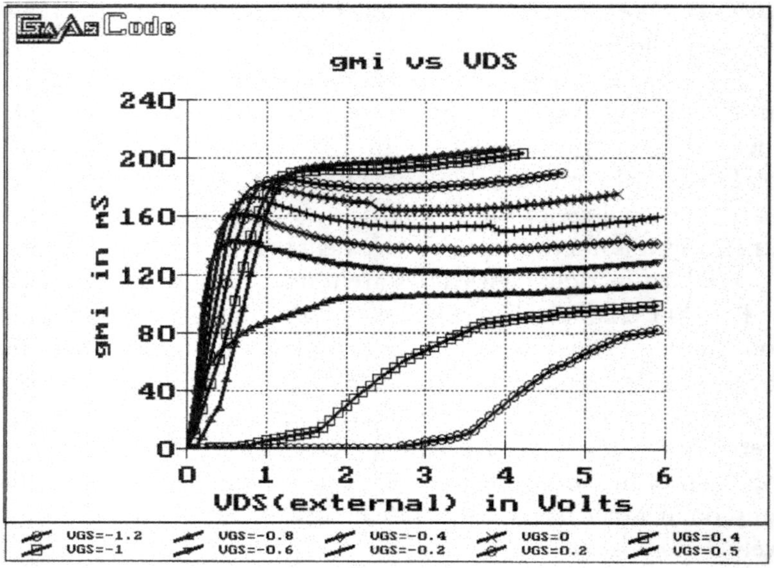

Figure 4.8 Simulated intrinsic transconductance.

The factual nature of the crossed-over C_{gc} means little in the quest to establish a valid nonlinear model. If we were to use it in the equivalent circuit model, it would immediately build in an irretrievable error. Faced with such an extreme distortion of the true element, the only solution would be to intervene with a physics-based scheme like the one used here.

Having used the simulator to develop insights into the problem of Figure 4.4, there is further guidance to be had generally, like the error arising from using other capacitive and resistive elements obtained by fitting the commonplace circuit model rather than the true element functions. The other elements which are affected by fitting the standard model to bias-dependent small-signal S2P are C_{gd}, C_{ds}, g_0, and R_i.

Figure 4.9 shows the form to be expected for the extracted gate-drain capacitance while Figure 4.10 shows the true nonlinear element, both as calculated by the simulator. Whereas Figure 4.9 has the familiar steep rise as V_{DS} is decreased towards 0V, that feature is nowhere near so pronounced in Figure 4.10.

Similar comments apply to the space-charge contribution to the drain-source capacitance in Figures 4.11 and 4.12.

Over the years, whether C_{ds} can be negative below the knee in the $I_D(V_{GS},V_{DS})$ characteristics, or whether it should unreservedly be negative, has been a topic of energetic conversation among device engineers. What Figure 4.11 suggests is that when fitting the commonplace equivalent circuit model to bias-dependent small-signal S-parameters, C_{ds} should indeed be allowed to go negative, even though such behavior is not expected when reasoning the problem from physical considerations alone (Figure 4.12). In a sense, therefore, both views have merit. Interestingly, the external C_{ds}, Figure 4.11, has much the same form as in Figure 3.26, in contrast to the true element, which is different below the knee. The space-charge component is very small, so C_{ds} can be regarded as constant to good approximation, dominated by the geometrical pad and bus capacitance.

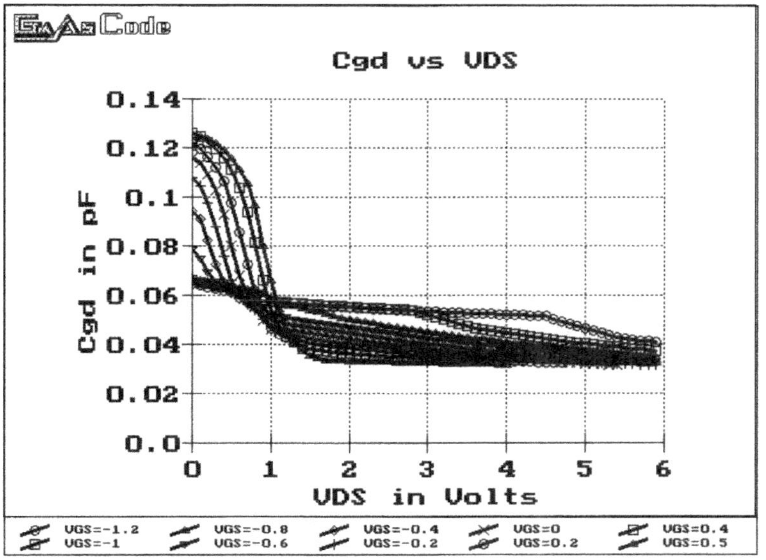

Figure 4.9 Simulation of what the fitted gate-drain capacitance would look like.

The form of the output conductance between what might be expected from fitting a bias-dependent linear model (Figure 4.13) and the intrinsic nonlinear element (Figure 4.14) is unchanged; only the value in the below-the-knee region is different.

Expressing the difference in resistance, in Figure 4.13, the smallest value of resistance that occurs at V_{DS} = 0V is 4Ω, whereas in Figure 4.14, it is about 1Ω. Caution in interpreting the difference is needed because in the limit of open channel and low drain-source field, the separation of the total drain-source resistance into R_s, R_d, and g_0 becomes somewhat ill-defined.

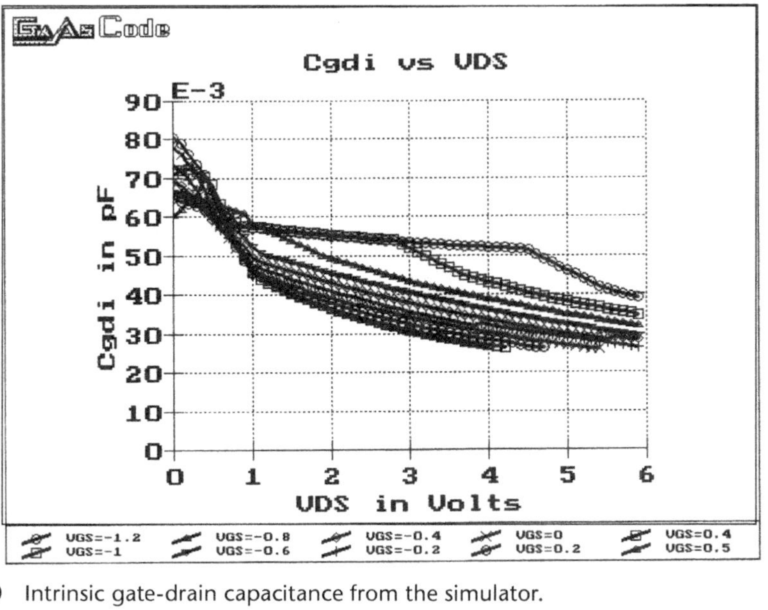

Figure 4.10 Intrinsic gate-drain capacitance from the simulator.

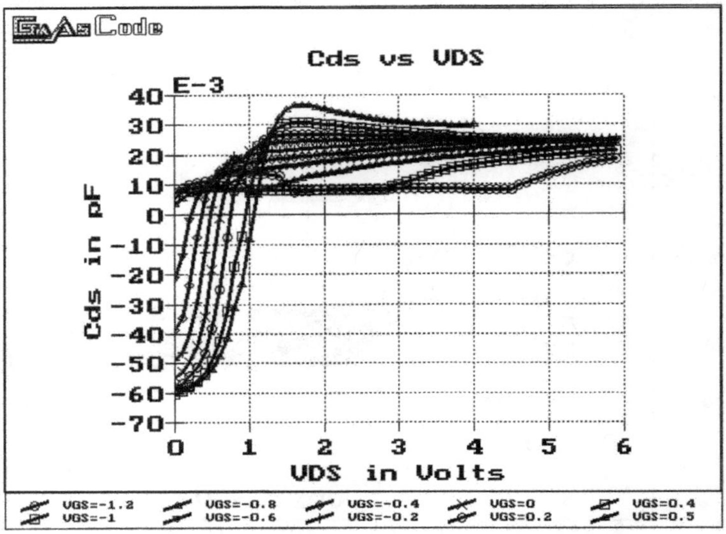

Figure 4.11 Simulated version of drain-source capacitance as would result from fitting the standard model to multibias S-parameters.

Figure 4.15 shows the simulated function for τ_{gm}. It exhibits a crossover at $V_{DS} = 2$V, a feature that has been seen before in practical devices where the delay was expressed in the equivalent measure of the space-charge-layer extension X, as in Figure 1.8.

The intrinsic channel resistance R_i in Figure 4.1 is an element whose very existence has been disputed by some workers who regard it as an artefact of the transconductance with a value of $1/5g_{m0}$, or $1/3g_{m0}$, or similar. Comparing Figure 4.8 with Figure 4.1 suggests that this view has a measure of validity, at least in the

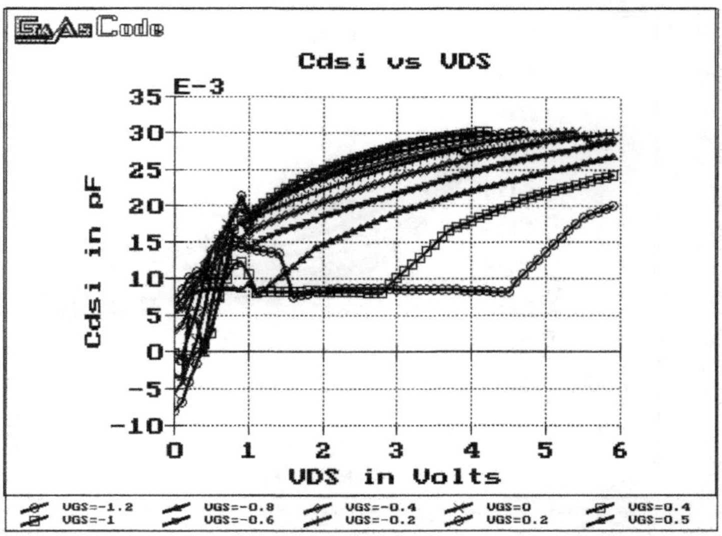

Figure 4.12 Intrinsic space-charge part of C_{ds} (simulated).

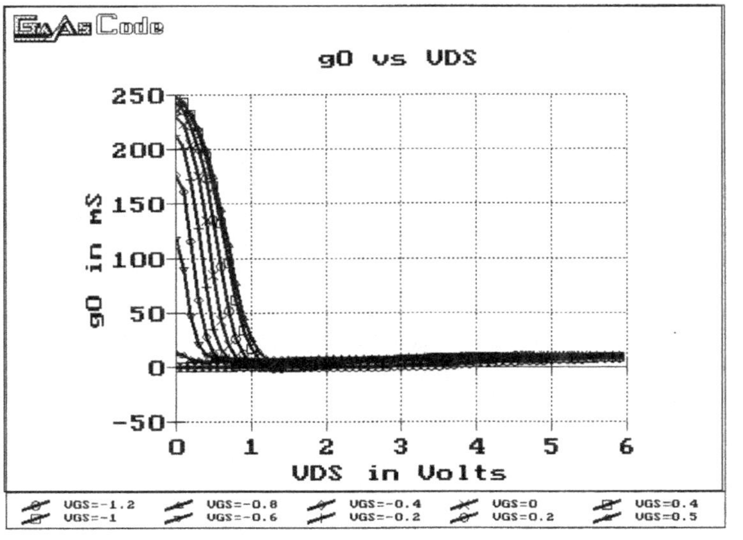

Figure 4.13 Output conductance as fitting the standard model would have it.

current saturation region. Other workers regard R_i as being the resistance of the tapered part of the channel under the source side of the gate where the drift velocity of the carriers is below v_{sat}. Changes in charge in the edge of the space-charge layer in this region therefore occur by a process akin to dielectric relaxation, albeit nonlinear, at a rate governed in approximate circuit terms by $C_{gc}R_i$. In this sense, R_i has a physically identifiable origin and purpose independent of that part of the active channel modeled by g_{m0}. Adopting the second of these two views, that R_i is identifiable independently of g_{m0}, leads to Figure 4.2 for the true or intrinsic

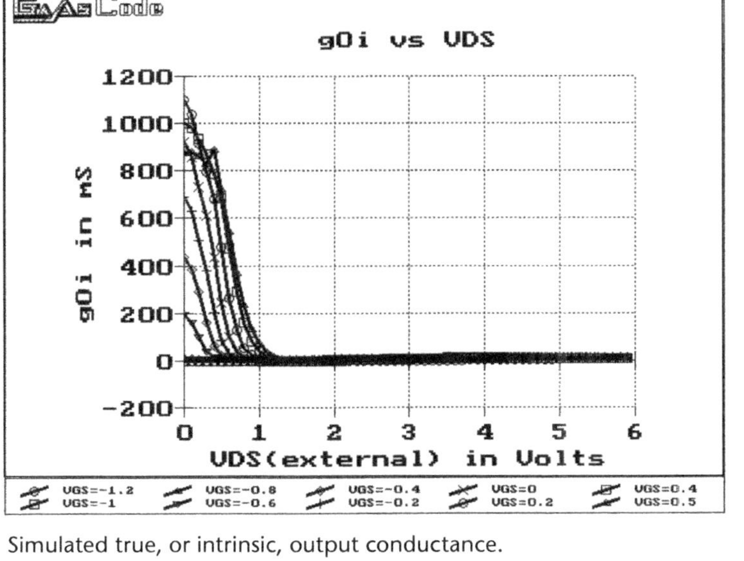

Figure 4.14 Simulated true, or intrinsic, output conductance.

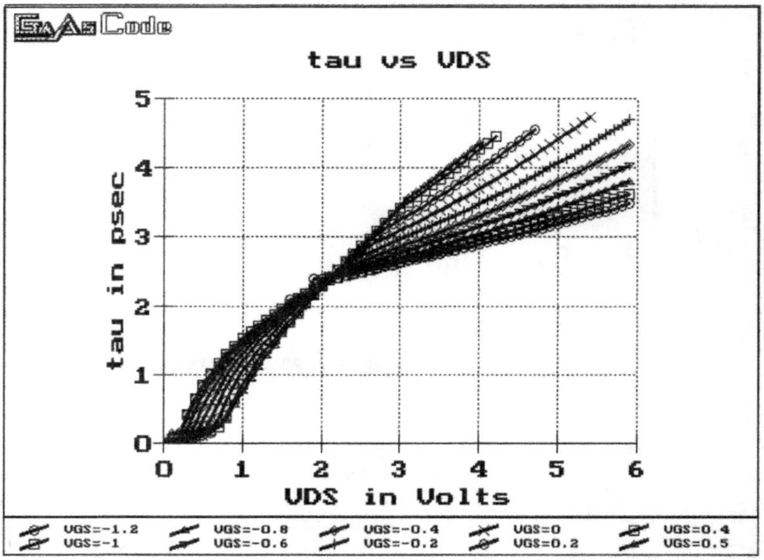

Figure 4.15 Simulated transconductance delay.

nonlinear element, and to Figure 4.1 for the element as would be fitted to measured S-parameters. Again, the shape or form is unchanged, but the values are different.

Finally, in many model extraction schemes, the source resistance R_s is assumed to be constant, while in others (as in this book), it is fitted unconstrained, in which event it often ends up negative and variable. Figure 4.16, which is the true nonlinear element for the PHEMT example in question, shows R_s to be variable but positive.

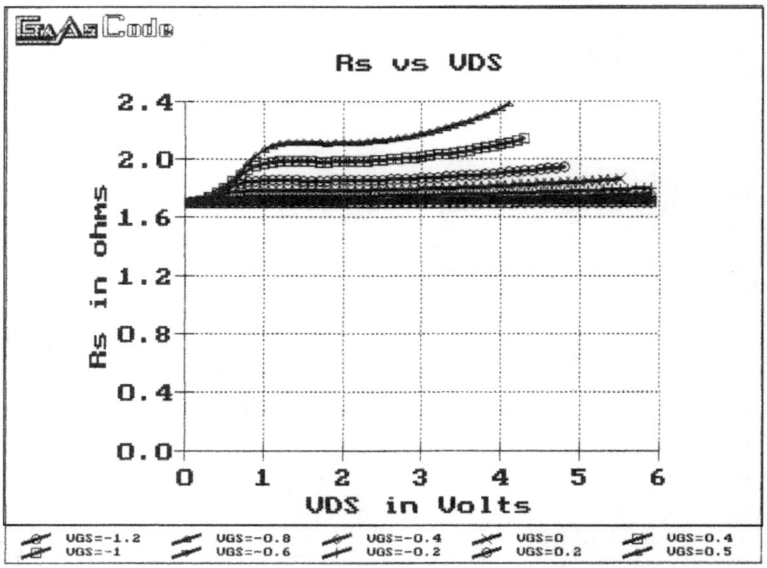

Figure 4.16 Simulated version of the intrinsic source resistance.

R_s is near enough constant only for low channel current. When the channel is wide open, the source resistance is larger, at first sight surprising perhaps. The physical requirement of current continuity, however, leads to high enough electric fields in the source-gate access region for electron transport to become nonlinear, manifest as nonlinearity in R_s itself.

In practical devices, this mechanism comes into play well before the ultimate limit outlined previously in connection with Figure 4.4. It is a self-evident disadvantage of this particular HEMT structure for large-signal applications, leading as it does to gross nonlinearity limiting the forward swing available. To alleviate the problem, the source-gate access region needs to have a higher sheet density of mobile carriers. Although no structural data were available for the device, the problem may have stemmed from the gate recess extending too far from the gate on the source side, and a tighter recess might have led to a different outcome.

For completeness, Figure 4.17 shows the drain characteristics calculated by the macro-cell solver. As no trapping or heating effects were accounted for, Figure 4.17 represents the dynamic, not static or dc, characteristics.

Here again is evidence of the R_s nonlinearity: the current compression in the topmost curves. Similar (although lesser) current compression is evident in Figure 3.5.

Before going on to see where the findings from the simulator lead, we note that in practical devices compression without crossover can occur in the elements of the standard equivalent circuit as evidenced by Figures 4.18 and 4.19, which show the fitted C_{gc} and g_{m0} for a 0.3 μm HFET. There are no peculiarities in the other elements, among them Figures 4.20 and 4.21, for R_{ds} and C_{gd}.

Returning to the implications of the simulations, the first contribution to the difference between the extracted and true nonlinear element functions is the topology of the standard equivalent circuit itself, the elements of which necessarily are discrete. It does not have enough elements in the right positions to model the distributed and

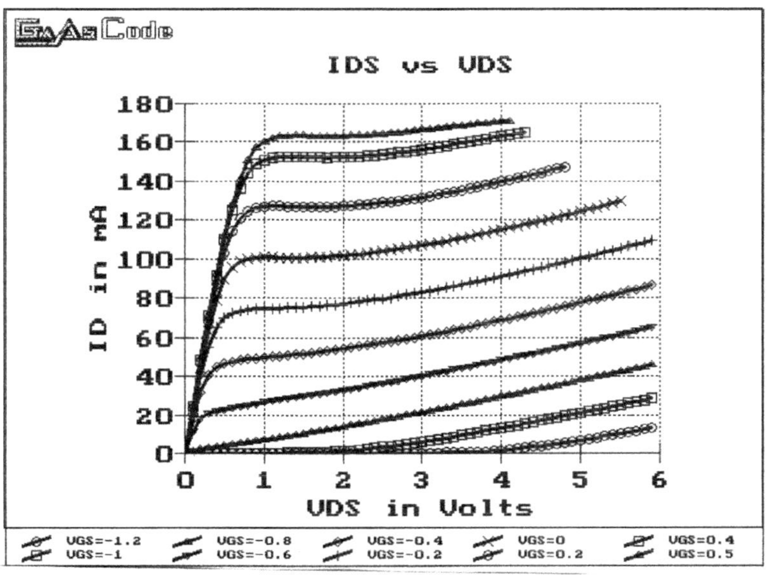

Figure 4.17 Dynamic drain characteristics (simulated).

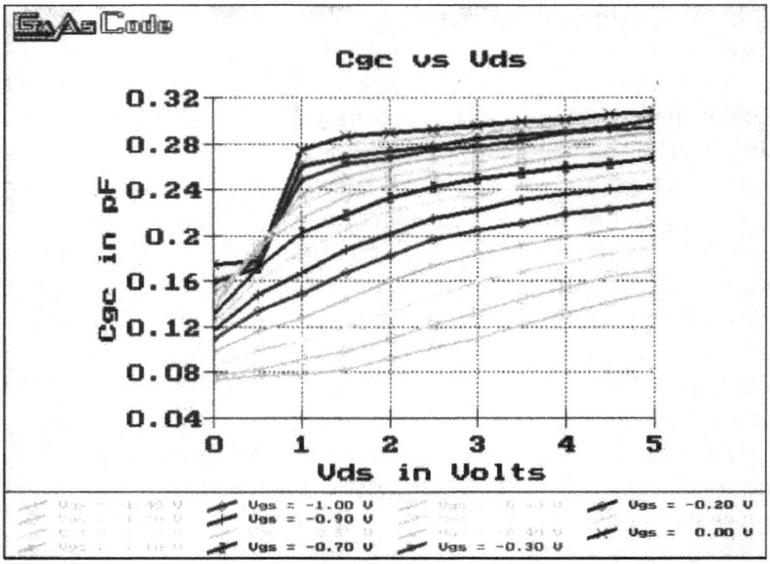

MATCOM 0.3μm FET

Figure 4.18 Fitted gate-channel capacitance C_{gc} for a 0.3 μm gate HFET.

interactive processes occurring within the device. The second contribution lies in the neglect of all nonlinear terms of the type $i\delta R$ and $v\delta C$ when fitting the standard model. Only terms of the type $R\delta i$ and $C\delta v$ are taken into account, which amounts to finding a bias-dependent linear circuit (extraction of a proper nonlinear model would require all terms to be accounted for simultaneously when fitting the circuit model to measured S-parameters). We now have to find a way of assessing the relative importance of these two factors.

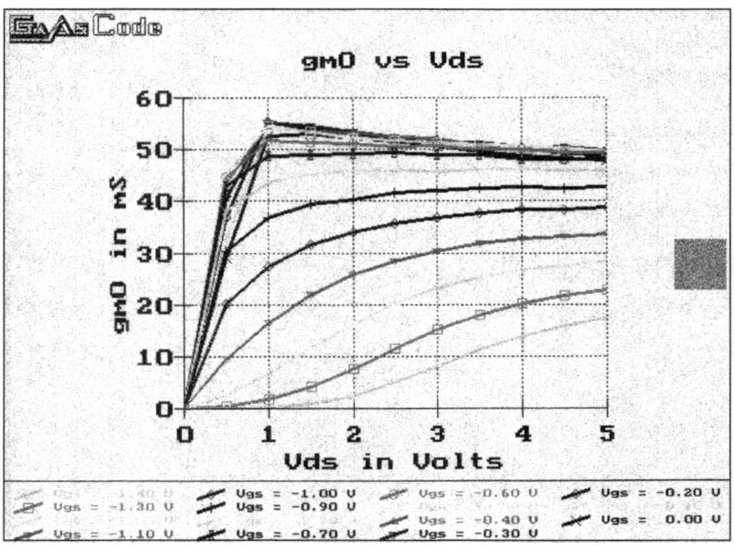

MATCOM 0.3μm FET

Figure 4.19 Fitted transconductance for a 0.3 μm HFET.

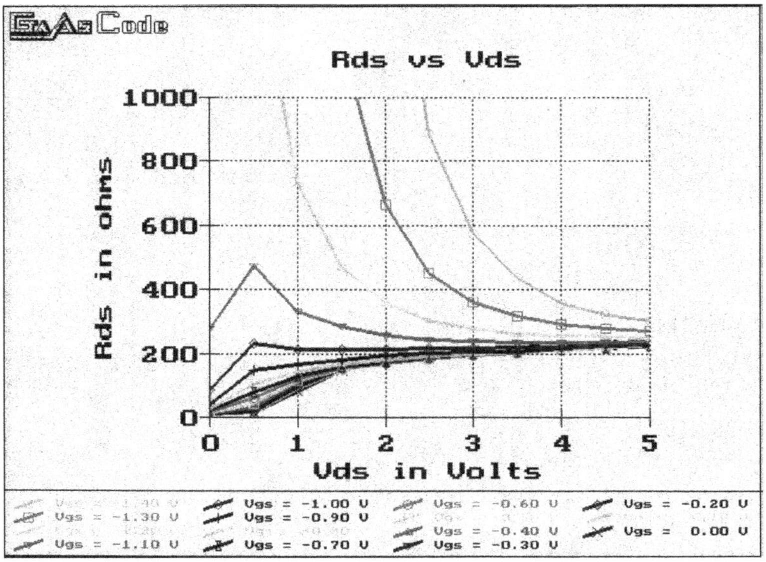

MATCOM 0.3μm FET

Figure 4.20 Output resistance fitted for a 0.3 μm HFET.

At the risk of being too abstract for a practically oriented text, there is value in reviewing in general terms the procedures involved in constructing a model where key nonlinear data are present and then become lost.

Consider that we have a 2-port, a box containing a real device and that we choose to measure the small-signal S-parameters at multiple bias points (Figure 4.22). We label the resultant set S2P-1.

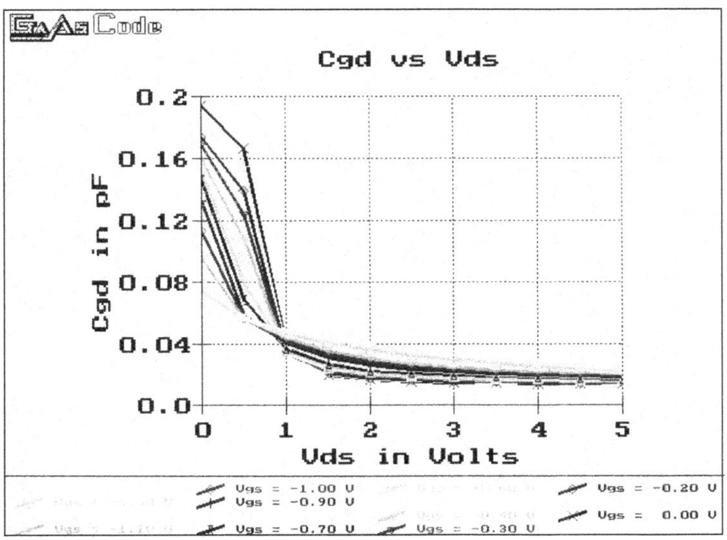

MATCOM 0.3μm FET

Figure 4.21 Fitted gate-drain capacitance for a 0.3 μm HFET.

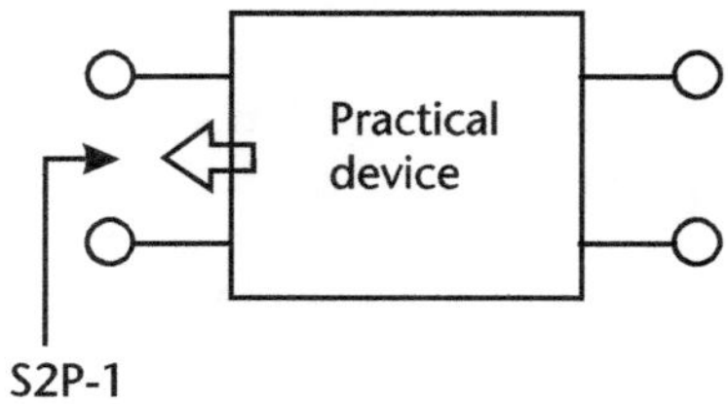

Figure 4.22 Concept of a practical device in a black box possessing a set of S-parameters S2P-1.

Nothing in this procedure has done anything to eliminate the effect on S2P-1 of nonlinear terms like $i\delta R$ and $v\delta C$. Even though we have elected to monitor only the local RF properties at each bias, all nonlinear processes have contributed to the result.

Suppose now that we have managed to develop an exact physical model of the practical device (the simulator we used above does not fit the exact requirement; it was indicative only). If the exact physical model is now substituted for the practical device in the box, under the same measurement conditions as before, we will see exactly the same set of S-parameters, S2P-1 (Figure 4.23), at the input port. Also, as before, the influence of all nonlinear processes is implicit in S2P-1.

In a sense, the physical model serves as a transform between a practical, physical device and its electrical properties.

Next, suppose we start with an empty box, but a box for which the S-parameters are known and are the set S2P-1. By inverting the transform, it should be possible to reconstruct what was in the box: that which was deconstructed by the forward transform (Figure 4.24). The input to the inverse transform is S2P-1, and the result is the reconstructed physical model that is now placed in the (empty) box.

Given that a physical model of sufficient fidelity to be of value in real-world circuit design does not exist, a different kind of inverse transform could be tried, one that is correlative with the exact physical model. The obvious contender is a circuit model wholly compliant and consistent with the physics of how the practical device operates (Figure 4.25).

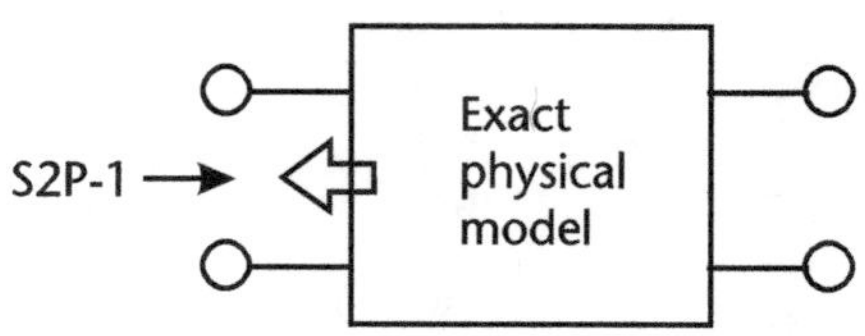

Figure 4.23 An exact physical model produces S-parameters, S2P-1.

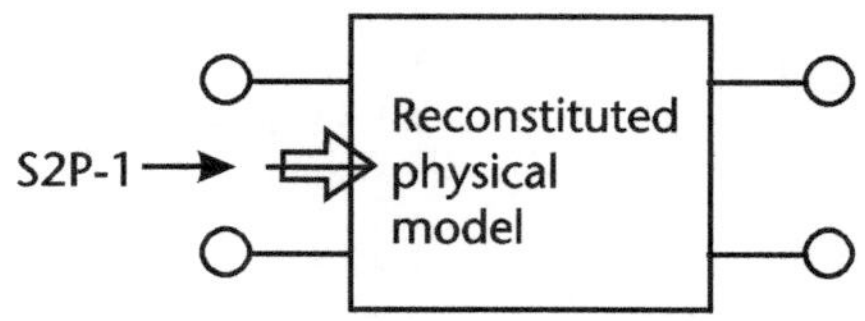

Figure 4.24 Reconstruction of the content of the box from S2P-1.

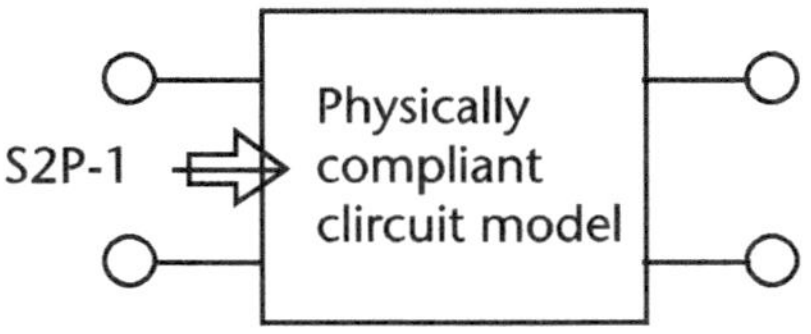

Figure 4.25 Substituting a physically compliant circuit model as a reconstruction in place of an exact physical model.

With S2P-1 as input, this new interpretive inverse transform places inside the box a description of the device that should perform exactly as did the physical model. If the new transform is a complete and true inverse, it will reconstruct the entire nonlinear properties of the device because it makes exactly the same use as before of the nonlinear data latent in the set S2P-1. If the inverse transform is incomplete, the reconstruction will be compromised. Neglecting the $i\delta R$ and $v\delta C$ terms amounts to an incomplete inverse transform.

Lastly, we take the case of an interpretive inverse transform, still of the circuit kind, but one that is not compliant with the physics (Figure 4.26). Despite being physically noncompliant, it nevertheless is capable of reproducing exactly the set S2P-1 (as in Section 3.2.2).

However, the difference now is that the box reproduces S2P-1 correctly but nothing else, with S2P-1 being properties local to each bias point. What the model retains of the nonlinear content in S2P-1 ends up in a disordered state, with the result that the nonlocal properties are compromised. In other words, the box contains a model that is in error for nonlocal or large-signal excitation, even though the fundamentals of nonlinearity were present in the S2P-1 from which it was derived.

In summary, the issue is not so much with the bias-dependent, small-signal source of practical data per se as with the way it is subsequently used. The $i\delta R$ and $v\delta C$ terms latent in S2P-1 become neglected only when fitting an equivalent circuit in the traditional manner, the consequences of which may be more severe for some topologies than others. An example of a circuit model less sensitive to error would be one with no wildly varying elements or, if such are unavoidable, two wildly varying elements that tend to cancel one another.

Experience gained through the present work suggests that for many HFETs the topology is the principal source of inaccuracy, the exceptions being devices like those in Figure 4.4, which are highly nonlinear.

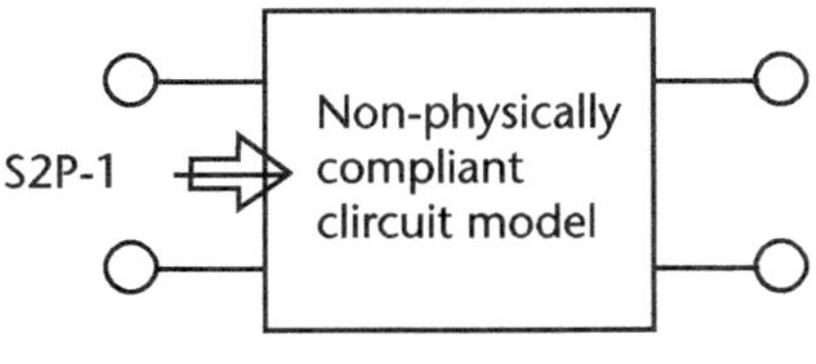

Figure 4.26 Reconstruction of a physically noncompliant model from S2P-1.

4.2.3 Consequences for Nonlinear Circuit Simulation

In this section, for the time being we overlook the fact that the standard model does not represent correctly the physical processes governing FET behavior and focus on the shortcoming in the last part of Section 4.2.2, that of neglecting nonlinear terms of the type $i\delta R$ and $v\delta C$ when fitting the model to measured data. Even if we assume the standard model to be an adequate approximation for nonlinear use (which it sometimes is), there is still fault with each element function arising from the manner in which we obtained it. For completeness, we next try to assess the magnitude of the error.

In the remainder of this section "the error" refers to the inaccuracy contributed by each element when the model is run in a nonlinear circuit simulator due to neglect of nonlinear terms of the type $i\delta R$ and $v\delta C$ in fitting the model to measured data.

Considering for simplicity only the input part of the equivalent circuit model (Figure 4.27), the objective of extraction is in essence to split i_G into i_{GC} and i_{GD}, to find the constituent parts of v_{GS} (being $i_G R_g$, v_C, v_{Ri}, and v_{Rs}), and then to find the element values, all as functions of V_{GS} and V_{DS}, which best fit the ratios (i_{GC}/v_C), (v_{Ri}/i_{GC}), (v_{Rs}/i_S), and so forth.

By deriving the model as a bias-dependent linear model, two categories of error are introduced that generate false terms when the model is run in a nonlinear circuit simulation:

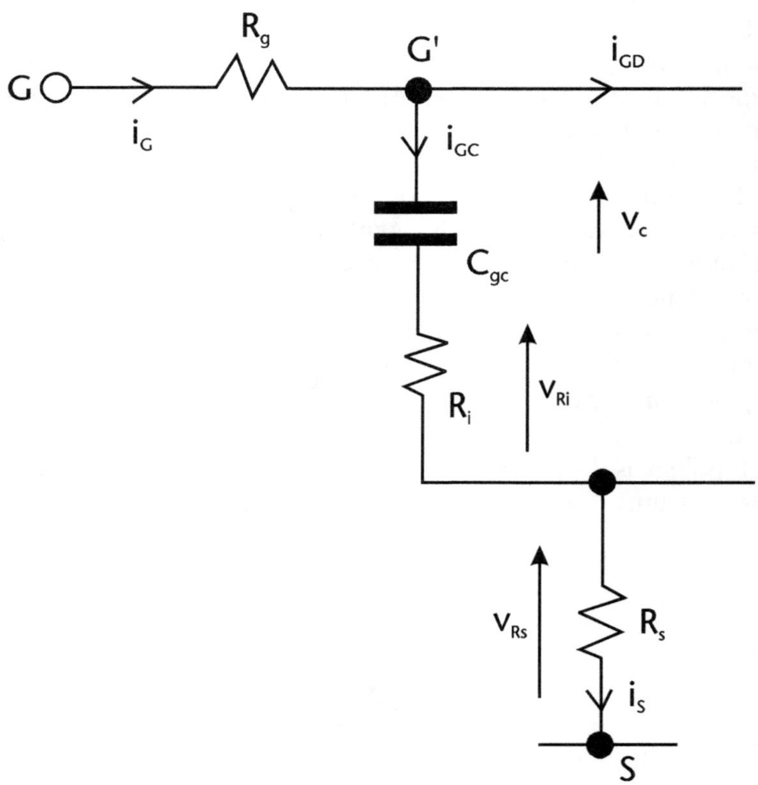

Figure 4.27 Input part of the standard circuit model.

1. Supposing the decomposition of i_G into i_{GC} and i_{GD}, and of v_{GS} into $i_G R_g$, v_C, v_{Ri} and v_{Rs} could be effected without error, C_{gc} should be determined from practical data by solving

$$i_{GC} = C_{gc}\frac{\partial v_C}{\partial t} + v_C\frac{\partial C_{gc}}{\partial t} \quad (4.1)$$

whereas, by treating the model elements as bias-dependent linear, C_{gc} is determined as

$$C_{gc} = \frac{i_{GC}}{\frac{\partial v_C}{\partial t}} \quad (4.2)$$

2. Given that, in step 1, improper element functions are fitted, and also that in practice the voltages and currents are found simultaneously with the circuit elements, there will be errors in i_{GC}, v_C, v_{Ri} .... as well. Although it is difficult to quantify these errors, that there is an error in principle is clear.

Recognizing that the same argument applies throughout the entire equivalent circuit, all the extracted element functions end up in error, although to a degree that differs from element to element as suggested by Section 4.2.2.

From this background, it is now possible to see what happens when a model extracted in the bias-dependent-linear approximation is lodged into a nonlinear circuit simulator. Denoting an extracted element by the superscript E and a true nonlinear element by T, from (4.1) and (4.2), we can write for the continuing example of C_{gc}:

$$C_{gc}^E = C_{gc}^T + v_C\frac{\partial C_{gc}^T}{\partial v_C} \quad (4.3)$$

When the simulator is run with C_{gc}^E, it gives rise to a change in the charge stored in this element of

$$\delta Q = \left(C_{gc}^E + v_C\frac{\partial C_{gc}^E}{\partial v_C}\right)\delta v_C \quad (4.4)$$

$$\delta Q = \left(C_{gc}^T + v_C\frac{\partial C_{gc}^T}{\partial v_C}\right)\delta v_C + \left(2\frac{\partial C_{gc}^T}{\partial v_C} + v_C\frac{\partial^2 C_{gc}^T}{\partial v_C^2}\right)v_C\delta v_C \quad (4.5)$$

What the simulator really needs is just the first term in brackets; the remainder is erroneous. Note that the unwanted terms still contribute when a small-signal simulation is done with a nonlinear (as opposed to linear) simulator.

Before attempting a rough estimate of the percentage error contributed by the second bracketed term, (4.3) offers an intriguing but untried possibility to recover the intrinsic or true element from the fitted or external element C_{gc}^E. For low frequencies where the capacitive reactance is large,

$$v_C \simeq V_{GS}$$

So (4.3) may be rewritten as

$$V_{GS}\frac{\partial C_{gc}^T}{\partial V_{GS}} + C_{gc}^T = C_{gc}^E\left(V_{GS}, V_{DS}\right) \tag{4.6}$$

At a given, fixed value of drain-source voltage, this is a first-order differential equation where the right side is known; it is the fitted element. A numerical solution would yield $C_{gc}^T(V_{GS})$ at the chosen V_{DS}. By repeating the calculation at all V_{DS} of interest, the result would be $C_{gc}^T(V_{GS},V_{DS})$, the true capacitance function.

On the face of it, an entirely similar differential equation with V_{DS} as the independent variable should apply:

$$V_{DS}\frac{\partial C_{gc}^T}{\partial V_{DS}} + C_{gc}^T = C_{gc}^E\left(V_{GS}, V_{DS}\right) \tag{4.7}$$

At each of a series of fixed values for V_{GS}, a solution for $C_{gc}^T(V_{DS})$ could be found numerically, again yielding a function $C_{gc}^T(V_{GS},V_{DS})$.

For the method to be viable, the two results would naturally have to be the same; otherwise, a solution in the two independent variables would have to be concocted instead of the split scheme.

Bearing in mind that here we are dealing only with methods for correcting the omission in the fitting process of terms like $i\delta R$ and $v\delta C$, attempting to reconstruct the true element in the foregoing manner would be worthwhile only in well-conditioned cases, as in Figure 3.21, for example. It would be pointless for cases like Figure 4.4 where extreme nonlinearity in the practical device has badly distorted C_{gc}^E, in other words, where the fault in C_{gc}^E does not arise from the forgotten nonlinear terms but from limitations associated with the standard model.

Returning now to the error arising from the second term in brackets in (4.5), the magnitude clearly depends on the shape or form of C_{gc}^T, presuming that we have some way of finding it. With that proviso, we can at least develop some idea by making use of Figure 4.6.

Staying with the gate-source voltage dependence for a start, for the topmost four curves, C_{gc}^T is only weakly dependent upon V_{GS} so, from (4.5) and again invoking $v_C \simeq V_{GS}$, little error can be expected in that region. Below, in the mid-range, the capacitance varies roughly as 0.1 pF per volt of V_{GS}. Picking the curve for $V_{GS} \simeq -0.4$V and the point where $C_{gc}^T = 0.2$ pF, the first bracketed term in (4.5), which is the true term the circuit simulator wants, comes to

$$(0.2 + 0.4 \times 0.1) = 0.24 \text{ pF}$$

and the error term comes to (the second derivative term is near enough zero)

$$2 \times 0.4 \times 0.1 = 0.08 \text{ pF}$$

Thus, the error in δQ is 33%.

Taking it that (4.7) for the V_{DS} dependence can be expanded similarly to (4.5), we begin by noting that for the uppermost curves in Figure 4.6 the capacitance varies approximately linearly as 0.02 pF per volt of V_{DS}. Then, focusing by way of example on the curve where $C^T_{gc} = 0.25$ pF at $V_{DS} \simeq 3$V, the correct term comes to

$$(0.25 + 3 \times 0.02) = 0.31 \text{ pF}$$

whereas the error term comes to (again with the second derivative near enough zero)

$$2 \times 3 \times 0.02 = 0.12 \text{ pF}$$

As a percentage, the error in the charge change δQ is therefore just under 40%.

These figures are illustrative only and may be summarized by the qualified statement: "... for a specific and restricted example, the charge change when C^E_{gc} is used rather than C^T_{gc} in a nonlinear circuit simulator is in error by several tens of percent."

However, it must be stressed that the outcome is wholly dependent upon the functional form of C^T_{gc}.

Looking beyond the example of gate-channel capacitance, we recall that similar considerations apply to the other equivalent circuit elements. There are no general rules for the magnitude of the error, and each element has to be treated on its own merits, device by device.

There is, however, an exception affecting g_{m0} and g_0. If fast-sampled pulsed $I(V)$ measurements are made to obtain the dynamic current generator function $i_D(v_{GS},v_{DS})$, and this measured function is used to replace g_{m0} and g_0, the replacements are the true RF-valid elements.

The error arising from ignoring the $i\delta R$ and $v\delta C$-type terms when fitting applies equally to the elements of the reformed model, which is the subject of Part II. Generally, these elements have forms that render them less susceptible to the error, at least judging by the examples given in Section 10.7. What we shall call the gate-channel branch capacitance, C_{IB}, on fundamental grounds can be expected to be flat with respect to V_{DS}, and its associated transcapacitance C_N has a similar form. The net gate-drain capacitance resembles Figure 4.10 and is slowly varying with respect to both V_{GS} and V_{DS}. A dynamic $i(v)$ current generator replaces g_{m0} and g_0. The resurgent τ appears fitted with no discernible error. That leaves the V_{GS} dependence of the aforementioned capacitive elements where the derivatives in (4.5) can be large.

It should be apparent that pulse-biased S-parameters do nothing to alleviate the error dealt with in this section, but they do correct another fault: they place the element functions on a correct voltage mesh rather than the distorted mesh, which results when the bias-dependent S-parameters are measured at dc. Section 10.2 covers the voltage net distortion.

4.3 Difficult Cases: Early SiC FET Example

Occasionally something crops up to tax whatever techniques we have for modeling devices. It would be welcome if such were outliers, but unfortunately it is not so. Difficult cases do arise and have to be dealt with on individual merit; each has its own quirks or foibles. For all that, an example serves to illustrate what we mean.

Figure 4.28 shows the measured and fitted S-parameters for an early SiC FET biased at V_{GS} = 0V, V_{DS} = 6V.

All the others in the multibias measured set showed similar unaccounted features. No clue was given as to their likely origin; the task put simply was: see what you can make of them.

Purely speculatively, the oscillatory nature of s_{22} suggests some kind of resonant behavior, the frequency being too high for the cause to be deep levels or traps.

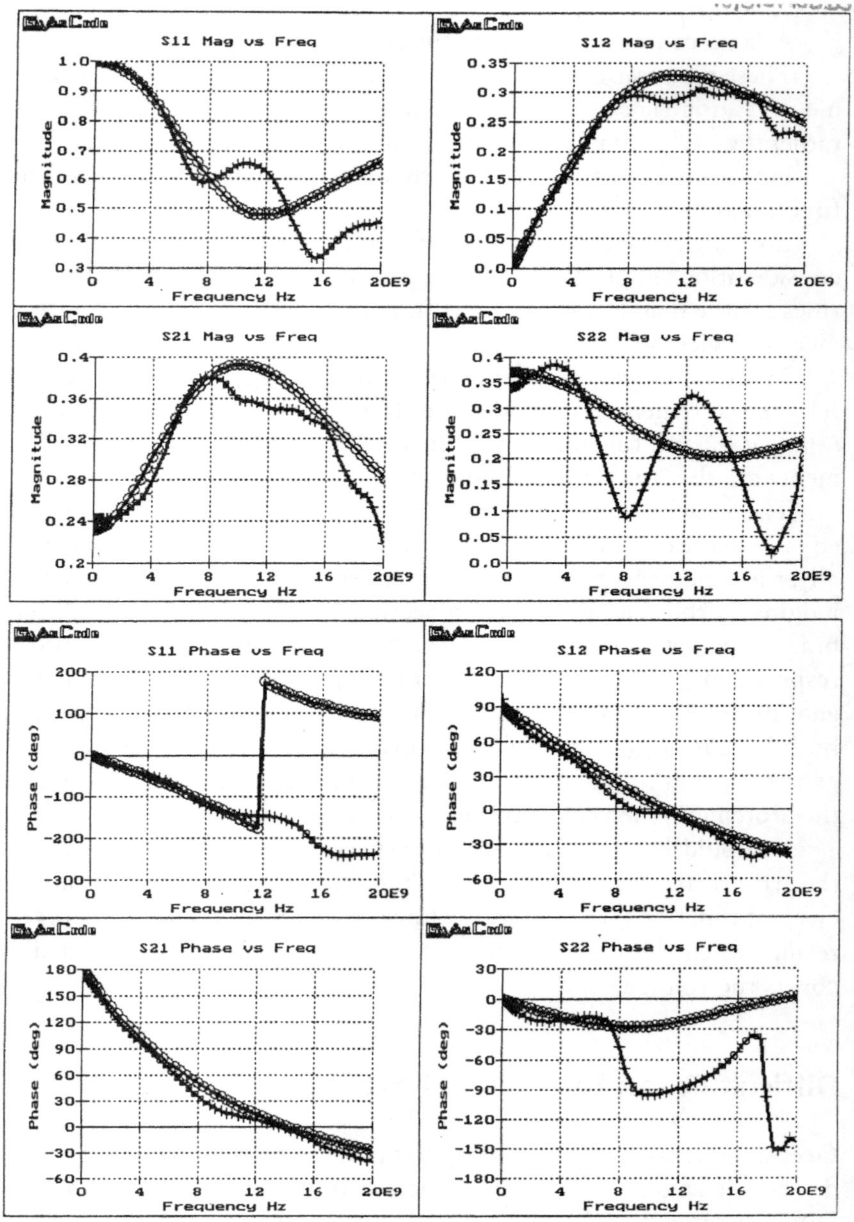

Figure 4.28 S-parameters at V_{GS} = 0V, V_{DS} = 6V, measured and fitted for a SiC FET.

As no information was provided concerning the device cross-sectional structure, the plan view layout, or the fixture in which it was mounted, again it was possible only to guess where the fault lay. One guess is that it was a packaged device and the cause was package parasitics. Irrespective, developing a package model as per Section 3.2.2 was not part of the plan.

A little background on the fitter is warranted to help to understand the results to follow. While not structured around physics-based algorithms directly, it nevertheless seeks relative values between elements that are physically believable or sensible. By this means, values found for an element or two that seem secure can help to constrain values for elements that are less certain. The commonplace model topology helps in this regard, as it has just about the minimum number of elements required to fit well the S-parameters of a real device; it does not have too many degrees of freedom. A circuit model with many more degrees of freedom would doubtless come close to fitting exactly the excursions in the measured S2P, but the result would be physically meaningless. To its credit, the standard model topology is close enough to the general manner in which HFETs operate physically to enable interpretation of the fitted elements for physical sensibleness after the fashion of the preceding sections.

There is no controversy about the current-voltage characteristics, which came with the measured S-parameter data (Figure 4.29).

Figure 4.30 shows the C_{gc} function that emerged from the fitting routine.

Despite the peculiarities in the measured S-parameters, the fitted element has an entirely reasonable-looking form. Although many SiC devices are constructed on a doped channel layer, the crowding together of the curves is more consistent with a 2-DEG channel than a bulk layer.

Figure 4.31 is the result for C_{gd}. Interestingly, the steep rise in C_{gd} towards V_{DS} = 0V is not in evidence for most of the curves. For reasons unknown, the shape of C_{gd} better resembles Figure 4.10 than Figure 4.9.

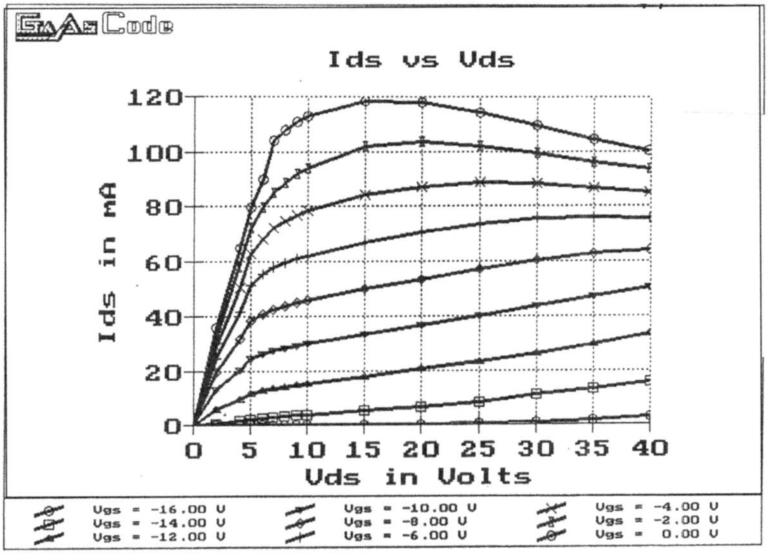

Figure 4.29 Measured drain characteristics for the SiC FET example.

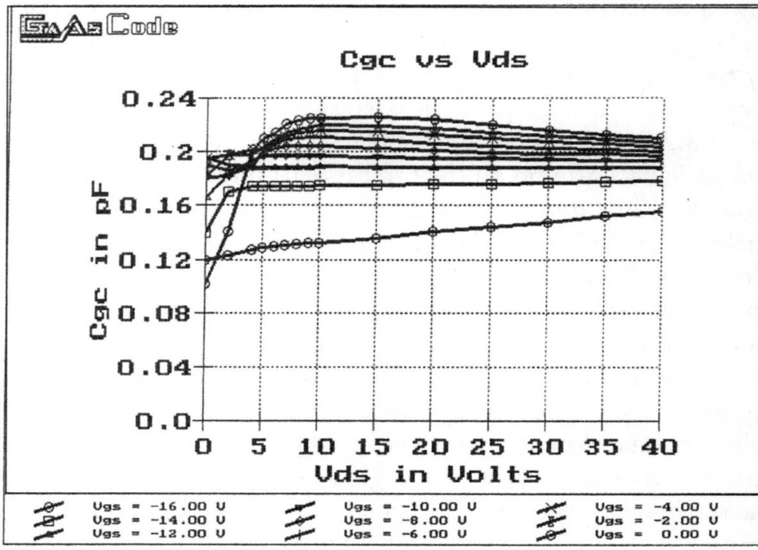

Figure 4.30 C_{gc} for the SiC FET.

Similarly, the form which eventuated for C_{ds} (Figure 4.32) is more akin to Figure 4.12 than Figure 4.11.

Bearing in mind that the general character of the fitted C_{gd} and C_{ds} near the origin has more to do with having to fit through the remainder of the equivalent circuit rather than physical behavior, we have no ready explanation for the results of Figures 4.31 and 4.32. It is worth noting, however, that the jagged fits in C_{ds} may be a manifestation of the uncertainty arising from the oscillatory nature of the measured s_{22}.

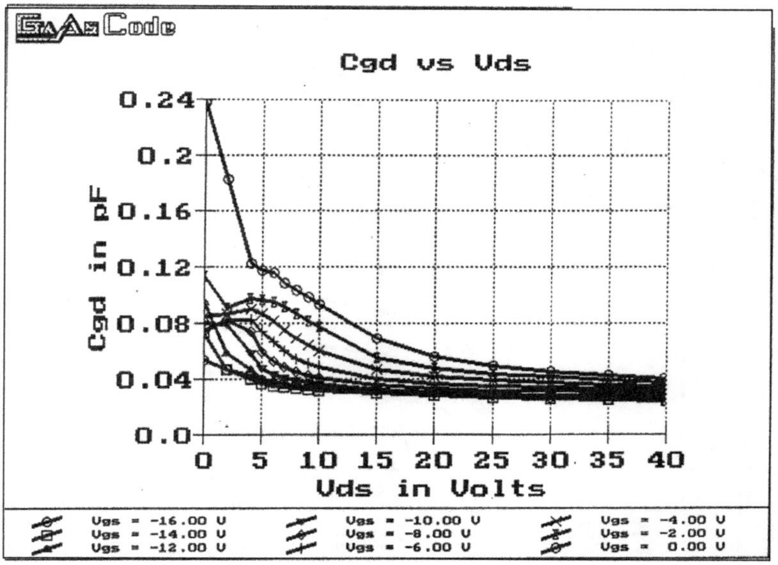

Figure 4.31 Fitted C_{gd} function for the SiC FET.

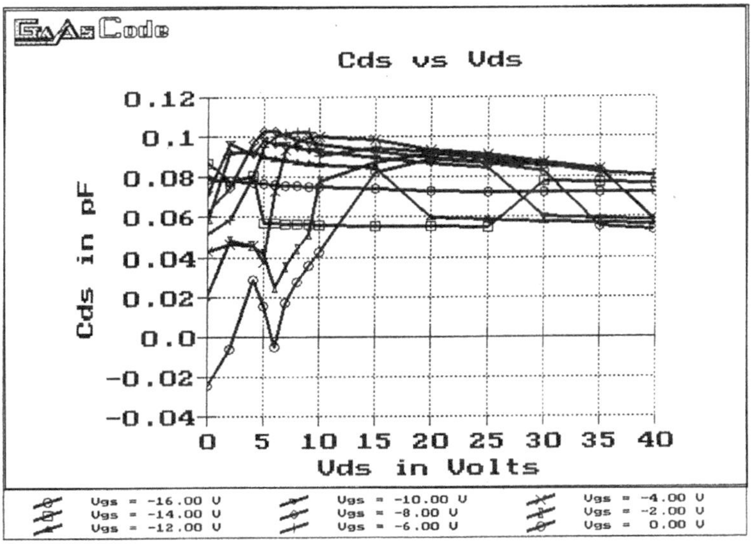

Figure 4.32 The fitted C_{ds} for the SiC FET.

Like the gate-channel capacitance in Figure 4.30, the transconductance of Figure 4.33 compresses at the most positive gate biases, which again suggests a 2-DEG (rather than bulk layer) channel. There too is rollover in the topmost curves: g_{m0} decreases as V_{GS} is increased positively.

The intrinsic channel resistance of Figure 4.34 is largest for the most positive V_{GS} (i.e., when the channel is open the widest), which is counterintuitive. Presuming this to be genuine and not an artefact, it implies the presence of electric fields, in the channel underneath the gate, large enough to degrade the carrier mobility.

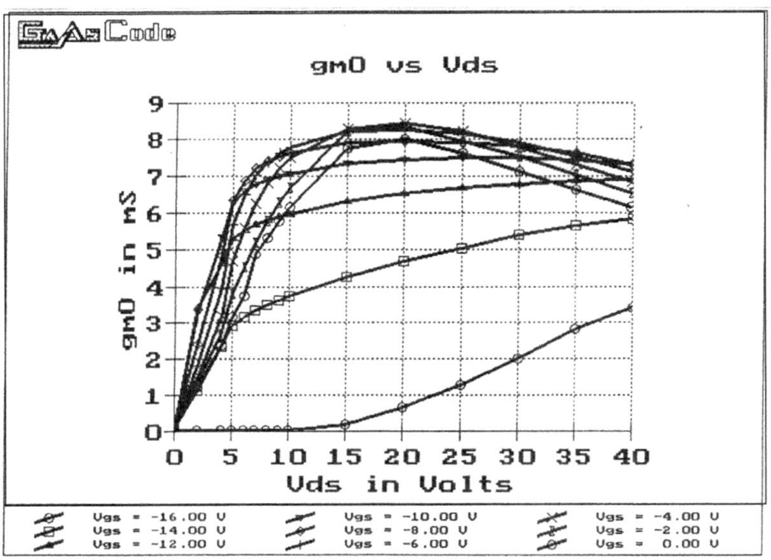

Figure 4.33 The fitted transconductance.

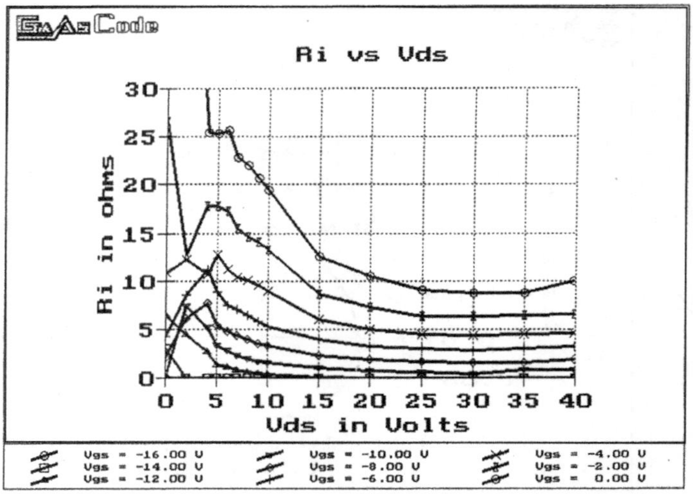

Figure 4.34 The intrinsic channel resistance for the example SiC FET.

There may be other interpretations, however. Suspicion is aroused by two of the curves lying on the horizontal axis, which, in turn, suggests an incorrect division of the total resistance in the gate-source branch between R_g, R_i, and R_s.

Without nasty surprises requiring awkward explanation, the drain-source resistance (Figure 4.35), behaves according to convention.

The only element in the equivalent circuit to be constrained during the process of fitting was the gate resistance (Figure 4.36). The 1Ω was pure guesswork on the "try it and see if it works" principle.

Finally, we come to the tell-tale element of the current generator delay τ.

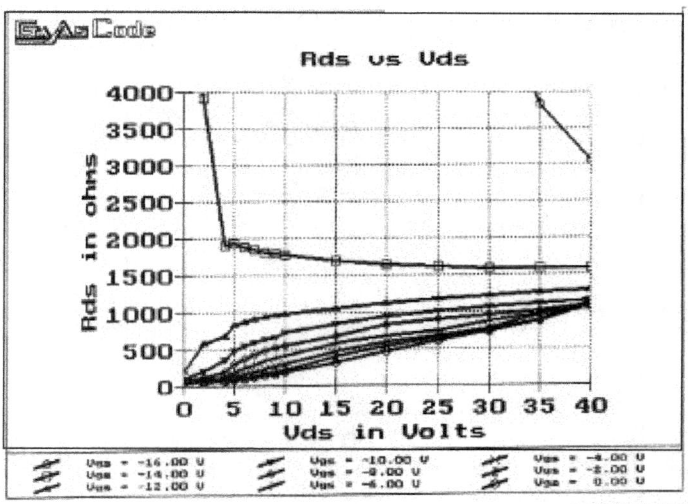

Figure 4.35 The output resistance.

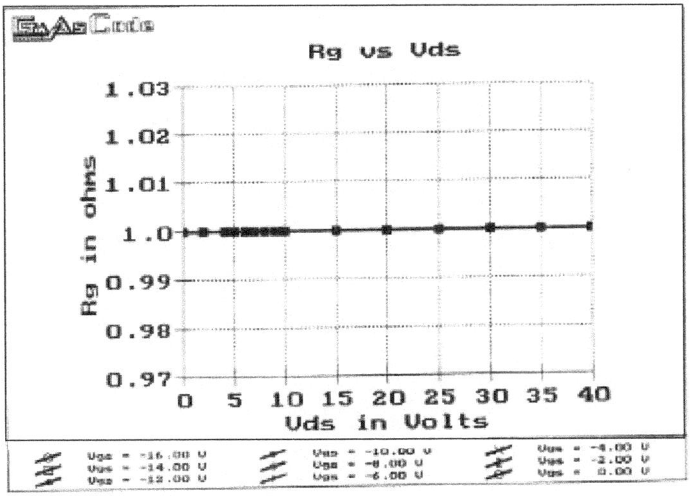

Figure 4.36 The arbitrary choice of 1Ω for the gate resistance.

In the 20V to 40V V_{DS} range, the ordering of the curves with respect to V_{GS} is unusual. Near pinch-off τ is relatively large, and it then decreases but increases again as the channel is opened up. One possibility lies with the channel doping profile.

τ is synonymous with X, the spatial extension of the depletion layer beyond the drain end of the gate. At a given value of V_{DS}, X depends upon the average density of positive ions through the depth of depletion: the larger the mean ion density, the smaller are X and τ. Consequently, the curve ordering in Figure 4.37 suggests a doping profile like that of a deep ion implant where the peak carrier concentration is in the middle of the channel.

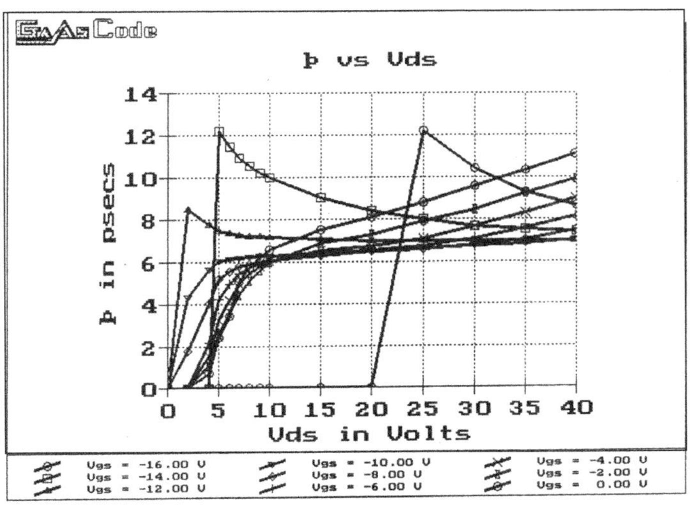

Figure 4.37 The multibias fit achieved for the delay τ in the current generator.

Overall, when faced with measurements like those in Figure 4.28, the difficulty is in knowing whether to spend time on it or assume that the whole proposition is not worth the effort. Much depends upon the fitting software coupled with whatever gave rise to distortion in the measurements in the first place. In the foregoing example, the fitter was able to see through the irregularity.

4.4 Improvements Towards a True Nonlinear Model

Before looking in greater detail at the problem of capacitance in the standard model, we briefly review what the findings so far mean in pursuit of a proper nonlinear model.

Even if the measurements are carefully done and without fault, two errors remain. Those errors are the topology of the standard equivalent circuit model itself and the neglect of nonlinear terms in the commonplace extraction procedure. Against this background, the possibilities for improvement are:

1. To retain the standard model topology but to include the missing terms of the type $i\delta R$ and $v\delta C$ generated by all elements;
2. To reform the equivalent circuit model so it conforms to the way that FETs and HEMTs work without compromise;
3. To underpin the equivalent circuit model with a physics-based model such that the circuit elements are related to physical variables that are then found by fitting the coupled physical and circuit model to measurements;
4. By measuring directly the large-signal conduction and charge functions.

In method 1, to find the element values at any (V_{GS},V_{DS}) point, it would be necessary to make use of practical data at (at least) four neighboring points to find the δR and δC magnitudes. Such an approach amounts to the same thing as the speculative method following (4.7) in Section 4.2.3. Like that variant, the suggestion of method 1 is untried. It involves a complexity well beyond that of day-to-day, bias-dependent, linear model extraction, which is difficult enough as it is.

Method 2 is the subject of Part II of this book.

Method 3 forms the basis of the BSIM4 MOSFET model [5] and is the direct counterpart of Figures 4.24 and 4.25. This method of nonlinear model extraction does what a physical device simulator does, only in reverse. How well it works depends on the efficacy of the physical model in the first place, and such models are notoriously difficult to construct. Our experience is that, for GaAs HFETs, a suitable simulator of adequate fidelity can be evolved [6], albeit in a rather protracted manner, whereas for a structure like a PHEMT, there are too many physical variables, mainly having to do with the material layers, which must be known with too great an accuracy for a simulator to be practicable.

A scheme conforming to method 4 has been promoted by Qi et al. [7], among others. While capable of giving right answers, the method takes to another level of complexity the instrumentation, calibration, and data retrieval required. The conduction and charge functions that result are those at the terminals of the device; they are not identified with specific physical mechanisms of device operation: they are black-box models. Thus, they are subject to the applicability reservations dealt with in Section 2.4. To cope with the kind of problem evinced by Figure 4.4, however, behavioral models may be the only way.

References

[1] Roblin, P., and H. Rohdin, *High-Speed Heterostructure Devices*, Chapter 16, Cambridge, U.K.: Cambridge University Press, 2002.

[2] Ladbrooke, P. H., *MMIC Design: GaAs FETs and HEMTs*, Norwood, MA: Artech House, 1989.

[3] Maas, S. A., *Nonlinear Microwave and RF Circuits*, Norwood, MA: Artech House, 1988.

[4] Parker, A. E., and D. J. Skellern, "A Realistic Large-Signal MESFET Model," *Asia Pacific Microwave Conference*, Tokyo, December 6–9, 1994.

[5] Aaen, P. H., J. A. Plá, and J. Wood, *Modeling and Characterization of RF and Microwave Power FETs*, Chapter 2, Cambridge, U.K.: Cambridge University Press, 2007.

[6] Ladbrooke, P. H., A. J. Hill, and J. P. Bridge, "Fast FET and HEMT Solvers for Microwave CAD," *International Journal of Microwave and Millimetre-Wave Computer-Aided Engineering*, Vol. 3, No. 1, 1993, pp. 37–60.

[7] Qi, H., J. Benedikt, and P. J. Tasker, "Nonlinear Data Utilization from Direct Data Lookup to Behavioural Modelling," *IEEE Transactions on Microwave Theory and Techniques*, Vol. MTT-57, No. 6, June 2009, pp. 1425–1432.

References

CHAPTER 5

The Capacitance Puzzle in the Standard Model

No matter how we choose to measure the $I(V)$ characteristics of a FET, either dc or pulsed, we have a pretty good idea not just of the current function itself but also what its derivatives g_{m0} and g_0 should look like. By comparison, how confident can we be about the form or shape of the capacitances, especially in light of the contradictions evident in Figures 4.9 and 4.10, and in Figures 4.11 and 4.12?

One thing is clear: when it comes to fitting multibias small-signal S-parameters, capacitances with a dependence on (V_{GS},V_{DS}) as in Figures 3.23 and 3.26 are the forms necessary to fit the measurements. What that means, to overstate the point, is that they are the forms that capacitances must have in the standard equivalent circuit model to account correctly for the small-signal properties of the device at the terminals. It means no more than that: it does not mean the same forms are in any sense physically correct, nor does it mean that are valid for use in a large-signal nonlinear model. So what, exactly, is their nature? Should it transpire that they are artefacts of the fitting scheme, their use in a nonlinear model would be hard to justify.

5.1 The Form of C_{gd} and C_{ds}: Fact or Artefact?

We saw from Figures 4.9 through 4.12 that the fitted forms of C_{gd} and C_{ds} in the commonplace equivalent circuit model exhibit steep changes as V_{DS} approaches 0V. Using a much simplified argument, we now demonstrate that such behavior is indeed more a property of the model than the actual device.

To begin with, we need to recall how the fitting routine works. When trying to find the value of any given element (in this case, C_{gd}), all the other elements are held fixed at their last iterated values while varying the element to be found to find the minimum error between the measured S2P and those of the model. Element by element, the whole iterative routine is repeated until the agreement improves no more.

Thus, the need is to construct a simple account of what goes on in the fitting process in the hope of exposing, in easily understood terms, why C_{gd} behaves as it does.

Figure 5.1 is a simplified version of the equivalent circuit when opened at the point where we wish to determine what kind of circuit element the fitter will fit there.

To determine the reactive element, we apply a test signal voltage v, work out from elementary network theory the current i, and then find the imaginary part of the ratio i/v. In the limit of small ω, which has no justification beyond seeking an algebraically simple result,

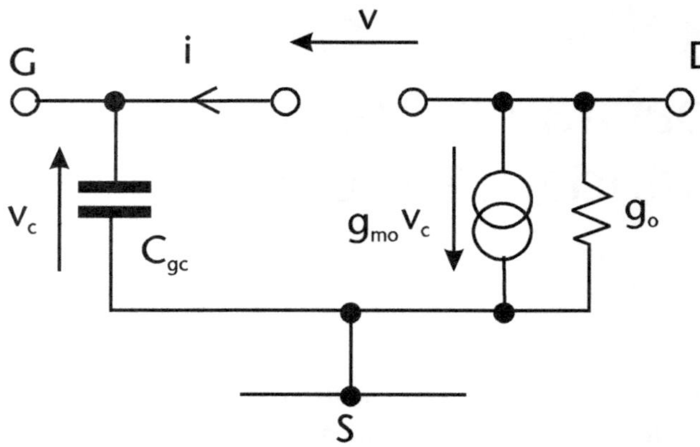

Figure 5.1 Minimal model for determining the form of C_{gd}.

$$\frac{i}{v} = \frac{j\omega C_{gc}}{1 + g_{m0}r_0}$$

where $r_0 = 1/g_0$. We call this ratio $j\omega C_{gd}$, so

$$C_{gd} = \frac{g_0 C_{gc}}{g_{m0} + g_0} \tag{5.1}$$

Equation (5.1) generates a function much like that in Figures 3.23, 4.9, and 4.21.

Herein lies an anomaly: to add a capacitance given by (5.1) to the terminals between G and D in Figure 5.1 amounts to double accounting. After all, Figure 5.1 already presents capacitance at those terminals of value C_{gd}, precisely because that is how (5.1) was derived. The only argument to justify the addition is that, if the gate and source were shorted in Figure 5.1, the circuit would present no capacitance at the open terminals, yet the real device would without question possess capacitance between the gate and drain. The fundamental problem is that the circuit model does not adequately represent the physical device; it does not have enough capacitive elements in the right places.

The inference is that C_{gd} given by (5.1) is an artefact in the following sense: it is what is required for the standard model to fit the measured S-parameters rather than being identifiable with anything physical in the FET. Irrespective, in accord with Figure 4.10 and as shown in Figure 6.39(b) and (6.15) of [1], a physical basis exists for the form of C_{gd} at, and above, the value of V_{DS} where the curves cross over and the device enters the saturated regime of operation. The principal doubt is the upsweep in the curves as $V_{DS} \to 0$V. Reservations apart, if an approximation to the fitted C_{gd} is required, as, for instance, in C from g, (5.1) has merit.

A similar low-level analysis can be attempted for C_{ds}. It results in the left side of (5.2).

$$\frac{C_{gd}C_{gc}}{C_{gc} + C_{gd}} \neq C_{ds} \tag{5.2}$$

The inequality sign is merely a way of indicating that this ratio of capacitances is not the answer to the shape of C_{ds} that we are seeking to explain, with the conclusion that the approach is in this case too simplistic. Once again, the main area of departure is below the knee: the left side of (5.2) suggests a positive-going upsweep like C_{gd}, whereas Figures 3.26 and 4.11 show an equally sharp negative-going feature. Which to believe? The answer is probably neither; there is not a simple way of diagnosing the problem. Instead, the best guide that we have is the simulator result of Figure 4.12, which suggests that the space-charge part of C_{ds} is small and close to being a constant, as in the upper reaches of Figure 3.26.

So far as C_{gd} and C_{ds} are concerned, therefore, there is good reason to believe that the functions extracted from measured S-parameters are not appropriate for a nonlinear model. More on the subject of these two capacitances is given in Section 15.4.2.

5.2 The Composition of C_{gc}

From Section 15.4,

$$C_{gc} = C_{IB} + g_{m0}\tau_{gm} - C_N \tag{5.3}$$

where C_{IB} is the gate-source branch capacitance in the reformulated equivalent circuit, g_{m0} is the transconductance, τ_{gm} is the pure delay governing the drain current, and C_N is the transcapacitance in the gate-drain branch of the reformed model.

Each of the three terms on the right side of (5.3) has a single, clearly identifiable, physical origin. By implication, C_{gc} is a composite element that we now seek to deconstruct.

Given that the process of model extraction yields C_{gc}, g_{m0}, and τ_{gm} as functions of (V_{GS},V_{DS}), using experimental data, we can form

$$C_{gc} - g_{m0}\tau_{gm} \tag{5.4}$$

to find what $(C_{IB} - C_N)$ looks like. Then, allowing the approximation

$$C_{IB} \simeq \frac{g_{m0}L_G}{v_{sat}} \tag{5.5}$$

we can gain an impression of what C_N alone looks like provided that we have a value for the gate length L_G.

Before proceeding with example deconstructions, it is as well to be aware of the character of $g_{m0}\tau_{gm}$. The product $g_{m0}\tau_{gm}$ is an out-of-the-ordinary circuit element that models charge storage by pure delay. It is not a capacitance. In the small-signal limit, it can be reduced to a capacitance, but in that limit only. For the time being, we may regard it as capacitance because here we are dealing with elements derived from small-signal S-parameter measurements. Under large-signal conditions, $g_{m0}\tau$ must be incorporated into a circuit model for what it is, an element founded in pure delay, a challenge that requires a different way of thinking.

From Figures 3.21, 3.20, and 3.28 for C_{gc}, g_{m0}, and τ_{gm}, respectively, we can find $(C_{IB} - C_N)$ for the 0.5 μm gate FET examined at length in Section 3.2.4. With $v_{sat} = 10^7$ cm s^{-1}, we can find C_{IB} and hence C_N.

Figure 5.2 shows the result for $V_{GS} = -0.2V$ (after allowing a geometric (or pad) capacitance of $C_{gcgeom} \simeq 0.05$ pF).

Three facts emerge:

1. C_N has a form similar to C_{IB}.
2. C_N is smaller than C_{IB} but is of comparable magnitude.
3. Both facets are broadly in accord with expectation as evolved in Chapter 15.

Even more revealing are the relative values of the components of capacitance making up C_{gc}. Taking $V_{DS} = 3V$ to illustrate the point, the space-charge part of C_{gc} is

$$(0.128 - 0.05) \text{ pF} = 0.078 \text{ pF}$$

So

$$\begin{aligned} C_{gc} &= C_{IB} + g_{m0}\tau_{gm} - C_N \\ 0.078 &= 0.115 + 0.035 - 0.072 \end{aligned}$$

Herein lies one of the keys to the failure of the standard model. $(C_{IB} + g_{m0}\tau_{gm})$ determines how much gate-source branch current flows in response to changes in V_{GS}, while C_N determines the current flowing in the gate-drain branch in response to those self-same changes in V_{GS}. Thus, we conclude that the transcapacitance current (flowing in the orthogonal branch) is half the branch current (flowing in the branch where the voltage was applied). It is what happens in the practical device. Yet in the standard model where the three component parts are consolidated into the single element C_{gc} in the gate-source branch, all the current flows in that branch. To stress the point, the standard model has large reactive currents flowing in the wrong branches.

All the curves of Figure 3.21 can be deconstructed similarly. To the extent that (5.5) is a reasonable approximation, at least in the v_{sat} regime, we should expect C_{IB} to resemble the form of g_{m0}. (Section 10.7 is given over to practical results for the elements of the reformed model, all elements being freely extracted without constraining relationships between them.)

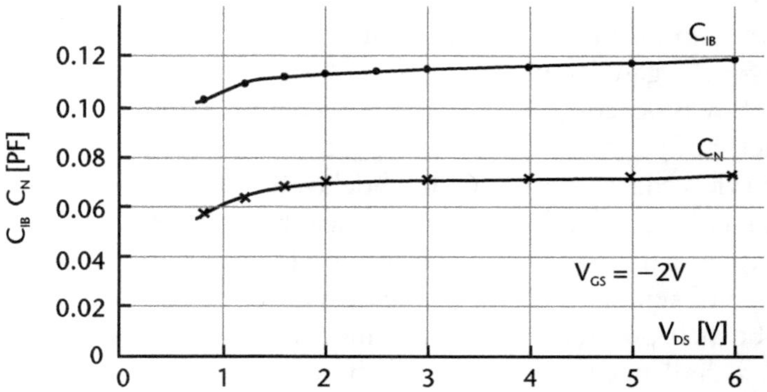

Figure 5.2 Components of C_{gc} resulting from deconstruction using practical data for $V_{GS} = -0.2V$ from Section 3.2.4.

To assess whether breaking down capacitance components in this way is more widely applicable, the SiC FET data in Section 4.4 provides an even stiffer test. Figure 5.3 results after taking data from Figures 4.30, 4.33, and 4.37 for $V_{GS} = -8$V (chosen arbitrarily).

Bearing in mind that no structural information was available for this device, some guesses had to be made to construct Figure 5.3. First, the crystal structure, which among other things determines the saturated drift velocity of electrons as needed for (5.5), was not known. At 300K, typical values quoted for the saturated drift velocity are in the range 0.8×10^7 cm s^{-1} to 2×10^7 cm s^{-1} [2]. Knowing no better, we assume $v_{sat} = 10^7$ cm s^{-1}.

Next, a value is required for the gate length. A rough estimate of what the gate length can be made from the measured delay τ. The relationship between τ_{gm} and X from Section 13.8 comes in handy:

$$\tau_{gm} = \frac{X}{2v_{sat}}$$

In Figure 4.37, τ_{gm} is in the range 6 ps to 8 ps, implying $X \simeq 1.5$ µm. As experience suggests X can be around L_G to $2L_G$ for large gate-drain voltages ("large" being a relative term depending upon the material and device type), we speculate that $L_G \simeq 1$ µm.

These, then, are the estimated values used in the deconstruction in Figure 5.3 (other values could be tried). The accompanying uncertainty in the magnitude of C_{IB} and C_N has to be borne in mind when making consequent deductions.

Irrespective, while C_N is not as large a proportion of C_{gc} as it was in Figure 5.2, it is still large; about a quarter of the current through C_{gc} is flowing in the wrong branch. In a model true to the actual device, it should be flowing in the gate-drain branch.

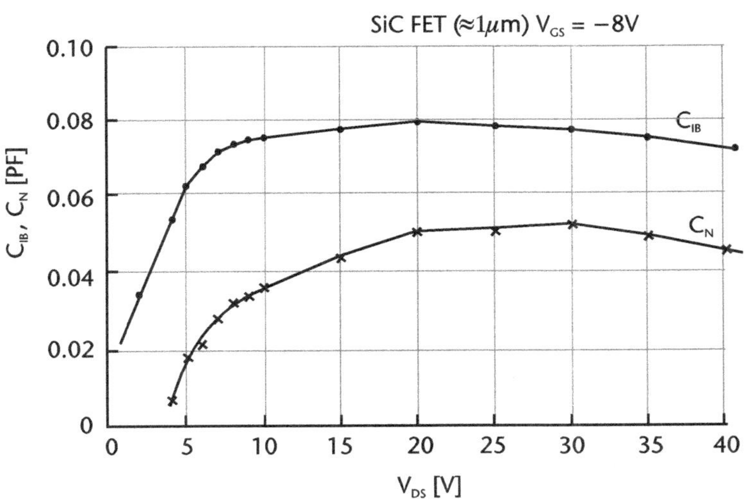

Figure 5.3 Example decomposition of the gate-channel capacitance for a SiC FET into branch capacitance and transcapacitance. The curves have been extended below the strict range of validity, which is the v_{sat} range.

5.3 C from g: Deriving Capacitance from Conductance

The work of the previous two sections suggests the possibility of constructing approximate capacitance functions for the standard model from the transconductance g_{m0}. Viewing capacitance functions in the round, the limits of common practice to recover C_{gc}, C_{gd}, and C_{ds} from measurements mean there are inaccuracies, as evidenced by the examples in Section 4.2.2. These inaccuracies create a little leeway that can be exploited: if, in the standard model, inaccuracies are hard to eliminate, can we do comparably well (or badly, depending upon how we perceive it) by pursuing a much simpler approach?

Elementary principles of device physics establish relationships between capacitance functions (being derivatives of the charge function $Q(v_{GS},v_{DS})$) and the conductance functions $g_{m0}(v_{GS},v_{DS})$ and $g_0(v_{GS},v_{DS})$ (being derivatives of the current function $i_D(v_{GS},v_{DS})$). From these generic relationships, useful approximations for the capacitance functions can be derived from pulsed $I(V)$ measurements. The resultant capacitances are large-signal elements and are bias-point dependent.

Some basic physical information about the device is needed to effect the recovery of capacitance. At its most basic, the metallurgical gate length needs to be known. For some variants of the method, the delay element τ also needs to be known, which requires network analyzer measurement and so partly defeats the object of a simple technique. Further, the method yields information about the space-charge component of capacitance only, so network analyzer measurements are still required to determine the geometric and parasitic components of capacitance.

Different degrees of sophistication exist, from basic to convoluted. An example of the latter can be found in [3]. Despite showing promise and offering intellectual interest, the more elaborate technique has not been pursued because it centers on the standard equivalent circuit, which has limitations for nonlinear use. Nevertheless, the more basic implementation, which requires the delay element τ, finds ready application in dissecting and understanding the composition of extracted element functions as in Section 5.2.

To summarize what was said there, the individually identifiable changes of charge in the depletion region in response to a change in gate-source voltage lead to the following expression for C_{gc}:

$$C_{gc} = C_{IB} + g_{m0}\tau - C_N \tag{5.6}$$

where C_{IB} is the gate-source branch capacitance of the reformed circuit model, g_{m0} is the transconductance, preferably derived from the dynamic $I(V)$ function, τ is the pure delay governing the drain current, and C_N is the transcapacitance in the gate-drain branch of the reformulated model.

In the reformed model of Part II, these elements are empirically derived, exposing the subtleties of each. For the lowest-level prediction scheme under discussion, the following approximations apply (in the saturation regime only):

$$\begin{aligned} C_{IB} &\simeq \frac{g_{m0} L_G}{v_{sat}} \\ C_N &\simeq C_{IB}\left(\frac{X}{L_G + X}\right) \end{aligned} \tag{5.7}$$

The expression for C_N is derived in Section 15.2.3. C_N has a form similar to C_{IB} but smaller. For instance, if $X = L_G$, then $C_N \simeq C_{IB}/2$, whereas for $X = 0.5L_G$, $C_N \simeq C_{IB}/3$.

Having thus found an approximate function for C_{gc}, we can find C_{gd} in the form that results from fitting the standard model to multibias S-parameters:

$$C_{gd} \simeq \left(\frac{g_0}{g_{m0} + g_0} \right) C_{gc}$$

(this relationship is derived in Section 5.1).

Expressed simply, these procedures reverse what was done in Sections 5.2 and 5.1. Instead of applying the equations to break down a practically derived $C_{gc}(V_{GS},V_{DS})$ element into its component parts by forming

$$C_{IB} - C_N = C_{gc} - g_{m0}\tau$$

the starting relationship

$$C_{gc} = C_{IB} + g_{m0}\tau - C_N$$

is used to form C_{gc}.

As a useful approximation, in some practical devices C_N can be regarded as only weakly dependent upon V_{DS}, as evidenced by Figure 5.2: it varies less than implied by (5.7) (it is, however, a strong function of V_{GS}). For a partial explanation, we can look to the derivation of (5.7) under the condition $V_{D'G}$ = constant, which differs from the horizontal axis in the empirical plot. Nevertheless, a few examples working the equations in reverse provide useful background to using them for quick C from g predictions.

5.4 Standard Model Capacitance in Review

For easy reference, we finally gather in one place the key issues emerging in the capacitance puzzle as it applies to or in the standard model.

Even for straightforward cases, the capacitive elements fitted in the standard model are distortions of the true elements required to model correctly the physical processes governing device behavior.

C_{gd} as fitted is, below the knee, an artefact of the model itself although, in the saturation regime, it bears a fair resemblance to the physically correct element. Just like C_{gc}, the lone element C_{gd} in the gate-drain branch places reactive current in the wrong branches of the model.

C_{ds} is more difficult to discern. In principle, the space-charge part of C_{ds} should show only weak dependence upon (V_{GS},V_{DS}) so, coupled with the geometric (or pad) capacitance, it should be near-enough constant. In typical extracted versions the expectation is borne out. Below the knee, however, the fitted C_{ds} often exhibits a sharp negative excursion. Once again, suspicion rests with it being a property without physical basis but one resulting from the standard mode topology itself.

The capacitance C_{gc} as fitted to small-signal S-parameters is, naturally enough, a small-signal element like the others. Here though, there is a difference. C_{gc} embodies a coupled delay-transconductance term that can only be regarded as a capacitance in the strict small-signal limit. It should not be forgotten that accepting it as a capacitance is an approximation; in truth, the term $(g_{m0}\tau)$ represents charge storage by pure delay, something of a wayward concept by mainstream standards. Regardless, a large fraction, typically a quarter to a half, of the reactive current to which C_{gc} gives rise should not be flowing in the gate-channel (or gate-source) branch at all: C_{gc} comprises a large transreactance current that should be flowing in the gate-drain branch instead.

Beyond the straightforward cases, other challenges now and again turn up. An example from which much was learned involved the PHEMT wherein a crossover occurred in the extracted C_{gc} near the open channel (Figure 4.4). The effect is not real insofar as the true element internal to the FET does not behave in that manner, but at the terminals C_{gc} has this characteristic due to extreme nonlinearity in R_s (in turn, due to the penetration of high electric fields into the gate-source region). The distortion in the (V_{GS},V_{DS}) dependence of C_{gc} arises from V_{GS} being determined by a nonlinear potential divider, internal to the FET, of which R_s is a part. If R_s were not itself distorted by the need to compensate for topological defects in the standard model (see the end of Section 4.1), its true form could be established in fitting. Then, in principle, V_{GS} could be corrected to unhook the crossover but, as it is, there is no chance so to do. Such instances are a problem without a clear solution. The restriction applies equally to the reformed model of Part II. Behavioral models apart, the only way so far envisaged to deal with problems of this nature is to construct a true physics-based model of the type used in Section 4.2.2. Not only is that is beyond the remit of this book but the degree to which such a simulator conforms to practice would have to be high for it to be adequate as the basis for correcting distorted elements. How many years of effort that would entail is anybody's guess.

References

[1] Ladbrooke, P. H., *MMIC Design: GaAs FETs and HEMTs*, Norwood, MA: Artech House, 1989.

[2] *Silicon Carbide (SiC)—Transport Properties in High Electric Fields*, NSM Archive, ww.iofe.ru.

[3] "Pulsed I(V) Measurement of Semiconductor Devices with Applications," Chapter 8, Appendix 8B, *Accent Optical Technologies Application Handbook*, 2004.

CHAPTER 6

Dynamic $I(V)$ Measurements

By way of introduction to pulsed measurement instruments, there is a riddle that will make sense only to those with the patience and persistence to read and understand Chapter 10: A curious fact is that realizing that you may not need one only comes from having one.

6.1 Development of a Desktop Pulsed $I(V)$ Instrument

With the availability of fast digital ICs including microprocessors, pulse generators, sampling gates, and memory, the prospect of being able to manufacture an instrument to measure the dynamic $I(V)$ characteristics of HFETs became a reality. If a small, portable, affordable instrument could be made available, it would open up such measurements to small design houses, contract engineers, and university researchers so that the technique would no longer be the sole province of large companies.

From humble beginnings, such a piece of equipment was developed and marketed for less than $10,000, including Application Notes and a protective, soft carry case (Figure 6.1).

Working in collaboration with another small company in the Cambridge Science Parks area, a single-channel instrument was developed for studying the rate dependence of gate-drain breakdown in GaAs FETs, HEMTs, and PHEMTs. Fifth Axiom was a small enterprise with special expertise in parallel computing, a natural for creating the hardware [1]. Although the alliance began with modest aims, it evolved through all phases of development and commercialization of the complete instrument and its eventual sale and technology transfer to Accent Optical Technologies of the United States. Continuing with the original personnel, Accent expanded the effort to develop higher-voltage versions, marketed internationally under the DiVA name.

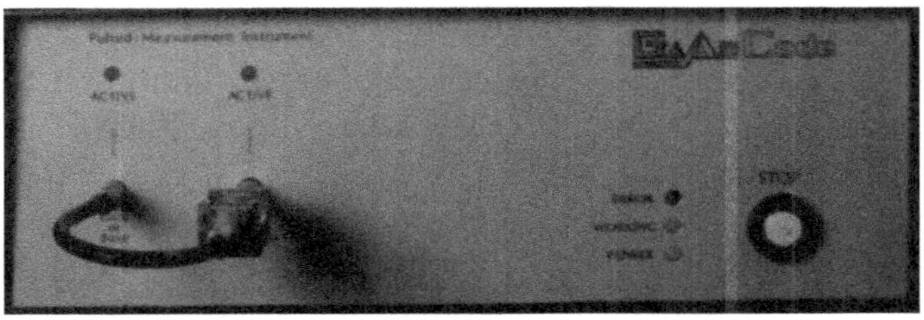

Figure 6.1 Frontal view of the compact pulsed $I(V)$ measurement instrument.

The impetus for the whole development, the gate-drain breakdown study, was research contracts let by a department of the U.K. Ministry of Defence. Following the work of Barton [2], it had become clear that pulsed measurements could reveal much about the role of deep levels, particularly surface states, in determining the breakdown characteristic of these devices.

Having engineered a single-channel instrument capable of setting a dc bias on the device and then applying fast pulses starting from that point, conceptually, at least, it was a short step to duplicate it to create a two-channel instrument. From a hardware point of view, further development was needed to synchronize the channels and, from a software point of view, a whole new operating system needed to be developed: the architecture of the instrument was what was termed "virtual," a somewhat obscure description for a piece of test equipment without knobs, switches, or dials but commanded and run entirely by a PC. The software provided full graphical processing and display facilities.

As is often the case, it soon came to light that the idea of a compact pulsed $I(V)$ measurement instrument was not new: at the University of Kent, Canterbury, England, Jastrzebski and coworkers Donarski and Barnaby had already developed a well-engineered, rack-mounted version [3]. Although not developed for commercialization, one of their instruments could be purchased for research use. Collaboration with them led to the application of the instrument in determining the physical origin of some rather obscure device characteristics [4].

Ideas evolve. From work done at Plessey Research Caswell in 1978 to 1979, it was clear that GaAs power FETs exhibited different $I(V)$ characteristics when measured on a Tektronix curve tracer depending upon the measurement conditions. The Tektronix curve tracer not only made provision for the voltages applied to the device under test to be true dc, it allowed the drain source to be driven by a voltage sweep at the mains frequency while the gate was pulsed, at a low duty ratio, with 80 μs or 300 μs pulses. With a duly historical air about it, Figure 6.2 is an original example. It raised all manner of questions: what was causing the loops in the dc characteristics, and what was going on to make the pulsed set so different from the dc set? These questions and their answers are as relevant today as they were then.

To explain the loops, we have to look at how the curve tracer works. A steady voltage is applied to the gate of the FET and the drain is swept slowly from $V_{DS} =$ 0V up to a maximum and then down again with a rectified sinusoidal waveform. Loops in the $I(V)$ characteristic could only develop if something had changed within the device between the upward and downward parts of the V_{DS} sweep. Suspicion rested on deep levels, or slow states, by virtue of the scale of time on which they operate, typically tens of microseconds in GaAs and up to units of seconds in the wider bandgap materials like GaN and SiC. It did not much matter where in the device they existed as long as the sites were accessible by mobile electrons. Some, if not all, electrons trapped on the upsweep would be held long enough to affect the size and shape of the depletion region on the downsweep. Given that the depletion layer governs how much current the channel can carry, here in principle was a mechanism with a chance of explaining the curve tracer's dc results. With this as the probable reason, the pulsed set of $I(V)$ characteristics seemed less obscure but still demanded a detailed explanation.

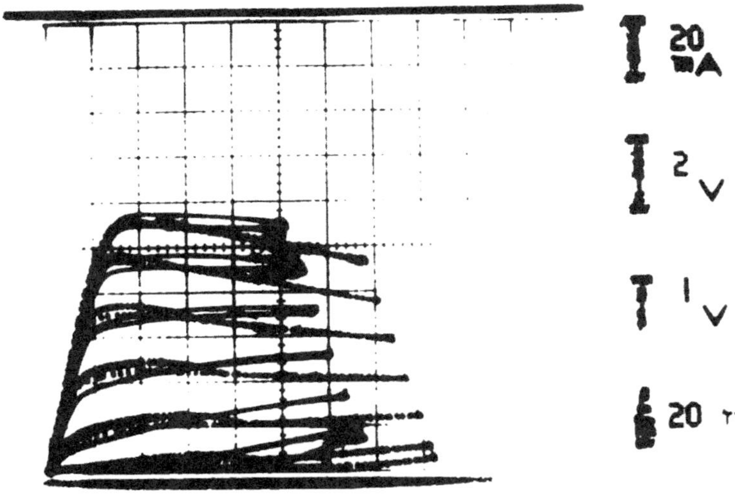

Figure 6.2 GaAs FET characteristics measured using the 80 μs pulsed drain facility on a standard curve tracer from the 1970s.

The problem rested there until Barton, who was aware of the work, took it to the next level and assembled what must have been one of the earliest two-channel, fast-pulse measurement systems [5].

Figure 6.3 shows Barton's experimental arrangement. Pulses of length up to 100 μs were applied at a repetition rate of 500 Hz. Importantly, the current could be sampled at times down to 100 ns into the pulse. Some results from his investigations were given in Figure 3.10.

To function as a fully fledged dynamic $I(V)$ characterization system, Barton's implementation needed the addition of a facility to set, anywhere on the $I(V)$ plane, a bias point from which both positive-going and negative-going pulses then emerge.

Not all pulsed measurement instruments are the same: there are variants for specific tasks. Depending upon the material from which the HFET is fabricated, pulses of different length and duty cycle (as well as magnitude) are required. What matters is whether the pulses are in accord with the response times of the memory effect: the characteristic capture and reemission times of deep levels or thermal time constants when self-heating is the prime concern.

The interpulse period is important: it must be long enough to allow the memory mechanism to reset fully to the (presumed) steady state in which it was before the pulse was applied. GaN and SiC FETs may require very different pulse properties from GaAs devices. Generally, in GaAs, a reset time of a few milliseconds is sufficiently long, but in the wide bandgap materials, several seconds may be nearer the mark.

At the other end of the timescale, the pulse rise time must be shorter than the time that it takes for the memory mechanism to begin to respond. Pulse rise times down to a few nanoseconds may theoretically be necessary in some cases but are practically difficult to maintain. Here it is necessary to distinguish between the intrinsic rise time of the pulse, as it appears at the output of the instrument, from

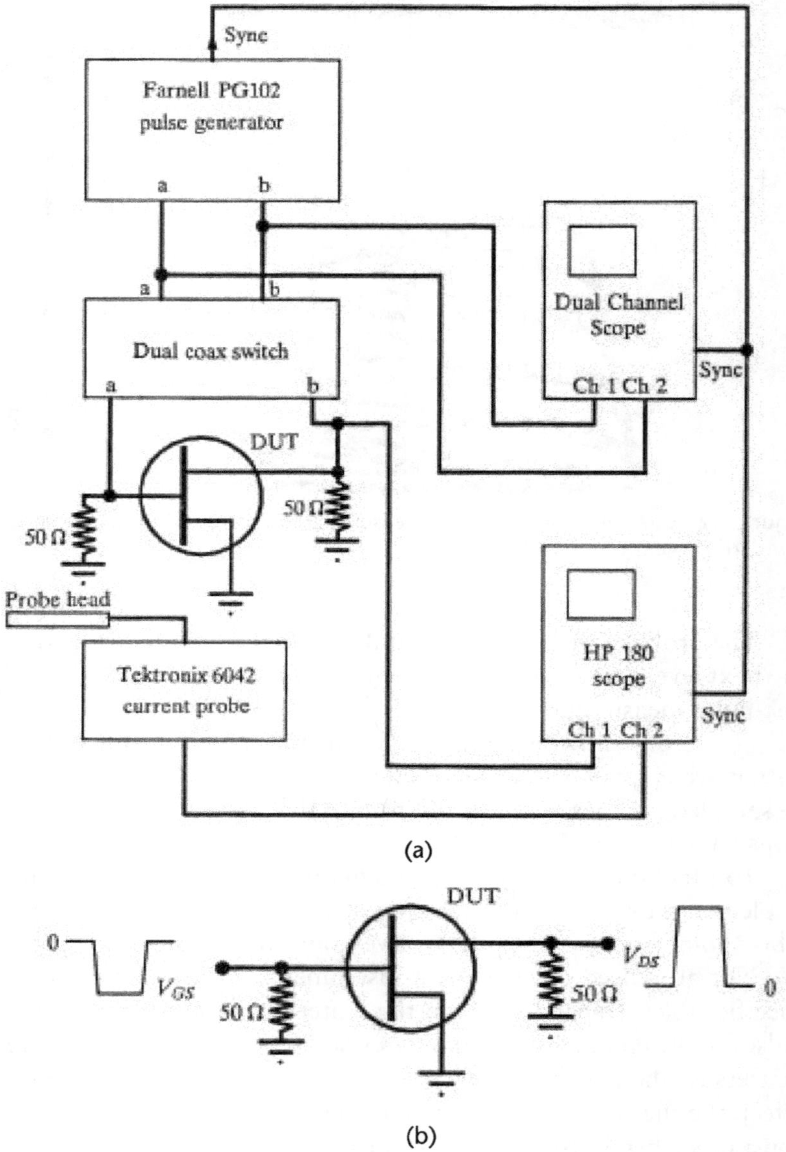

Figure 6.3 Barton's fast-pulse measurement system. (*From:* [5]. Reproduced with permission.)

how it appears at the device terminals after degradation by cables, device mounts, and parasitic coupling to surrounding objects. In Figure 6.1, for instance, at bottom left of the front panel can be seen a small microwave jig with a short connecting cable, all of which keeps the parasitics to a minimum and preserves the integrity of the pulse. If long cables are unavoidable, the only solution is to remove the final-stage pulse generation circuits and amplifiers from the instrument proper and integrate them into remote heads at the device end of the cables [6]. These and other topics, including RF stability of the device under test, are covered in [7].

6.2 Operation and Utilization

A simple diagram that summarizes the operation of a pulsed $I(V)$ measurement instrument is given in Figure 6.4. In the diagram, the arrows depict the trajectories of the combined gate-source and gate-drain voltage excursions executed by the synchronized pulses.

With regard to timing, crucially the FET's response has to be sampled at a time short enough after the pulse is applied for the device's response to be the same as it would be at RF. Thus, if everything that is known about the device suggests a pulse of 200 ns should be fast enough, in practice, that translates into sampling the response after 200 ns. The pulse can be longer, even much longer, providing no undesired consequence follows (overheating of the device is one possibility). Irrespective, it is easier to refer to the test condition as a 200 ns pulse.

The significance of the dc bias is that it establishes a steady state for the lattice temperature due to self-heating, and similarly for the charge held in bulk deep levels, interface states, and surface states (the trap nature of all three being the same, the different labels signifying only their physical location). In the parlance of the cognoscenti, the single word "state" is invoked to refer to this preset condition. How the device responds to high-frequency signals, both large and small, depends upon the state in which the FET is set by the bias. Change the bias, the state of the FET is changed, and the RF behavior is changed too.

By now, it should be apparent that circuit designers face a real problem; somehow they have to contend with devices that do not have a single set of microwave characteristics, but as many different sets as bias points as could arise in their chosen circuit.

To deal in a systematic manner with this reality, we now write the dependence of the drain current as

$$i_D = i_D\left(v_{GS}, v_{DS}, V_{GS}, V_{DS}\right)$$

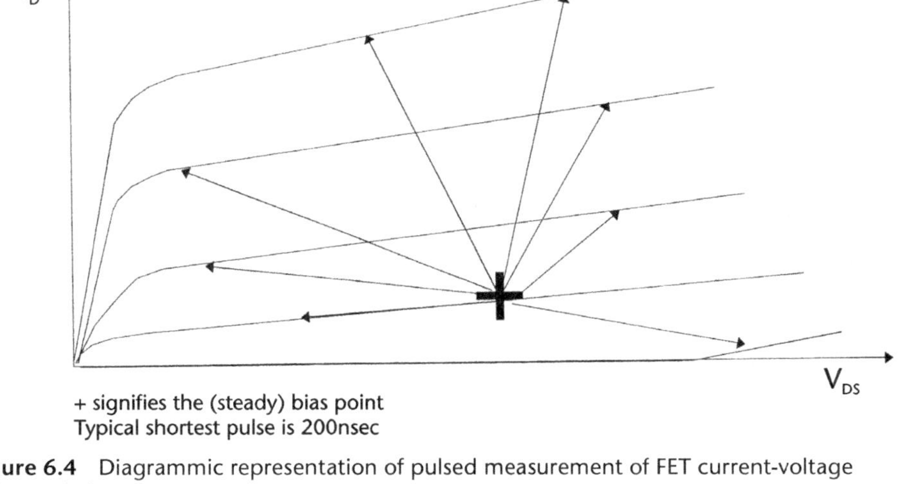

Figure 6.4 Diagrammic representation of pulsed measurement of FET current-voltage characteristics.

where V_{GS},V_{DS} signify the dc condition and v_{GS},v_{DS} represent the large-signal excursions. Provided that the pulse length and the interpulse period are set appropriately to the material type (GaN, SiC, GaAs-AlGaAs), a pulsed measurement instrument will measure the $i_D(v_{GS},v_{DS},V_{GS},V_{DS})$ characteristics along which the FET will work at RF. Therefore, the pulsed measurement instrument provides an essential piece of information required for a nonlinear model: the high-frequency, current-voltage characteristic function $i_D(v_{GS},v_{DS},V_{GS},V_{DS})$.

On the face of it, however, this is only half the information required for a nonlinear model: the other half, the charge storage characteristic, is not measured. In the event, the large-signal current-voltage characteristic has value well beyond the apparent half. Its derivatives

$$g_{m0}\left(v_{GS},v_{DS},V_{GS},V_{DS}\right)$$

and

$$g_0\left(v_{GS},v_{DS},V_{GS},V_{DS}\right)$$

are true RF conductance functions from which key features of how charge is stored in the FET automatically follow, as will be dealt with in Parts II and III. Application of the g_{m0} function to forming large-signal capacitances under the C's from g's approximation scheme has already been mentioned in Section 5.3.

As reference to Figure 3.1 reveals, a dynamic $i_D(v_{GS1},v_{DS},V_{GS},V_{DS})$ curve crosses its corresponding $I_D(V_{GS1},V_{DS})$ dc curve at some point, both being measured for the same value of gate-source voltage, which is an instantaneous value in the case of the dynamic curve but a steady value in the dc case. Figure 6.5 is a sketch of this condition.

If we now focus on a pulse-to point (v_{GS1},v_{DS1}), which lies above the bias point indicated by X, the instantaneous current response i_{D1} lies above the long-term stable current i_{D1}, which is the dc value. Hence, the current transient should look something like that in Figure 6.6.

Pursuing the idea logically, if successive pulsed measurements were made while decreasing v_{DS} from v_{DS1} to the crossing point, the amplitude of the transient would gradually decrease until, upon arriving at the crossing point, it would disappear altogether. Then, if v_{DS1} were decreased still further (i.e., to a values below the crossing point), the sense of the transient would invert as in Figure 6.7.

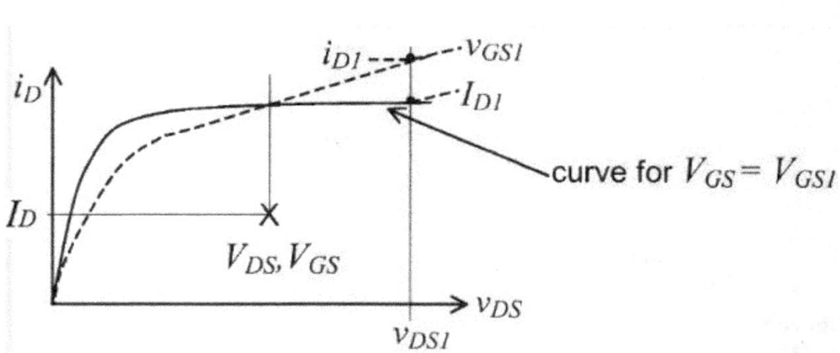

Figure 6.5 Sketch illustrating the crossing of dc and dynamic curves.

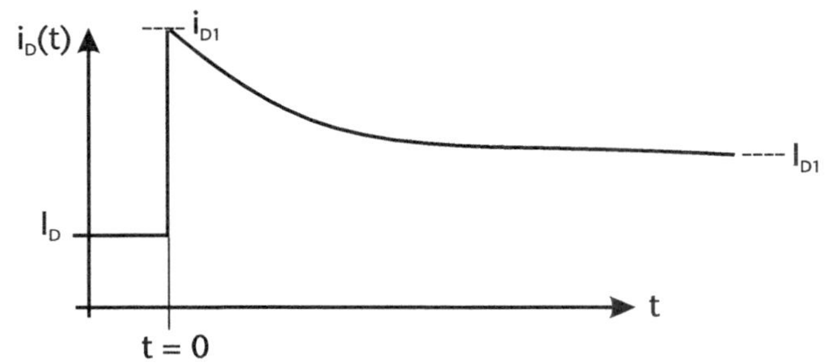

Figure 6.6 Idealized form of drain current transient with decay from a higher value to a lower value.

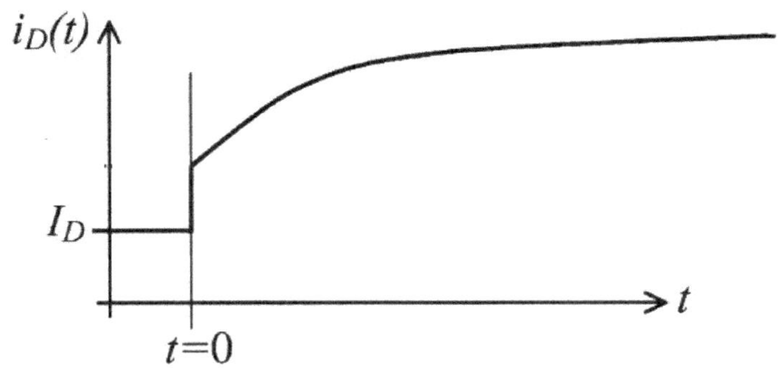

Figure 6.7 Inverted transient where the drain current rises from a lower value to a higher value.

Based on these simple deductions, even in the ideal case of a monotonic evolution in time of the drain current, transients of different duration, amplitude, and sense can be expected. Apart from all else, this fact is something that has to be accommodated in the simulation of complete circuits.

6.3 Memory and Other Effects

Any physical process involving delay has the potential to affect what is happening now by what has happened in the past and is, in a strict sense, a memory effect. According to that definition, even the current generator delay that we are calling τ, which operates on a picosecond scale of time, could be labeled as a memory effect. In the context of pulsed measurements, the term is used to refer to slow phenomena: charge trapping and self-heating.

In general, slow-rate mechanisms are active simultaneously; they interact, so the task of unraveling what is going on becomes difficult. Specifically, the characteristic times for capture and emission of charge carriers by deep levels depend upon the lattice temperature, so self-heating alters the rates. Whatever the interplay, the outcome is manifest in the current-voltage characteristics of the FET and their rate dependence, which usually is where the enquiry begins.

A pulsed measurement facility can help to unravel the contribution made by each of the competing mechanisms. An example illustrates the point.

In some GaAs power FETs, near open channel the dc drain current droops, meaning that it decreases with increases in drain-source voltage in a manner reminiscent of Figure 6.2. A logical and oft-invoked assumption is that self-heating is the cause. Reasonably enough, if the device temperature rises due to the dc power dissipated, the saturated drift velocity of the electrons carrying current through the channel decreases [8], so the current decreases. What the assumption leaves unanswered is any role that deep levels may play.

Jastrzebski used the Kent pulsed $I(V)$ measurement system to separate the contributions of the two slow-rate mechanisms [9]. It showed that the main cause of current droop is not self-heating but charge exchange with deep levels, and that a degree of droop would persist even if self-heating could be eliminated completely. Tracing the likely source back to technology tends to be speculative. Taking the risk pointed to surface states in the gate recess as the likely culprit, in turn determined by the offset of the recess edge from the gate edge on the drain side and the surface treatment overlaying it.

In contrast to the simple forms of Figures 6.6 and 6.7, transients of mixed origin can be complex. All manner of time variations can be observed depending not only upon the material and technology that went into the device but also upon the bias point and the pulse-to point. Figure 6.8 is one example.

Figure 6.9 is a sketch of even less understandable responses. Should memory events like this affect adversely the RF behavior of a particular type of circuit, the

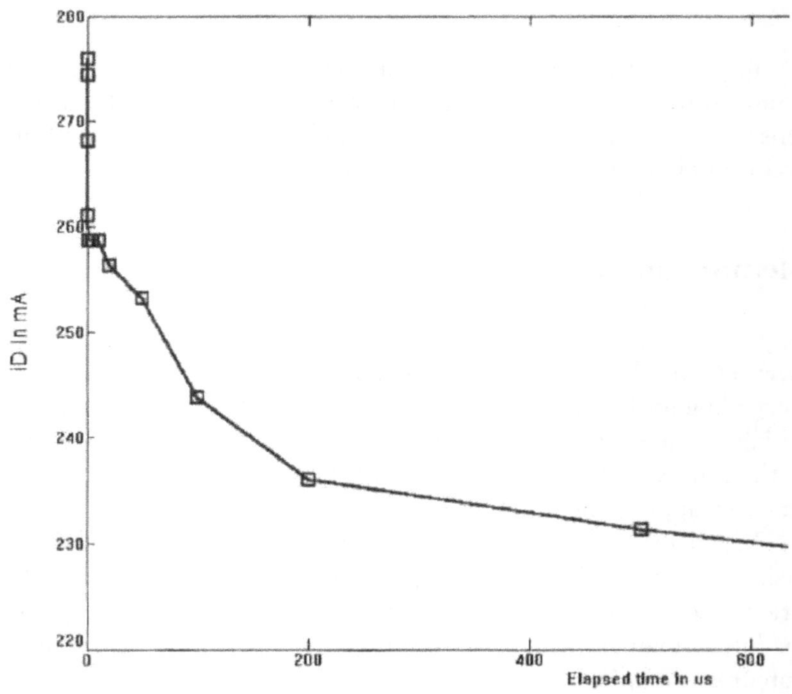

Figure 6.8 An example of a transient resulting from the interplay of several slow-rate mechanisms.

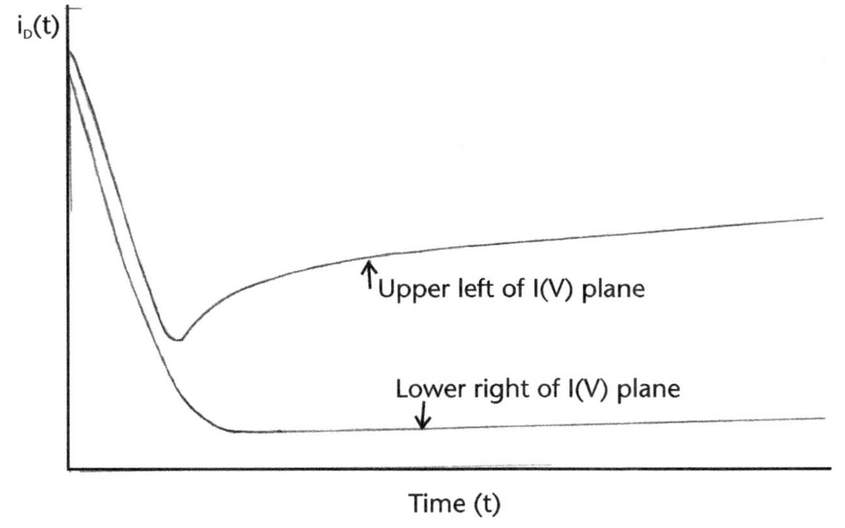

Figure 6.9 Transient responses of the same device when pulsed to different areas of the $I(V)$ characteristics.

best solution might be to seek another device. Whether there is a way to account for such complicated slow-rate effects in a circuit simulator is not known.

Fortunately, not all transients are of a compound nature. Figure 6.10 shows a measured transient from a relatively clean GaAs HFET cell. Transients such as this foster hope of incorporating memory effects into a circuit simulator manageably, which, after all, is the purpose of the whole exercise.

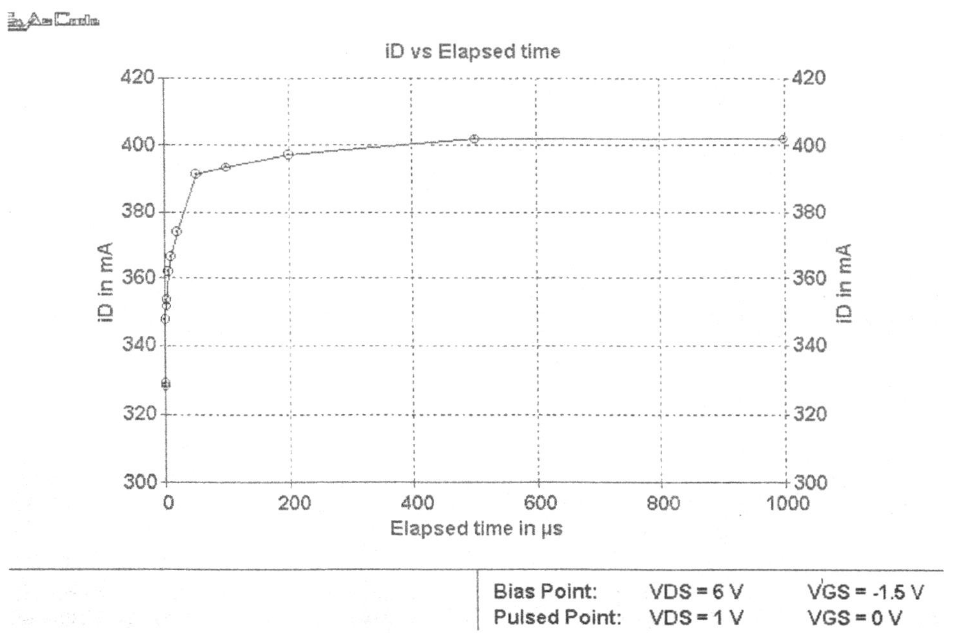

Figure 6.10 A clean drain current transient measured for a GaAs HFET cell.

In addition to the foregoing uses, a pulsed $I(V)$ measurement instrument has application in trying to work out why devices behave the way they do. Occasionally, unexpected features show up in the current-voltage characteristics of FETs and HEMTs, more often than not in power FETs. Kinks occurring at drain-source voltages of 6V to 8V were, at the time, a hitherto unseen phenomenon [10]. In this case, the role of the instrument, as important as it was, was limited to making measurements at drain voltages beyond the point where dc would have resulted in overheating. Apart from revealing the kinks clearly in the 200 ns pulsed characteristics, however, no hint as to their origin stood out from the measurements. A physics-based simulator of the type described in Chapter 16 showed the effect to be consistent with the space-charge layer reaching the gate recess edge nearest the drain.

6.4 Contrariness as a Positive

It is often said that there are two sides to every story; the dispersion problem is no exception.

There are two ways of interpreting the difference between the dc $I(V)$ and pulsed (or instantaneous) $i(v)$ curves. The first is that, for given voltages ($v_{GS1} = V_{GS1}$) and ($v_{DS1} = V_{DS1}$), the value of the drain current is different (except at the crossing point, where it is the same). The second inverts cause and effect and relies on reasoning helped by Figure 6.11. It hinges on the premise that for a given current the voltages must be different.

In the figure, the heavier curves depict the dc and dynamic curves as before (Figure 6.5). We now focus on a general value of dc drain current I_{DP}, removed from the crossing point for clarity. Just to the left of where the vertical line at V_{DSP} crosses the dc characteristic is a faint curve that represents a pulsed measurement made at a gate voltage v_{GSP-}. This curve intersects the horizontal line I_{DP}. In other words, at the intersection the instantaneous current equals the dc. Similarly, just to the right of where the vertical line at V_{DSP} crosses the dc characteristic is a faint curve that represents a pulsed measurement made at a gate voltage v_{GSP+}. This curve

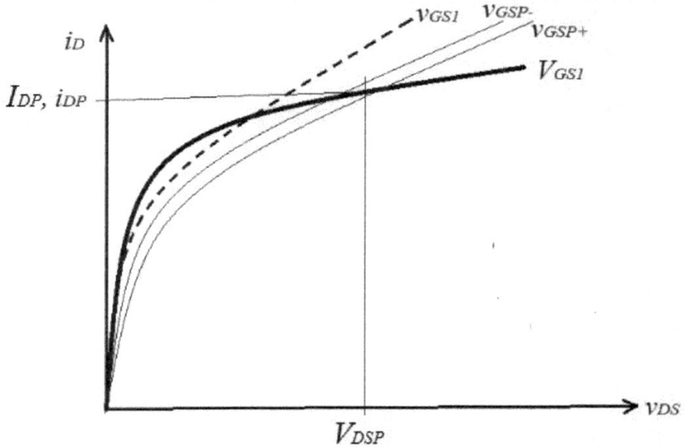

Figure 6.11 Leading to the concept of a distorted voltage mesh.

too intersects the horizontal line through I_{DP}, so, once again, at the intersection the instantaneous current equals the dc. We now have two dynamic curves for which the equality $i_{DP} = I_{DP}$ applies. (A little thought leads to the realization that there is a continuum of such curves in the vicinity of the point I_{DP},V_{DSP} that will satisfy the condition.)

Continuing with the argument, Figure 6.11 tells us that any given value of drain current, for instance $i_{DP} = I_{DP}$, occurs at values of (v_{GS},v_{DS}) (being the pulse-to voltages), which are different from (V_{GS},V_{DS}) (the dc voltages). In the figure, the particular current of interest results for dc voltages of (v_{GS1},v_{DSP}), whereas for the left thin curve, the required voltages are $v_{GS} = v_{GS-}$ and $v_{DS} < V_{DSP}$, while for the right thin curve, the requirement is for $v_{GS} = v_{GS+}$ and $v_{DS} > V_{DSP}$.

Each of the pairs of voltages (v_{GS},v_{DS}) and (V_{GS},V_{DS}) can be viewed as belonging to their own 2-D mesh or net, meaning there are two meshes: a dynamic voltage net and a dc voltage net. Thus, the single drain current $i_{DP} = I_{DP}$ occurs at different pairs of voltages on the two meshes.

In the dynamic case, the voltage excursions occur too fast for any charge trapped in deep levels to change, so the state of the device remains unchanged. We might say that the dynamic mesh is clean or unchanged or uncorrupted. However, when the voltages change slowly, as at dc, the state of the device alters: the size and shape of the depletion region change, which, in turn, changes the current. Turning that around the other way, to achieve a dynamic current equal to the dc, we need values of RF gate-source voltage and drain-source voltage that are different from the dc pair: we need to transform the dc voltage mesh to get back to the dynamic mesh. By so doing, we render the principal features of the space-charge layer, d and X, the same in both dc and the (new) dynamic cases. We will have systematically undistorted the dc net to get back to the dynamic mesh, reversing the distortion to the former caused by memory effects.

Abstracting the essential point, dispersion originates mainly from the space-charge conditions in the active region of the FET being altered by trapped charge, which is equivalent to the applied voltages being altered. On this basis, we can regard the dc mesh as being a distortion of the dynamic voltage mesh.

In summary, the first of the two views takes the voltages as reference and concludes that the current is modified, whereas the second takes the current as reference and concludes that the voltage mesh is modified, in a manner better described as distorted. Of these, the second is closer to the physics; it is a valuable concept in deciding how to handle the dispersion problem in a nonlinear model for a circuit simulator, a subject which is taken up in Section 10.2.

6.5 Contemporary Instrumentation

Equipment for making pulsed $I(V)$ measurements has come a long way. With the advent of the newer materials GaN and SiC for high-power HEMTs and FETs, so the demand for higher-voltage and higher-current test equipment has increased too. A key factor in configuring fast-pulse instrumentation is avoiding degradation of the outgoing pulse and the return response by the cables connecting the device under test to the instrument.

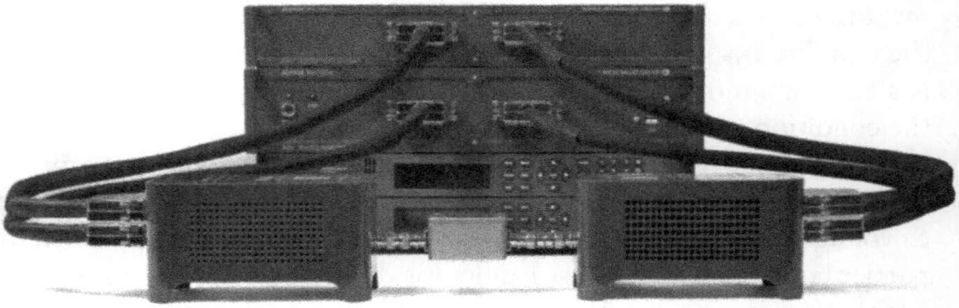

Figure 6.12 Auriga fifth generation pulsed measurement instrumentation. (Courtesy of Focus Microwaves.)

As surprising as it may seem at first given the comparatively slow (relative to microwave timescales) rise and fall times of the pulses, cable capacitance and inductance have a dramatic effect, leading to the time-domain counterpart of a microwave problem. The solution is to take the pulse generators and sensing circuitry out of the instrument proper and incorporate them in remote heads.

Focus-Auriga has a range of instruments now in their fifth generation: the remote heads are the units in the foreground of Figure 6.12 [11]. Taking the mid-range PHD 600-5 instrument as an example, the specification includes the following impressive combination:

- Maximum voltage: 600V;
- Maximum pulsed current: 5A;
- Minimum output rise and fall time: 55 ns;
- Minimum pulse length: 200 ns.

A capability of this order represents a major advance over an early-generation model like Figure 6.1.

References

[1] Goodship, N., and D. Battison, "Hardware Development for a Two-Channel Pulse-Measurement Instrument," 5TH Axiom Ltd., Internal Technical Memorandum, Cambridge, April 1, 1996.

[2] Barton, T. M., et al., "Narrow Pulse Measurement of Drain Characteristics of GaAs MESFETs," *Electron. Lett.*, Vol. 23, No. 13, June 18, 1987, pp. 686–687.

[3] Jastrzebski, A. K., and R. J. Donarski, "Pulsed I-V Measurement System for Temperature and Dispersion Characterisation of Microwave FETs," *20th ARMMS Meeting Digest*, Nottingham, U.K., March 1994, pp. 95–100.

[4] Ladbrooke, P. H., et al., "Mechanism of Drain Current Droop in GaAs MESFETs," *Electron. Lett.*, Vol. 31, No. 21, October 12, 1995, pp. 1875–1876.

[5] Barton, T. M., "Characterisation of the Physical Behaviour of GaAs MESFETS," Ph.D. Thesis, University of Leeds, May 1988.

[6] "Introduction to Pulsed IV Measurements," Focus Microwaves, https://youtub.be/iwefqSYqflk.

[7] Ladbrooke, P. H., *Pulsed I(V) Measurement of Semiconductor Devices with Applications*, Accent Optical Technologies, 2004.

[8] Ruch, J. G., and W. Fawcett, "Temperature Dependence of the Transport Properties of Gallium Arsenide Determined by a Monte Carlo Method," *J. Appl. Phys.*, Vol. 41, No. 9, August 1970, pp. 3843–3849.

[9] Jastrzebski, A. K., "Characterisation and Modelling of Temperature and Dispersion Effects in Power MESFETs," *Proc. 24th European Microwave Conf.*, Cannes, France, September 1994, pp. 1319–1324.

[10] Ladbrooke, P. H., and J. P. Bridge, "Benign Mechanism Giving Rise to Kinks in GaAs MESFET and HEMT I(V) Characteristics," *Electron. Lett.*, Vol. 31, No. 22, October 26, 1995, pp. 1947–1948.

[11] Focus Microwave Product Brochure, https://focus-microwaves.com/wp-content/uploads/2019/09/Product-Brochure AU5-2019-1.pdf.

Part II

CHAPTER 7

Reformulating the Circuit Model

7.1 Introduction

The commonplace equivalent circuit model, frequently referred to as the standard model, needs reformulating for the design of high-specification nonlinear circuits. The reason is simple: as will be shown at the conclusion of Chapter 15, the standard model is a purely static representation of an active FET or HEMT. It is necessary to explain what this means.

When a device is operating in the steady state, meaning that the terminal voltages and currents are time-independent, the charge stored in the depletion region is similarly unchanging. In part, the commonplace model is composed of capacitors, or capacitive elements, which account for charge storage in the various parts of the depletion region. In the steady state where the depletion region is unchanging or static, the model is a correct representation of where and how much charge is stored, but that is where it ends.

As soon as the terminal voltages become time-varying, as they must for the device to have any practical purpose, the standard model has charge flows occurring in the wrong branches of the equivalent circuit. Expressed in circuit terms, the charging currents for the capacitive elements are wrongly placed. The model is dynamically flawed, and no amount of tinkering with equations to fit the empirically derived current and capacitances is ever going to correct it. (An interesting fact is that identification of what is amiss with a particular theory, view or convention is often made at near enough the same time by individuals working separately and unknown to each other. That said, it is likely that David Root (then of Hewlett Packard) was among the first to realize the true nature of the standard model.) What has to replace it, therefore, is a dynamically correct equivalent circuit model.

Truth or criticism apart, whichever way we care to look at it, the standard model has proven value. The big question is where.

A circuit operating at a single frequency well below the current cutoff frequency (f_T) of the devices within it and with a relatively simple performance specification of gain and power output can be well enough designed using the standard model. However, the risk is that good correspondence between theory and practice under these conditions can give the impression, falsely, that the approach has accuracy on its side. A better term would be adequacy.

The supposition of adequacy fails in what might be called high-demand applications. For example, a power amplifier handling a 40 Gbps data stream might be operating at or above $f_T/2$ of its devices, the compromise being between a long gate for power and a short gate for the frequency response. For such an application, a dynamically sound model is a prerequisite for good design.

Contrary to common belief, the static nature of the standard model renders it no more accurate for small-signal use than for large-signal use. As mentioned in several places in this book, for example Section 3.2.3, the following applies: if, when fitting the model of Figure 3.19 to measured S-parameters the minor elements L_s, R_s, R_d, L_d are fixed at sensible physical values, the fit is not as good as if the minor elements are unconstrained or freely fitted, in which event they (the minor elements) take on seemingly nonphysical values.

Generally, the contributions to physically absurd values are several. For instance, negative values of R_s arise in part from the wrong current being forced through the source [1], while negative values of R_d are thought to be associated with charging delay in a negative capacitance in the gate-drain branch (see Section 15.1 for an account of negative capacitance), all very convoluted and to a degree qualitative. So as not to lose sight of the important point, however, the standard model is subject to the same kind of restrictions for small-signal use as for large-signal use.

When adopted as the basis for nonlinear design, the standard model is normally drawn as in Figure 2.1(a), were the symbol resembling an elongated angular S signifies a nonlinear component. Variations can be encountered, mainly C_{ds} also being treated as nonlinear, and the inclusion of a resistor R_{gd} in series with C_{gd}. There are good arguments for both.

Revealing the limitations of the common approach requires performance measures more stringent than simple gain and output power. We now enter the domain of the circuit and subsystem expert, a labyrinthine repository of knowledge and terms like third-order intermodulation (IM3), adjacent channel power (ACP), orthogonal frequency division multiplexing (OFDM), and similar, which, for the device engineer, can be perplexing when it comes to knowing what kinds of tests of a device model are appropriate to serve them. Among the possibilities for general tests are harmonic generation and phase tracking. Phase fidelity even at modest frequencies would appear to qualify as a stringent test, because for the phase to be right, just about everything else has to be right too (AM-to-PM conversion comes to mind). Be that true or a fallacy, phase has been regarded in this work as a test of the reformed model's worth.

7.2 The Core

In the physical device, the current in the channel beneath the gate in region I responds near enough instantly (Section 14.1) to any change in terminal voltage (say, gate voltage). Before the electrons of which this current is comprised can reach the drain, however, they have to pass through region II-III where their motion is limited to the saturated drift velocity, v_{sat}. As a result, the current change emerges delayed by a time τ (Section 13.8).

This vital distinction is lost in the standard equivalent circuit model. Reference to Figure 2.1(a) makes immediately clear that the same current as flows in the drain is forced into the source too. Overlaying the corrected circuit equivalent on Figure 7.1 results in Figure 1.9.

Now, with Figure 7.2, the idea that charge storage in region II-III of the space-charge layer should be, or even can be, represented by capacitance is redundant. In its place, charge is stored by pure delay.

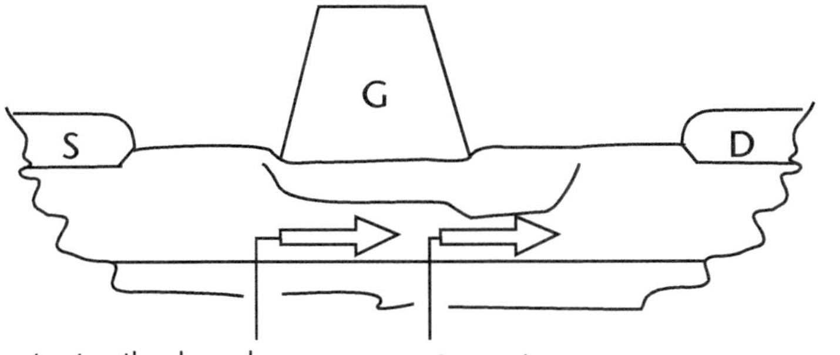

Figure 7.1 The current emerging at the drain end of the channel is delayed by a factor τ compared with the current at the source end.

Perhaps the best way to demonstrate the redundancy of capacitance principle is to recapitulate from Section 6.4.1 of [2]. Figure 7.3 is reproduced from [2, Figure 6.17].

In Figure 7.3, the combination of $C_{II\text{-}III}$ and the current generator resulted from consideration of region II-III, leaving the question: Where should node A be connected? Under a large-signal regime, the correct connection is to S′ via a current generator; the approximation of a short circuit to connect A to S′, just about workable for small-signal use, no longer holds. Because $C_{II\text{-}III}$ is located in the tapping branch between the two current generators, the current through it is

$$I(t) - I(t-\tau) \overset{\text{def}}{=} \delta I(t,\tau)$$

so all the presence of $C_{II\text{-}III}$ does is to fix the voltage at the tapping point. The last step is to realize that the voltage at the tapping point between the two generators is inconsequential to circuit behavior; it can be whatever it likes. Therefore, $C_{II\text{-}III}$ has no effect and is redundant.

Elimination of $C_{II\text{-}III}$ as the store of charge in region II-III necessitates acceptance of an unfamiliar concept, charge storage by pure delay (Section 14.4) where

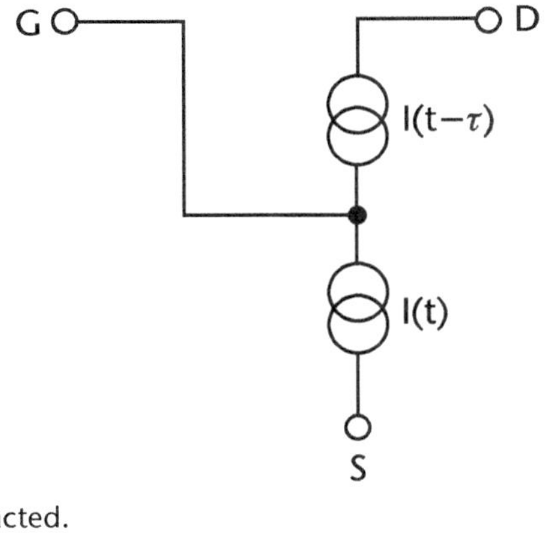

Figure 7.2 The core abstracted.

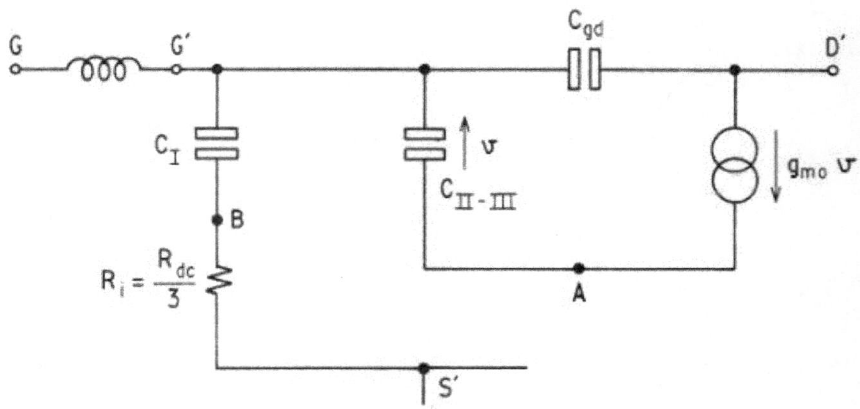

Figure 7.3 Partial equivalent circuit. (*From:* [2, Figure 6.17]. Reproduced with permission.)

the charge flows are given by $\delta I(t,\tau)$. FETs and HEMTs fundamentally are time-domain devices, so the mechanism is best thought of that way.

7.3 Charge Flows When V_{GS} Changes

To continue constructing the reformed model, we have to add circuit elements to describe charge changes which occur when the terminal voltages change. We deal with them one at a time, beginning with V_{GS}.

Figure 7.4 shows the bare cross-section of a device within, marked separately, the three distinct parts of the depletion region where there are changes following a change in V_{GS}. They are subscripted as being either primary or secondary to reflect being either a cause or an effect.

In response to the V_{GS} stimulus, region I responds near enough instantaneously (Section 14.1), followed by transmission of the change through region II-III at a controlled rate determined by v_{sat} (Section 13.8). Transmission from one end of region II-III to the other takes a time τ governed by the length of the region, which,

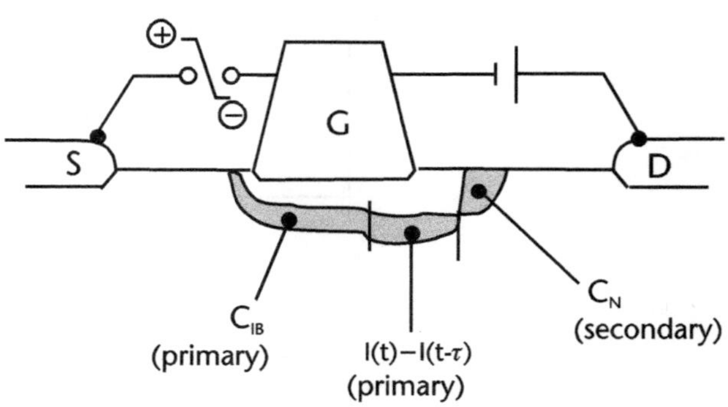

Figure 7.4 Three distinct regions where there is a change in stored charge in response to a change in gate-source voltage.

in turn, is voltage-dependent. By then, the depletion edge everywhere has settled at its new position, simply referred to as d.

Establishing the new d is a primary change that, as a secondary response, induces a change in the space-charge layer extension X. Relationships between the distribution of charge in the depletion layer and the potentials across it establish an interdependence of d and X (Section 13.6). Coupling of d and X imparts to the depletion a characteristic akin to breathing.

The breathing character results in dimensional changes to d and X that are in opposition to one another: if d increases, X decreases. In turn, that means if the increase in d results in the storage of more charge and therefore positive capacitance, the associated decrease in X results in negative capacitance (Section 15.1). In the present context, negative capacitance merely signifies a charge change in opposition to the primary charge change, so the net is smaller. It is a negative nevertheless.

In Figure 7.4, the three distinct areas of charge change are labeled with their equivalent circuit symbols, (Section 15.2.2), (Section 15.2.3), and (Section 14.4, more easily understood from the derivative account of Section 15.2.4). For an addition to the core model, these elements interconnect as in Figure 7.5.

C_N represents a charge change in the branch of the equivalent circuit that is at right angles to, or orthogonal to, the branch across which the applied voltage exists. It is a transcapacitance and is negative.

C_{IB} models a charge change in the branch across which the voltage is applied. It is designated a branch capacitance and is positive.

The tricky element is $\delta I(t,\tau) = I(t) - I(t - \tau)$, which establishes storage of charge by pure delay (Section 14.4) and can only loosely be referred to as a type of capacitance. As it leads to an increase in stored charge like C_{IB}, it similarly constitutes a positive charge storage term. $\delta I(t,\tau)$ is the charge flow in the tapping branch of the core.

Finally, adding Figure 7.6 to the core of Figure 7.2, we end up with Figure 7.6.

7.4 Charge Flows When V_{DG} Changes

Next, we look at the circuit components needed to account for what happens when the gate-drain voltage is time-varying.

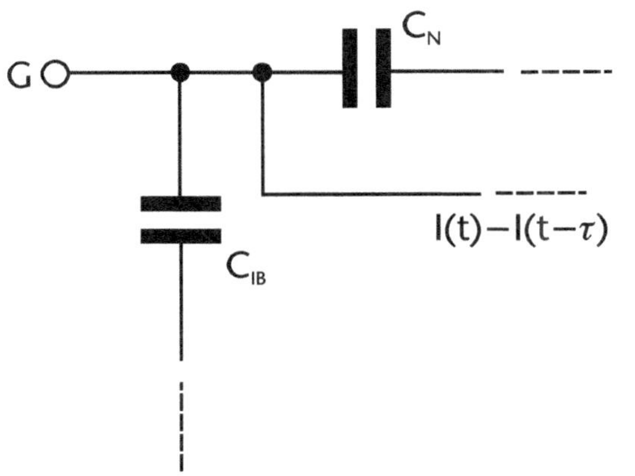

Figure 7.5 Circuit equivalent of Figure 7.4. C_N is a negative capacitance.

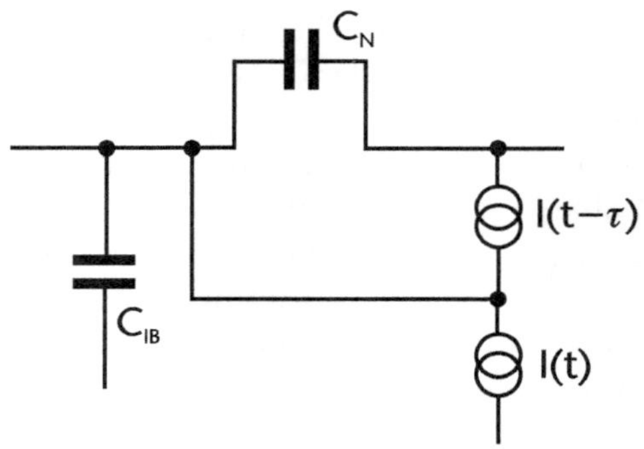

Figure 7.6 The core now including charge storage elements deriving from a change in gate-source voltage.

Similar to Figure 7.4, Figure 7.7 shows a bare cross-section of a device with the distinctly identifiable sites of charge change marked, but this time due to changes in V_{DG}. Again, there are three.

Here, there is a single primary charge change: a branch charge represented by a capacitance C_{gdB}, which is positive (Section 15.3.2). It is designated a branch charge because the change takes place in the same part of the device across which the voltage stimulus is applied.

By virtue of the $d \leftrightarrow X$ coupling, the secondary response to the increase in X is a decrease in d. Thus, the charge change in the d edge is opposite to that in the X edge, thereby contributing negative capacitance or capacitive-type terms. As is by now obvious, the part of the device or branch of the circuit where these secondary changes occur is the other, or orthogonal, branch. Hence, C_M is a transcapacitance (Section 15.3.3), occasionally written $-C_M$ as an aide-mémoire.

Translated into a circuit equivalent, we arrive at Figure 7.8.

Once again the less familiar contribution is that provided by the tapping branch of the core, $\delta I(t,\tau) = I(t) - I(t - \tau)$. This charge flow term now stores a charge of polarity opposite to that of the main or primary charge so it too counts as a negative. It is a transcapacitance only in a loose sense, but attempts at an exact description end up with something clumsy. For example, Section 15.3.4, which deals with this mechanism, is titled, "Orthogonal Branch Charge Storage by Pure Delay."

Figure 7.9 puts together the contributions thus far. It should be stressed that the current generator tapping branch has two roles, not one. For V_{GC} changes, it

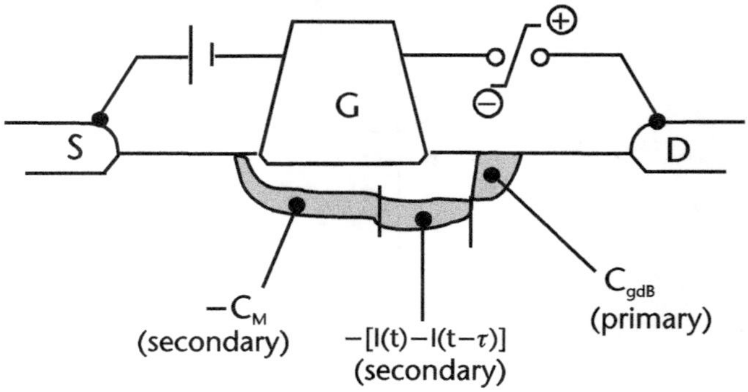

Figure 7.7 Three distinct sites of charge change deriving from a change in gate-drain voltage.

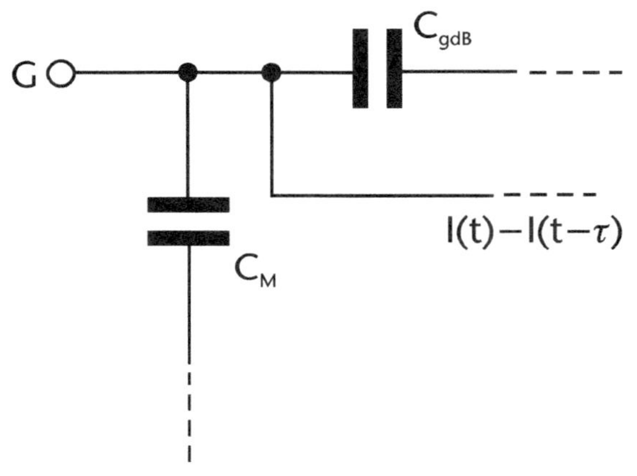

Figure 7.8 Circuit equivalent of Figure 7.7. C_M is a negative capacitance.

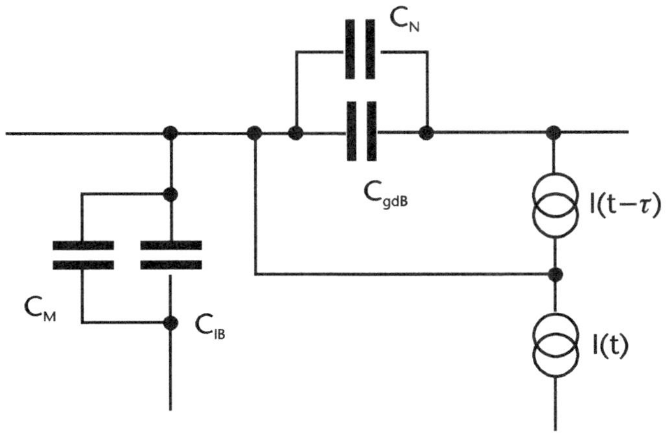

Figure 7.9 The new model so far.

acts like a positive branch charge, whereas for V_{DG} changes it acts like a negative transcapacitance, both of them in the gate-channel branch of the network.

7.5 Resistive and Ancillary Elements

Taking Figure 7.9 and adding resistances completes the reformulated model, Figure 7.10. In keeping with our general notation, R_g,R_s,R_d have been assigned lowercase subscripts to indicate that they are now linear (or constant), while R_I,R_{GD} are uppercase subscripted to denote a nonlinear element.

As the claim is that the model is now compliant with the physics of the HFET operation, the resistances R_g, R_d, and R_s should take on physically sensible values, if not constant values. Although all three are indeed treated as constant in the results of Chapter 11, there are minor considerations which may dictate otherwise. We deal with each in turn.

R_g, the resistance of the gate metal strips, interconnecting pads and busbars, can generally be regarded as a constant.

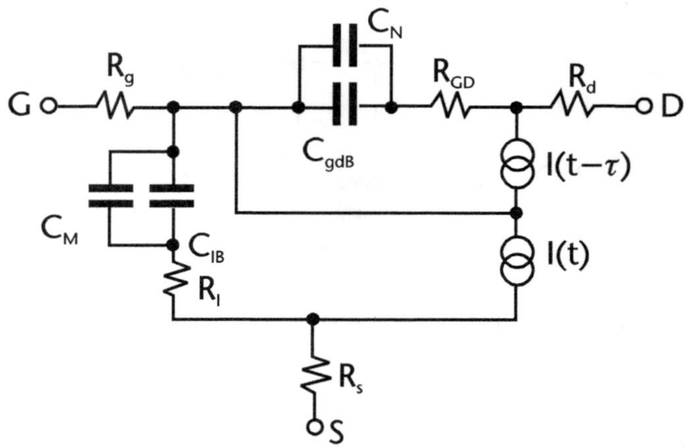

Figure 7.10 Configuration of the complete reformulated model.

The drain series resistance R_d may be modulated to a small degree by the movement of the X-edge of the space-charge layer in response to signal excursions. R_d is made up of metallization resistance, contact resistance, and the bulk resistance of the conducting layer between the drain electrode and the X-edge of the depletion. It is likely that modulation of the latter is a small effect.

Apart from extreme nonlinearity due to strong electric fields penetrating into the source-gate space (as in the PHEMT dissected in Section 4.2.2), which should not arise in a well-designed device anyway, R_s should be constant for practical purposes.

The series resistance R_{GD} in the gate-drain branch is included to set a charging time for C_{gdB}. With this arrangement, R_{GD} also sets the charging time for C_N, which strictly is incorrect. As C_N models a secondary charge adjustment coupled to the primary response of the d-edge in the channel, the time response of C_N is dictated by the settling time of the edge at d, which is τ. Incorporation of this effect requires an extra resistor and current generator, but these are minor additions that detract from the main theme. Irrespective, all the results in Chapter 11 included them.

Finally, the branch that is the junction between the two current generators expresses correctly the time evolution of the charging of region II-III. Consequently, R_I is free to express a single physical mechanism: the charging of region I alone. As discussed in Section 14.5, charging and discharging of region I is ill-defined to the point where the safest bet is to rely on whatever $R_I(V_{GS},V_{DS})$ function emerges from fits to practical data. Figure 4.2 can be taken as a rough guide. That said, R_I can be difficult to extract from measurements if there is noise present, leading to scattered data points so the functional form becomes indistinct.

Insofar as other elements are concerned, note that in the reformed equivalent circuit model there is no nonlinear capacitance C_{ds}. The omission needs to be emphasized.

The notion that there should be such an element is a hangover from the commonplace model. There is no space-charge component of drain-source capacitance in the reformed model because, as discussed in Section 15.4.2 the appearance of C_{ds} in the standard model is likely an artefact of that particular circuit configuration and given by

$$C_{ds} = \left(C_M + g_0\tau_{gm}\right) = g_0\tau_T$$

whereas the bracketed term is already present in the reformed model. Therefore, any addition of capacitance in the drain-source position should consist of a geometric component only.

Taking Figure 3.17 as a general disposition of ancillary capacitances and inductances, not just for a package, any element with a fixed value could be added to Figure 7.10 as circumstances demand. There are no hard and fast rules.

7.6 Voltage Dependence of the Elements

Example element functions are given in Section 10.7. As noted there, the continuous lines through the data points are equation fits in accordance with the methods of Chapter 9. In this form, the model is suitable for implementation as a user model in a circuit simulator.

Although there are two current generators in Figure 7.10, one is merely a delayed version of the other. In that sense, there is but one current function, yet two are needed, one dc and the other RF, the second being dependent upon whereabouts the device biases in any particular circuit.

The dc versus RF current problem was handled by the user model; thus:

1. For the dc solution, the two generators become one: the measured dc set of characteristics as, for example, in the upper frame of Figure 10.7.
2. Having found the bias point, the RF $I(V)$ reconstruction scheme of Section 10.2 is invoked to create the requisite dynamic function, as, for example, in the lower frames of Figures 10.7 and 10.8.
3. Simulation then resumes for the microwave solution.

Dealing with bias-dependent RF characteristics properly makes the writing of a user model obligatory. It has the merit of avoiding makeshift arrangements like Figure 2.1(b).

7.7 Reduction in the Static State to the Standard Model

If the following two statements are correct:

- The reformed model is compliant with the way FETs and HEMTs operate both dynamically and statically.
- The standard model is correct only statically.

then the reformed model should revert to the commonplace model in the static (strictly steady-state) limit.

We will now put it to the test.

First, with no time-dependent signals, large or small, the two current generators in Figure 7.10 become one and the tapping branch at the junction between them vanishes. We now have a circuit configuration that looks much like the standard model apart from the number of capacitive elements in the gate-channel and gate-drain branches. By then, however, the tapping branch has stored in region II-III a charge in of $g_{m0}\tau$ resulting from past V_{GS} changes and a charge of $g_0\tau$ resulting from past V_{DG} changes (Sections 15.2.4 and 15.3.4, respectively, give these more easily understood derivative accounts, the word derivative referring to g_{m0} and g_0, rather than $I(t)$).

Similarly, as a result of V_{GS} changes, charge will have been stored in both C_{IB} and in C_N. The charge in C_N partially offsets that stored in C_{IB}; C_N is a negative transcapacitance. Gathering together the means of storing charge in these three sites due to V_{GS} changes:

$$C_{IB} + g_{m0}\tau_{gm} - C_N$$

whereas in the commonplace model the same charge is taken to be stored in just a single element, C_{gc}. So we conclude

$$C_{gc} = C_{IB} + g_{m0}\tau_{gm} - C_N$$

which is (15.16). Hence, the same total charge is stored in both models but the sites of the individual contributions are kept in one model and lost in the other. In essence, the standard model rolls everything into one, thereby losing the detailed location of each contribution. Statically, however, the result is the same.

A comparable argument can be put forward for V_{DG} changes. In the reformed model, charge storage occurs in three locations within the depletion layer, namely, the X edge, region II–III and region I, represented by

$$C_{gdB} - g_0\tau_{gm} - C_M$$

whereas in the standard model the total is either lumped into one as C_{gd} (in this case, $C_{gd} = C_{gdB} - g_0\tau_{gm} - C_M$, which is (15.19)) or is split between C_{gdB} and C_{ds} depending upon how a fitter deals with C_{ds} as discussed in Section 15.4.2. Either way, in the static case, the total charge stored is the same; it is just that the individual site information is lost in the standard model.

The outcome is that the revised circuit model reduces to the commonplace or standard model in the time-invariant state: statically, they are equivalent. Under time-dependent conditions, however, the standard model has the charge flows occurring in the wrong branches. Dynamically, it is incorrect.

7.8 Previously Published Versions

Until now, no meaningful account of the reformulation and the background to it has been given. Instead, only a minimal suggestion was made that new ideas originating in device physics were in the offing. To round off this chapter, therefore, we give two versions of the reformed model as they appeared popular journals.

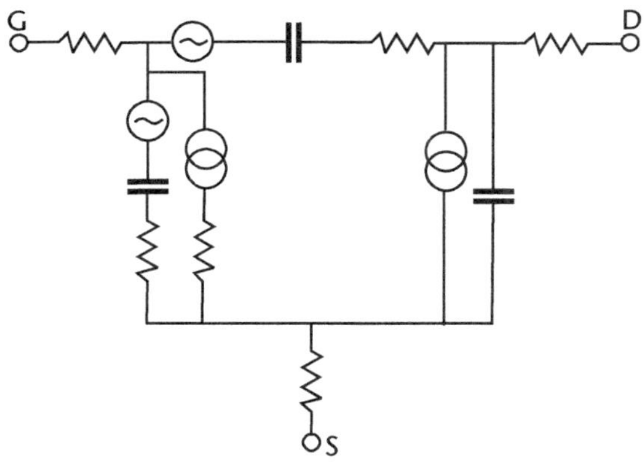

Figure 7.11 Variant from *Microwave Engineering Europe*, May 1995.

The first, shown in Figure 7.11, applied several simple network transformations to Figure 7.10.

- The series connection of two current generators was transformed into a single drain-source generator and a gate-source generator.
- Transreactance currents attributable to C_M and C_N were transformed into voltage generators.
- C_{ds} was shown explicitly.

No further detail was given, the purpose simply being to state that work with a reformulated configuration was under way as part of a nonlinear circuit design and analysis effort.

Figure 7.12 shows the second variant. Here, four current generators were shown to draw a distinction between dc and RF current-voltage characteristics. Also:

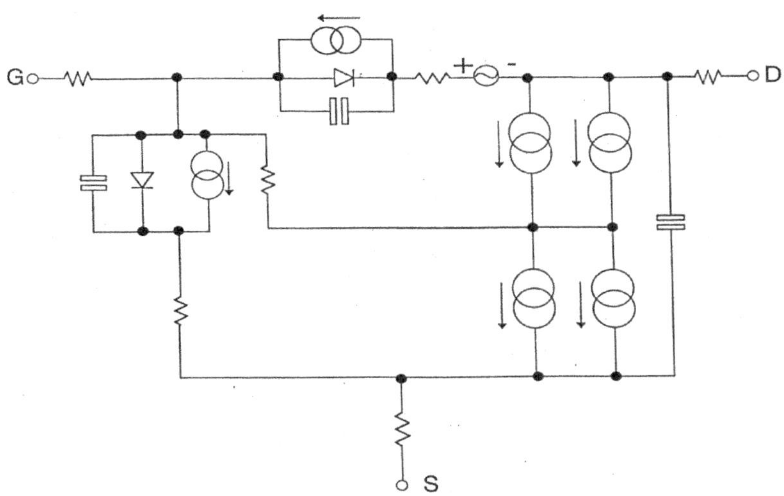

Topology of the physics-derived nonlinear model

Figure 7.12 Variant published in *Microwave Engineering Europe*, May 1998. The original figure caption is also shown.

- Diodes were included as shorthand for the gate conduction characteristics as in Figure 3.7.
- Transreactance currents attributable to C_M and C_N were shown as current generators.
- Ancillary components of a resistor in the tapping branch and a voltage generator in the gate-drain branch were included to correct the charging time of C_N as mentioned briefly in Section 7.5.
- Again, C_{ds} was included.

References

[1] Ladbrooke, P. H., A. J. Hill, and J. P. Bridge, "Negative R_s and R_d in GaAs FET and HEMT Equivalent Circuits," *Electron. Lett.*, Vol. 26, No. 10, May 1, 1990, pp. 680–681.

[2] Ladbrooke, P. H., *MMIC Design: GaAs FETs and HEMTs*, Norwood, MA: Artech House, 1989.

CHAPTER 8

The Importance and Utility of τ

Due to the central part played by τ in understanding and describing how FETs and HEMTs operate, at least according to the prescription of this book, collecting the important points in a succinct chapter is useful for ready reference.

8.1 Nature and Origin

τ represents the delayed response of the drain current to any voltage change at the terminals (Sections 14.1 and 15.2.4). By comparison, the source current responds instantly (Section 13.8). τ is a measure of pure delay and is distinctly different from a *CR* type delay (Section 14.4). For instance, a step voltage applied to the gate of a device for which $\tau = 2$ ps produces no change whatsoever in the drain current for 2 ps. After 2 ps, the drain current changes and remains at that value thereafter.

τ represents the time taken by a change in source current to propagate through the part of the channel where the current carriers are moving at their saturated drift velocity (Section 13.8), thence to appear at the drain. Pure delay has long been a topic of automatic control system theory where it is referred to as "transportation lag" [1].

8.2 Pivotal Role in the Reformed Model

Via the tapping point between the drain current generator and the source current generator, τ introduces into the gate-channel branch of the equivalent circuit:

- Charge storage, akin to a branch capacitance, deriving from changes in V_{GS} (Section 15.2.4);
- Charge storage, deriving from changes in V_{DG}, loosely described as a type of transcapacitance (Section 15.3.4).

In neither case is the storage of charge properly attributable to a capacitive element; charge is stored by pure delay, which is fundamentally different (Section 14.4).

In a nonlinear model, the drain current $I(t - \tau)$ must be treated as a large-signal, time-domain entity without compromise; approximating the delay by small-signal $\omega\tau$ type terms is flawed (Section 14.4).

8.3 Inclusion in Circuit Simulators

In the past, the facilities typically provided for coding a user model for a circuit simulator have made it rather less than easy to incorporate pure delay as a function $\tau(V_{GS}.V_{DS})$. Mostly, at best, provision has been made for a single-valued delay. Expedients to tax one's ingenuity may have to be conjured even to get close to a fair representation of τ in the simulator (Chapter 12). How much of the benefit of the reformed model stands to be lost if proper inclusion of τ proves elusive depends upon the type of circuit and the performance specification.

8.4 $X(\tau)$ as a Staple of Device Operation

τ is synonymous with X, the extension of the space-charge layer into the gate-drain region beyond the edge of the gate (Sections 13.3 and 13.8). Two-dimensional electric field arising in this region is a determinant of FET and HEMT properties, both at dc and RF. The shape of the $I(V)$ characteristics (Section 13.7), the output conductance (Section 13.7), and the capacitive elements (Chapter 15) of the equivalent circuit model are all dependent to the first order upon $X \leftrightarrow \tau$.

8.5 A Repository of Information on Device Technology

In conjunction with other equivalent circuit elements, τ allows information to be retrieved on the material and structural makeup of a device for which multibias S-parameter data are available (Section 1.2). Recovery of technological data from microwave measurements is known as reverse modeling [2]. By itself, the detailed dependence of τ upon $(V_{GS}.V_{DS})$ is indicative of the channel type, whether uniformly doped, containing doping gradients (Section 13.8) or a 2-D electron gas (2 DEG) (Section 1.2, Figure 1.5).

References

[1] Truxal, J. G., *Automatic Feedback Control System Synthesis*, Chapter 9, New York: McGraw-Hill, 1955.

[2] Ladbrooke, P. H., *MMIC Design: GaAs FETs and HEMTs*, Chapter 8, Norwood, MA: Artech House, 1989.

CHAPTER 9

Extraction

9.1 Introduction

If just one statement from this book proves memorable, it should be this:

Extraction of a model is as important as the model itself.

No matter how good intrinsically a model may be, it can be ruined as a designer's tool by poor extraction. As noted elsewhere, for some reason the art of model fitting and extraction has featured little in the literature compared with the models themselves, leading to an imbalance of knowledge. It is hard to resist the feeling that models generally might have been more sympathetically received had they been published as "Matched Empirical Model and Extraction" schemes, whimsically "MEME schemes."

It is interesting to speculate why this is so. One possibility is that the importance of extraction has long been understood in some organizations, so how to do it has been withheld as proprietary information. Conversely, experience may have shown it to be easy, almost trivial, so the obviousness of it has precluded publication.

Historically, for the group behind the findings of this book, devising and proving a scheme for fitting the standard model to multibias S-parameters took several years. The starting point was knowing that numerical searching with optimization methods could produce acceptable fits, but the resultant element values were often nonsense; they could not possibly be physically correct. Such physically implausible results naturally led to the idea that, somehow, device physics would have to be used as a constraint.

The scheme that was eventually settled is described in the next section as a precursor to outlining how it was extended to the reformulated model topology. This routine, for many years available as a commercial product (the package was marketed as FETlink by GaAsCode Ltd.), was used for all the results wherever they appear.

9.2 Obtaining the Element Functions

We first describe the basis of the fitter for the standard model and then look at how it can be adapted to obtain the elements of the reformulated model.

9.2.1 Obtaining the Standard Model Element Functions: The Fitter

For better or worse, the source data for constructing nonlinear models of FETs and HEMTs based on the commonplace model continue to be multibias, small-signal S-parameters. Such persistence is understandable: they require the minimum of costly test instrumentation (a vector network analyzer (VNA)), are easily understood, are the least difficult from which to extract a model, and, by and large, do a reasonable job for circuit analysis and design purposes. Against this workability, the finer points of theoretical inadequacy are easily dismissed as having too little consequence for practical engineers. All the empirically derived modeling of this book is based on this hard-to-dislodge tradition.

Before launching into the detail of how the fitter works, it will do little harm to be reminded that all the precursory steps must be carried out to high standard if the end result (the model) is to be the best its underlying approximations allow. Centrally, these are VNA calibration, test FET structure design, and de-embedding.

The basic approach of FETLink and its variants begins with noting that, at any frequency point, we have eight S-parameters (or Y-parameters or Z-parameters), which, in theory, give us eight pieces of information. In practice, this is not always true in the sense that s_{11}, for instance, is approximately 1 (unity) at low frequency because the input capacitance looks like an open circuit.

Using the symbol Y to signify admittances and Z to signify impedances, the topology of the equivalent circuit looks like a triangle of Y's with three series Z's, namely, $(R_g + j\omega L_g)$, $(R_d + j\omega L_d)$, $(R_s + j\omega L_s)$.

The problem is split into those two groups, noting that if we know the Z's, then we can remove them, leaving the real and imaginary parts of the Y's, which are easily obtained from the Y parameters. If we know the Y's, then, given the measurements, it is not too difficult to estimate the real and imaginary parts of the Z's. What emerges from this basic picture is the idea of a back-and-forth iterative technique.

We start by assuming that at low frequencies the parasitics have only a minor effect. In practice, this can amount to a coarse approximation when it comes to R_s, which obviously has an effect at all frequencies. Allowing for the fact that L_s is generally dominant at high frequencies, however, a rough guess for R_s in the first place is not too bad. Having only a minor effect does not mean that the parasitics are negligible, only that approximate values (but physically reasonable values) suffice to begin with.

FETLink then divides the frequency range into two: low frequencies (typically the first few points, or up to quarter of the range depending upon the total number of points), and high frequencies (the last few points or a quarter of the range similarly).

There follows an iterative process, looping around the two frequency bands. Starting with the low-frequency data, guessing initial values for the Z's, say, $R_s = R_g = R_d = 1$, $L_g = L_s = L_d = 0$, allows the Y's to be calculated. With these first-pass values for the Y's, from the high frequencies we find improved values for L_g, L_s, L_d, R_s, R_d (R_g is generally fixed). The new estimates of the Z's then give better estimates for the Y's at the lower frequencies, and so on.

Iteration is continued until all the elements converge to settled values, whereupon C_{gc}, R_i, C_{gd}, g_{m0}, g_0 are obtained from the Y's.

But what of $\tau(v_{GS},v_{DS})$, the parameter that we have promoted as all-important? Exactly how τ is extracted changed, if only slightly, from version to version as its relevance to nonlinear modeling became apparent. Initially, it was included in the low-frequency elements as it is part of the Y's, but in the later variants FETLink LS ("Large Signal") and FETLink NT ("New Topology") the data range for extracting τ was extended to intermediate frequencies.

Thus, the fitter:

1. Uses to advantage the knowledge of fitting to physical devices, so that we can assume rough relative orders of magnitude for elements.
2. Treats groups of elements as sets rather than attempting to optimize or fit values of individual elements in an uncorrelated fashion.
3. Uses (1) to partition the frequencies into ranges where different behaviors dominate, in the process allowing the values of key elements to be calculated given starting estimates for the values of second-tier elements. Designating the latter as second tier reflects the fact that they have only a second-order effect, greatly reducing the consequences of uncertainty in the estimated values.

Such a modus operandi also meant the time to fit to a set of S-parameters (as in Section 3.2.2, for example) was typically a few seconds.

In summary, the FETLink fitter for the standard FEMT model used physics in the sense that it relied on the values of elements being physically sensible in proportion to each other. When presented with a set of S-parameters calculated from the standard model with prescribed values for all the elements, it would recover the element values with which we started.

For measurements on real devices, it worked well (at a guess, as a commercial product, it probably fitted thousands of FETs and HEMTs from institutions worldwide). From the beginning, it featured basic reverse modeling algorithms to recover, from the fits once done, basic structural data such as the metallurgical gate length, which could be weighed against the known structural makeup of test devices, both FETs and HEMTs. Proof-positive assessment of this kind meant the basis of the fitting routine, as described above, was sufficiently sound not to have to resort to more direct methods of physics to implement a fitter.

9.2.2 Fitting the New Topology Model

Rather than starting afresh and devising a new fitting scheme, augmenting the existing fitter for initial trials of the reformed topology was the obvious route. While not originally intended to be permanent, it became so. In addition to covering the main points of the algebra, the following account goes into the thinking behind it.

Broadly, the principal difference between the new topology and the standard model is the charge storage elements, including charge storage by pure delay. For present purposes, we might refer to it as the charge storage core or simply the core for ease. The problem thus reduces to finding the capacitive element values in the new core from those in the old core. Step-by-step:

1. Fits done with the standard equivalent circuit fitter include the parasitic elements R_s,R_d and L_s,L_g,L_d, with R_g being preset and fixed in value. To prevent L_s being fitted as negative and other nonphysical element values, it is necessary to fix R_s and L_s (along the way accepting the inevitability of fits not up to the standard of Section 3.2.2).
2. The y-parameters of the core of both the new topology and the standard topology circuits are analyzed and the terms of lower order in ω are compared (we use the shorthand LIM/($\omega \rightarrow 0$)). Concentrating on low frequency is justified by capacitive reactance outweighing resistive effects. The comparison

gives us expressions for the new topology circuit elements in term of the standard topology elements except for R_I and R_{GD}.

3. Given the core element values for the new topology obtained in LIM/($\omega \rightarrow 0$), analysis of the high-frequency z-parameters yields a matrix of time constants from which R_I and R_{GD} can be retrieved.

We will refer to these step numbers as the analysis unfolds.

As stated in Section 3.2.3, due to topological errors a free fit of the standard model may result in unphysical values of R_s, notably negative values.

Because the new topology is (supposedly) free of these errors, R_s should be near enough constant and independent of operating point while retaining good fits. Even though imposing a fixed value of source resistance on the standard model may lead to poorer fits, it is a fact with which we simply have to live. With that proviso, the necessity for R_s to be the same in both models for (2) above means that the analysis has to start with the circuit of Figure 9.1.

Y represents the y-parameters of the charge storage core of one model or the other (new topology or standard topology), shorn of R_s, which is now treated as an element external to the core. We will write y_{ij} for the y-parameters of the core. The enveloping block Y^e represents the corresponding admittance parameters of the whole, including R_s, which we shall call Y_{ij}.

Knowing Y for the standard model from the standard fit, our task is to find the capacitive element values in the new topology core that produce the same Y^e. The first step is to write down the Y_{ij} in terms of the y_{ij}.

$$
\begin{aligned}
Y_{11} &= \frac{y_{11} + R_S \Delta y}{1 + R_S \Sigma y} \\
Y_{12} &= \frac{y_{12} - R_S \Delta y}{1 + R_S \Sigma y} \\
Y_{21} &= \frac{y_{21} - R_S \Delta y}{1 + R_S \Sigma y} \\
Y_{22} &= \frac{y_{22} + R_S \Delta y}{1 + R_S \Sigma y} \\
\Delta y &= \left(y_{11} y_{22} - y_{12} y_{21}\right) \\
\Sigma y &= y_{11} + y_{12} + y_{21} + y_{22}
\end{aligned}
\tag{9.1}
$$

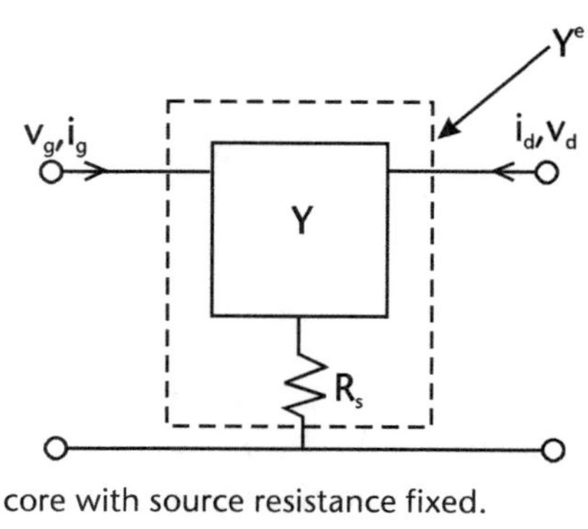

Figure 9.1 Charge storage core with source resistance fixed.

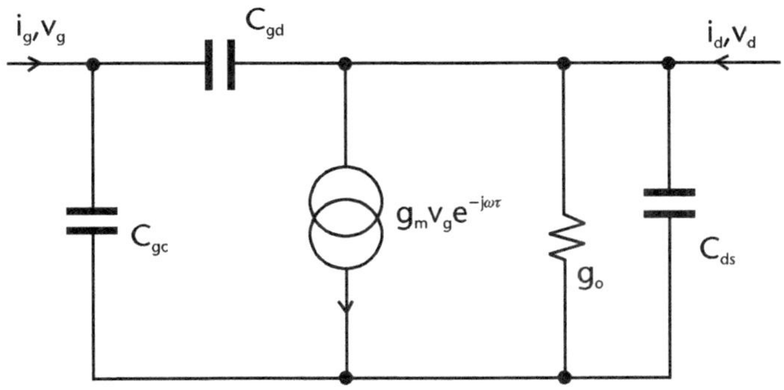

Figure 9.2 Charge storage core of the standard model.

We calculate the y_{ij} for the charge storage core of the standard model from the fitted values of C_{gc},C_{gd},C_{ds} using Figure 9.2.

In $\dfrac{\text{LIM}}{\omega \to 0}$:

$$\begin{aligned} y_{11} &= j\omega\left(C_{gc} + C_{gd}\right) \\ y_{12} &= -j\omega C_{gd} \\ y_{21} &\simeq g_m - j\omega\left(g_m\tau + C_{gd}\right) \\ y_{22} &= g_0 + j\omega\left(C_{ds} + C_{gd}\right) \end{aligned} \tag{9.2}$$

Thus, by substituting the known values of C_{gc},C_{gd},C_{ds} into these expressions for the y_{ij} of the standard model, the values of the Y_{ij} can be determined.

We now need comparable expressions for the charge storage core of the new topology circuit, Figure 9.3, with C_{ds} added as a geometric-only capacitance as per Section 12.2.1.

In $\dfrac{\text{LIM}}{\omega \to 0}$:

$$\begin{aligned} y_{11} &= j\omega\left(C_{IB} - C_N + g_m\tau + C_{gdB} - C_M - g_0\tau\right) \\ y_{12} &= -j\omega\left(C_{gdB} - C_M - g_0\tau\right) \\ y_{21} &= \left(g_m - g_0\right) + j\omega\left(C_N - g_m\tau - C_{gdB} + g_0\tau\right) \\ y_{22} &= g_0 + j\omega\left(C_{gdB} - g_0\tau + C_{ds}\right) \end{aligned} \tag{9.3}$$

By substituting these y_{ij} into the right sides of (9.1), and given that the left sides have values already determined from the standard model core, we arrive at a set of simultaneous equations for C_{IB}, C_N, C_{gdB}, and C_M.

To solve the equations for the four capacitances, the values of g_m, g_0, and τ must be known. Comparing the conduction current elements in Figures 9.2 and 9.3 suggests that g_m,g_0 should be the same in both circuits. That leaves τ. As the

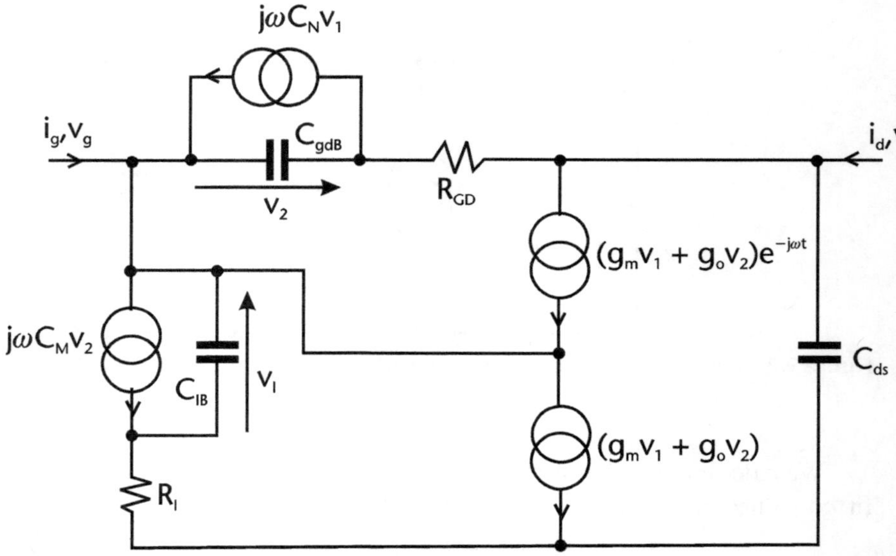

Figure 9.3 Charge storage core of the new topology circuit. C_{ds} is added as a geometric-only component.

delay element is similarly a property associated with the conduction current, it can be assumed to be the same for the two topologies.

It remains to deal with step 3.

To recover values for R_I and R_{GD}, we need data for frequencies high enough for the capacitive reactance no longer to swamp the resistances.

In matrix form, for the circuit of Figure 9.3:

$$\left[\frac{v_g}{v_d - v_g}\right] = \begin{bmatrix} 1 + j\omega R_I C_{IB} & j\omega R_I C_M \\ j\omega R_{GD} C_N & 1 + j\omega R_{GD} C_{gdB} \end{bmatrix} \left[\frac{v_1}{v_2}\right]$$

In whatever higher-frequency interval we choose, we can obtain y-parameters from the measured S-parameters to relate the terminal currents to the terminal voltages:

$$\left[\frac{i_g}{i_d}\right] = \begin{bmatrix} y_{11} & y_{12} \\ y_{21} & y_{22} \end{bmatrix}^{\text{meas}} \left[\frac{v_g}{v_d}\right]$$

where we have added the rider to the matrix as an aide-mémoire.

If we now describe the high-frequency core in terms of core admittance parameters, thus:

$$Y_i \triangleq j\omega\left(C_{IB} - C_N\right) + \left(1 - e^{-j\omega\tau}\right) g_m$$

$$Y_{gd} \triangleq j\omega\left(C_{gdB} - C_M\right) - \left(1 - e^{-j\omega\tau}\right) g_0$$

$$Y_{gm} \triangleq g_m e^{-j\omega\tau} + j\omega C_N$$

$$Y_{ds} \triangleq g_0 e^{-j\omega\tau} + j\omega C_{gdB}$$

it can be shown that

$$\begin{bmatrix} Y_i & -Y_{gd} \\ Y_{gm} & Y_{ds} \end{bmatrix} = \begin{bmatrix} y_{11} + y_{12} & y_{12} \\ y_{21} + y_{22} - j\omega C_{ds} & y_{22} - j\omega C_{ds} \end{bmatrix} \begin{bmatrix} 1 + j\omega\tau_1 & j\omega\tau_2 \\ j\omega\tau_3 & 1 + j\omega\tau_4 \end{bmatrix}$$

with

$$\begin{aligned} \tau_1 &= R_I C_{IB} \\ \tau_2 &= R_I C_M \\ \tau_3 &= R_{GD} C_N \\ \tau_4 &= R_{GD} C_{gdB} \end{aligned}$$

Rearranging:

$$\begin{bmatrix} 1 + j\omega\tau_1 & j\omega\tau_2 \\ j\omega\tau_3 & 1 + j\omega\tau_4 \end{bmatrix} = \begin{bmatrix} y_{11} + y_{12} & y_{12} \\ y_{21} + y_{22} - j\omega C_{ds} & y_{22} - j\omega C_{ds} \end{bmatrix}^{-1} \begin{bmatrix} Y_i & -Y_{gd} \\ Y_{gm} & Y_{ds} \end{bmatrix}$$

We now have an equation giving the time constants in terms of the measured y-parameters and the core admittance parameters, the values of the core parameters being calculable from the numerical values for C_{IB}, C_N, C_{gdB}, and C_M found from step 2.

The matrix product on the right is calculated over a range of frequencies and the real and imaginary parts divided by ω. If the parasitics R_g, R_s, R_d, L_g, L_s, L_d, C_{ds}, have been estimated correctly, the real parts should be close to 1 or 0 and the values of τ_1, τ_2, τ_3, τ_4, should be invariant with frequency. Experience suggests the key elements to have right are R_s and L_s.

The last step is to recover R_I and R_{GD}. By averaging the τ's over the top one-fifth of the measurement frequency range, the resistances are obtained as follows:

$$\begin{aligned} R_I &= \frac{(\tau_1 + \tau_2)_{\text{average}}}{(C_{IB} + C_M)} \\ R_{GD} &= \frac{(\tau_3 + \tau_4)_{\text{average}}}{(C_{gdB} + C_N)} \end{aligned}$$

9.3 Curve Fitting

In this section, without survey, rationale, or comparison with other methods, we simply give the equation evolved to fit all empirical data in this work, regardless of nature or origin, in the expectation that it might be useful to others. For those wanting a survey of past methods, Chapter 7 in [1] is a good place to start.

The function is:

$$f(x) = \alpha_1 \exp - \left(\beta_1 (x - \gamma_1)\right)^2 + \alpha_2 \exp - \left(\beta_2 (x - \gamma_2)\right)^2 + a + be^{-cx} + mx$$

A double Gaussian basis was conceived to cater for source data comprising a hump and a dip, as in Figure 9.4 for example.

Overall, the fit varies between excellent and adequate for data representation to a circuit simulator, the exception being the fourth curve down.

Gaussians have the merit of being smoothly varying functions with no tendency to wild excursion. For fitting to the nonlinear functions that describe FET and HEMT behavior, which similarly are smoothly varying without tendency to wild excursion, they are well suited to fitting and interpolation over the measured data range. Outside the measured data range, they should extrapolate smoothly, although that does not mean that the extrapolation is valid.

All characteristics and behavioral properties of a nonlinear FEMT are functions of the two voltage variables, V_{GS} and V_{DS}. Fitting to the V_{DS} variation and to the V_{GS} variation is done separately, so each element is presented to the circuit simulator in two parts.

The new topology element functions of Section 10.7 show the general quality of fit. Over the measured data range, there is little to complain about. As to the extrapolations of Figures 10.23 through 10.27, it is not possible to say whether they are valid, useful, or simply errant. Most look plausible, except the upper frame in Figure 10.25. The plain truth is that any extrapolation scheme can be thrown into disarray by noise on the measurements, especially in the end-of-range points, which then wrecks the derivatives of the data function upon which extrapolation depends.

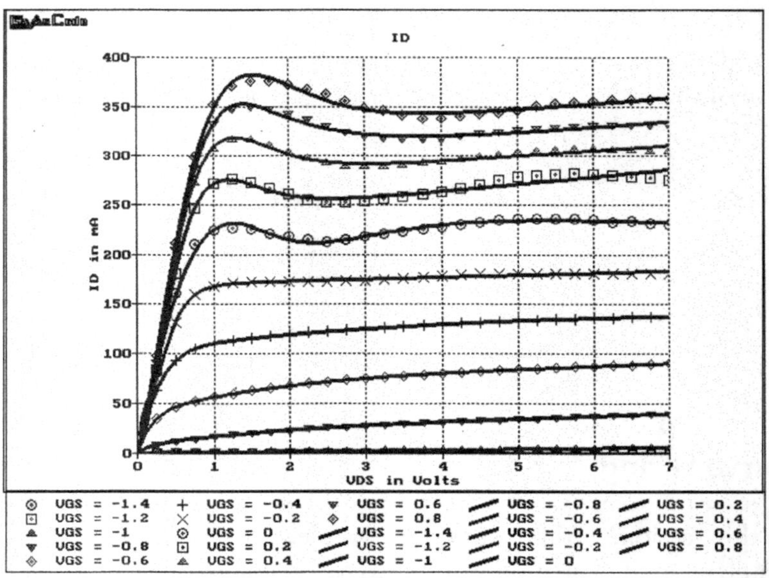

Figure 9.4 Example empirical data with a hump-and-dip character.

No doubt many of those who at one time or another have wrestled with the curve-fitting problem have their own favorite way of finding the parameter values α_1, β_1, γ_1, … . Ours consisted of setting trial values for

$$\beta_1, \gamma_1, \beta_2, \gamma_2, c$$

whereupon the exercise reduces to a linear problem allowing least-squares determination of

$$\alpha_1, \alpha_2, a, b, m$$

Precautions are few:

- If β_1 or β_2 is zero, it creates a term α_1 or α_2, which cannot be distinguished by the optimizer from the a term.
- Similarly, if $\beta_1 = \beta_2$ and $\gamma_1 = \gamma_2$, or if $c = 0$, we again have redundancy.

Measures to avoid these pitfalls are self-evident.

Reference

[1] Aaen, P. H., J. A. Plá, and J. Wood, *Modelling and Characterization of RF and Microwave Power FETs*, Cambridge, U.K.: Cambridge University Press, Chapter 7, 2007.

CHAPTER 10

Obtaining the Current and Capacitance Functions

10.1 Current Functions from Pulsed *I*(*V*) Measurements

In Section 3.1.1, practical examples were given of the difference between the dc characteristics and the set of $I(V)$ curves along which a FET or HEMT will work at RF. Further, the RF (or dynamic) characteristics depend upon the bias point about which they are measured, a striking instance being Figure 10.1.

A universally applicable FET or HEMT model must therefore contain all possible RF sets of characteristics $i_D(v_{GS},v_{DS},V_{GS},V_{DS})$, functionally written thus to remind us that the dynamic curves depend upon both the bias voltages (V_{GS},V_{DS}) and the signal voltages (v_{GS},v_{DS}). Herewith a sizeable measurement, data management, and programming task when it comes to incorporating the model in a circuit simulator. There has to be another way.

A proposal frequently seen is that it might be possible to reconstruct a dynamic $I(V)$ simile by integrating the transconductance or output conductance derived from small-signal S-parameters, thus:

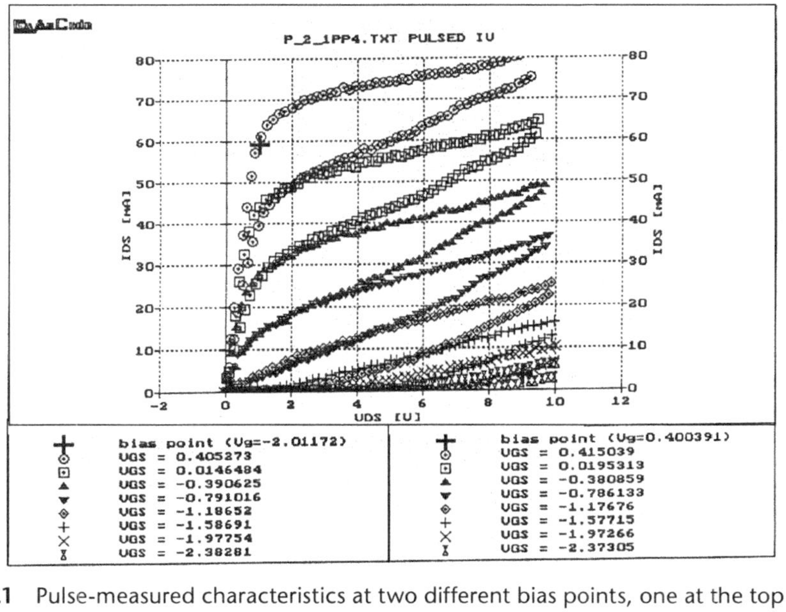

Figure 10.1 Pulse-measured characteristics at two different bias points, one at the top left and one at the bottom right (visible as + symbols).

$$i_D\left(V_{GSP},V_{DSP}\right) \leftarrow \int_{V_P}^{V_{GSP}} g_{m0}\left(V_{GS},V_{DSP}\right)dV_{GS}$$

or equivalently

$$i_D\left(V_{GSP},V_{DSP}\right) \leftarrow \int_{0}^{V_{DSP}} g_0\left(V_{GSP},V_{DS}\right)dV_{DS}$$

such that, by choosing pairs of (V_{GSP},V_{DSP}) to cover the entire current-voltage plane, the RF characteristics could be synthesized.

There are two principal drawbacks. The first drawback, for the procedure to be valid, the two integrals must produce exactly the same answer. Among the large number of devices studied in the course of this book, it is hard to recall the condition ever being true. The left-pointing arrows in the foregoing two equations have been purposely substituted for equality signs to act as an aide-mémoire of this inconvenient fact. In the parlance of the cognoscenti, g_{m0} and g_0 are said to form a nonconservative pair.

A second drawback is that, even if the two integrals did yield the same answer, the resultant curve set would apply only to a starting (or bias) point of V_{DS} = 0V. However, it is not an impasse, but rather a complicating factor in that the lower bound for integration could instead be any arbitrary bias point (V_{GSB},V_{DSB}). The job would then be to show that conservation held at all conceivable bias points.

Therefore, one of the major challenges in developing a model for coding in circuit simulators is to find a means of reconstructing bias-dependent dynamic $i_D(v_{GS},v_{DS},V_{GS},V_{DS})$ curves from S-parameters measured at multiple bias points, the basic data resource of equivalent circuit models. If that could be done, the need for a large database of pulse-measured characteristics could be dispensed with; the required set of dynamic characteristics could be reconstructed in the circuit simulator once the bias point had been found. The next section describes just such a method.

10.2 Dynamic *I*(*V*) Reconstructor

The precursor to the following development is Section 6.4. Expressed at its simplest, Section 6.4 proposed that there are two ways of looking at the fact that dynamic $I(V)$ curves differ from the dc curves. First, the voltage grid, composed of (V_{GS},V_{DS}) coordinates, is immutable and the $I(V)$ curves are distorted. Second, the complete converse: the $I(V)$ values are correct and it is the voltage grid that is distorted. Reconstructing bias-dependent dynamic $I(V)$ characteristics hinges on the second of these two views.

To follow the argument, we need to be acquainted with the idea from Part III that the voltages applied to the device fundamentally set two physical parameters: the depletion region depth d and extension X. To make progress, we further assume that dispersion principally arises from charge exchange with surface states, leading to changes that are representative of the changes that would result from charge exchange with deep levels anywhere, for example, at the interface between the N-channel layer and the buffer. (That is not to understate the importance of self-heating as a potential source of dispersion; it merely reflects our own experience. It also simplifies the task of reasoning out the reconstruction scheme.)

The underlying problem is that the dimensions of the depletion region, d and X, are different for given values of RF voltages (v_{GS},v_{DS}) compared with the values of d and X resulting from dc voltages (V_{GS},V_{DS}) having the same values. In other words, for given applied voltages, surface charge alters the values of d and X to values as if the applied voltage themselves had been changed in the absence of surface charge. Pictorially, the proposition is as shown in Figure 10.2.

In Figure 10.2(a, b), the voltages are the same, but the space-charge layer dimensions are different, whereas in Figure 10.2(b, c), the ds and Xs are the same, but the voltages are different. In short, via the intermediate step of Figure 10.2(b), it is as if the device, bare of surface charge, has different voltages applied to it. Put another way, the surface charge has the effect of distorting the voltage grid. Exactly how the space-charge layer changes depends upon location on the $I(V)$ plane, but for present purposes such details are unimportant. All that matters is the principle that, within limits, trapped charge alters d and X in a manner akin to having different terminal voltages and no trapped charge. Thus, Figure 10.2 is only indicative.

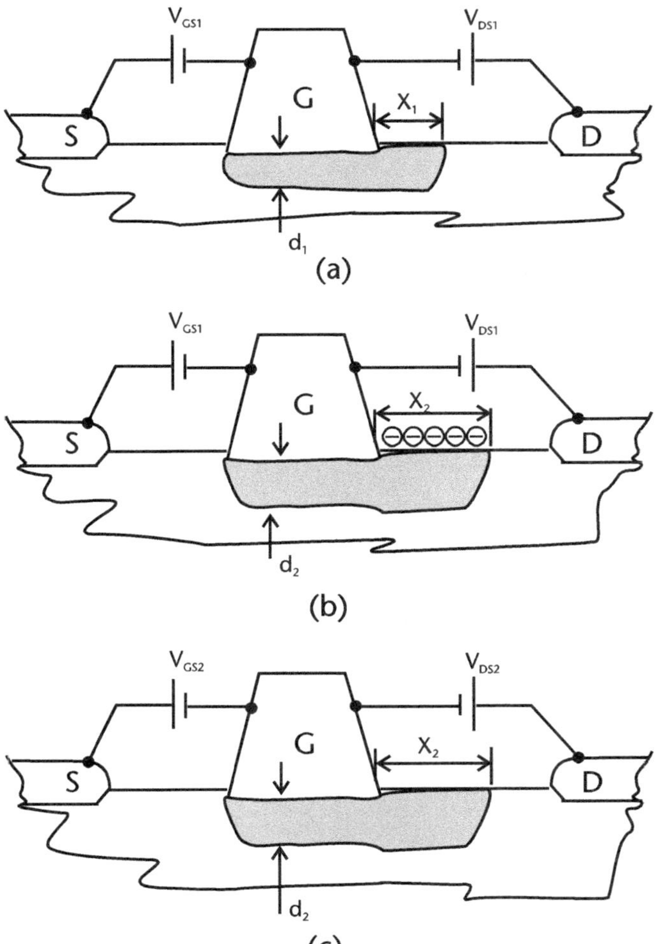

Figure 10.2 Illustrating the broad idea that surface charge has an effect equivalent to different applied voltages.

Going back to the integral inequality of the previous section, namely:

$$\int_0^{V_{GSP}} g_{m0}\left(V_{GS}, V_{DSP}\right) dV_{GS} \neq \int_0^{V_{DSP}} g_0\left(V_{GSP}, V_{DS}\right) dV_{DS}$$

the values of g_{m0} and g_0 are correct, but the values of V_{GS} and V_{DS} at which they occur are wrong. That is, g_{m0} and g_0 are nonconservative because the voltage grid or mesh is distorted; the foreground data component, consisting of g_{m0} and g_0 values, is the immutable component.

A simple diagrammatic parallel helps to dispel the evident strangeness of this proposition. Figure 10.3 shows an object overlaid with a gauge system, meaning a system of reference by which we gauge the relative positions of data or features: the grid is linear and square, the clock face is obviously circular although from the angle of view it appears elliptical. Now consider the interpretations possible if the grid were distorted as in Figure 10.4.

The first and obvious interpretation is that the object is still as it was, it has the correct shape and detail, but the grid is distorted, a fact well known to us, as we distorted it. Yet suppose the grid had become distorted without us knowing it and we tried to reconstruct the clock face from the data (recorded from Figure 10.4 at each mesh point) while assuming the grid still to be linear and square. We would end up with a contorted rendition of the clock face. The salient point is that, despite the distorted appearance and the assumption that led to it, the foreground object, the clock face, is correct; it is unchanged, but the means to measure it or position it spatially (the gauge system) has become corrupted.

To relate to the dispersion problem, all we need to do is substitute (g_{m0}, g_0) for the clock face and label the mesh with voltages instead of plain numbers. The foreground object, now (g_{m0}, g_0), is correct, whereas the grid, which in the FET case is (V_{GS},V_{DS}), is wrong.

Thus far, all we have managed to do is mount a qualitative argument to pinpoint the problem. The solution entails finding a way to undistort the (V_{GS},V_{DS}) coordinate system. Numerical optimization is one way, whereby the two voltages are

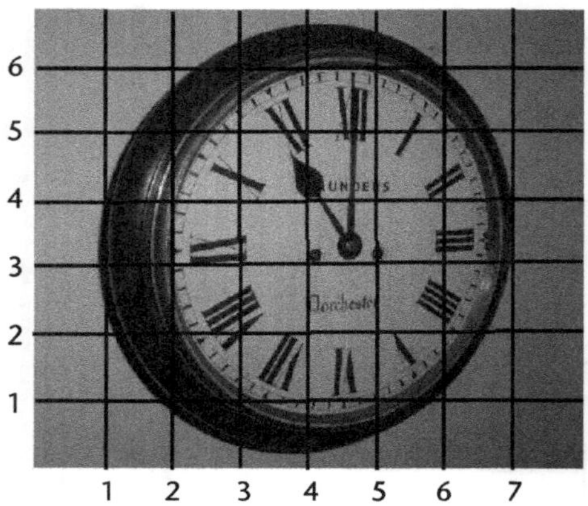

Figure 10.3 An object, in this case an antique wall clock, can be categorized by recording its features (wood, brass, black paint, metallic hands ...) at each point of a coordinate grid.

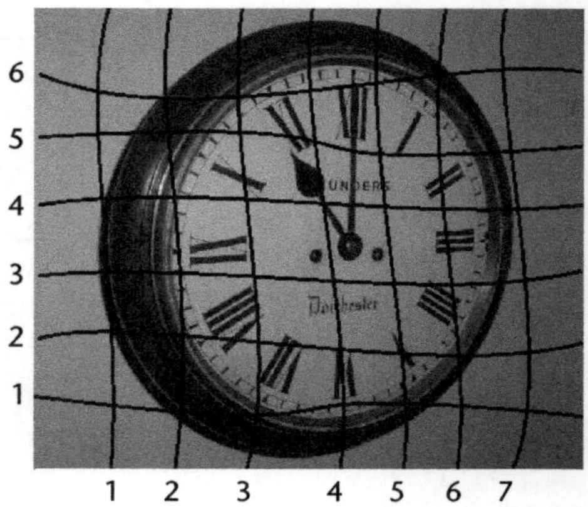

Figure 10.4 The object overlaid with a distorted grid.

manipulated together to render (g_{m0}, g_0) a conservative pair. A different approach has proved workable.

In what follows, we need to be clear from where the values of (g_{m0}, g_0) are to come. The answer is that they are to come from fitting to the commonplace model because:

- The raw data we are trying to undistort is the dc set of $I(V)$ characteristics.
- As the standard model is purely a static model, the set of $I(V)$ curves to which it naturally gives rise is the dc set.
- Primarily, the errors in charge movements under dynamic (i.e., RF) conditions arise from the incorrect placement of the capacitive elements in the static model; the conductance elements are in physically correct positions and so can be retrieved error-free (a relative term) by fitting to multibias S-parameters at low frequencies.

We start by identifying the line or trajectory of points where the dc and RF characteristic curves cross one another, as clearly visible in Figures 3.1 and 3.3, for example. One such point is the bias point, hereafter identified by the capital subscript B. At the point B, I_{DB},g_{m0B},g_{0B} are all correct values (Figure 10.5).

Another certainty is that there are other dc curves separated from one another by δV_{GS}.

To begin reconstruction of the dynamic curve set, the first step is to form a current

$$i_{D1} = I_{DB} + g_{m0B}\delta V_{GS} + g_{0B}\delta V_{DS}$$

where δV_{GS} is known (it is the next curve up) and δV_{DS} is adjusted to make $i_{D1} = I_{D1}$. The logic is this: i_{D1} amounts to a single value of RF current, and we know it must lie on the dc curve corresponding to $V_{GSB} + \delta V_{GS}$. This point is marked (1) on the figure.

As a second step, the foregoing procedure is repeated for all curves above and below the bias point, moving from one curve to the next.

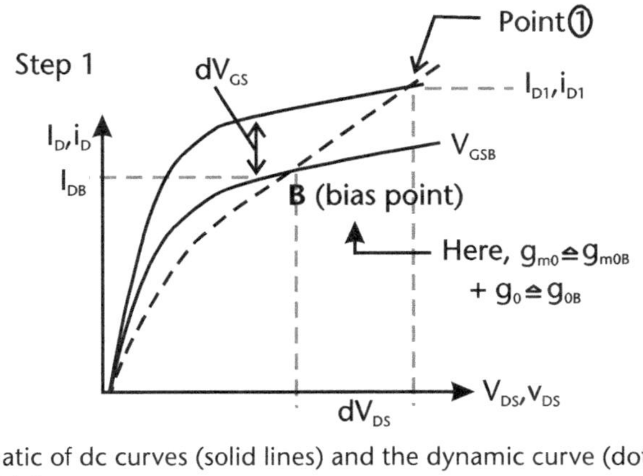

Figure 10.5 Schematic of dc curves (solid lines) and the dynamic curve (dotted) passing through the bias point B.

To review, all we have achieved so far is the limited determination of the crossing points where the RF curves intersect the dc curves. As yet, there is no shape to the $i_D(v_{GS},v_{DS})$ curves; all we have found is one point on each curve. The remaining task is to find a way of determining all the other points needed to define each complete $i_D(v_{GS},v_{DS})$ curve.

The third step therefore comes down to finding a way of shifting specific values of (g_{m0}, g_0) off the dc voltage grid to voltages (v_{GS},v_{DS}) that are different from the dc voltages (V_{GS},V_{DS}) at which those specific (g_{m0}, g_0) values occur. An algorithm that achieves that end to good approximation, while at the same time being relatively quick to implement, is as follows. The justification for it lies in the results achieved.

The method proceeds by interpolating between curves of g_0 on the (V_{GS},V_{DS}) coordinate grid to arrive at a $g_0(v_{GS},v_{DS})$ such that the point given by the integral

$$\int_0^{V_{DSB}} g_0\left(v_{GS}, v_{DS}\right) dv_{DS}$$

passes through $i_D = I_{DB}$, the dc bias current. By evaluating the generalization of this integral

$$\int_0^{v_{DSP}} g_0\left(v_{GS}, v_{DS}\right) dv_{DS}$$

for a range of v_{DSP}, we generate point by point a complete RF curve that passes through the origin $i_D = 0$. Practice shows the curve replicates to very good approximation the pulse-measured dynamic curve, which passes through the bias point.

We have now reconstructed the dotted curve in Figure 10.5.

To reconstruct all the other RF curves, the above procedure is repeated with V_{DSB} replaced by the previously identified crossing points, which we will now call V_{DSX}. Construction of each curve begins by interpolating between curves of g_0 on the (V_{GS},V_{DS}) coordinate grid to arrive at a $g_0(v_{GS},v_{DS})$ such that the point given by the integral

$$\int_0^{V_{DSX}} g_0\left(v_{GS}, v_{DS}\right) dv_{DS}$$

passes through $i_D = I_{DX}$, the dc current at the crossing point. Figure 10.6 summarizes the interpolation-by-trial routine.

Once the interpolated function $g_0(v_{GS},v_{DS})$ has been thus found to satisfy the crossing-point requirement, evaluating the integral

$$\int_0^{v_{DSP}} g_0\left(v_{GS}, v_{DS}\right) dv_{DS}$$

for a range of v_{DSP} generates a complete RF curve that replicates to very good approximation the pulse-measured dynamic curve, which passes through the chosen crossing point.

We have now reconstructed the whole set of dynamic (or RF) curves. Figure 10.7 is a typical result, in this case for a 0.5 μm ion-implanted foundry FET.

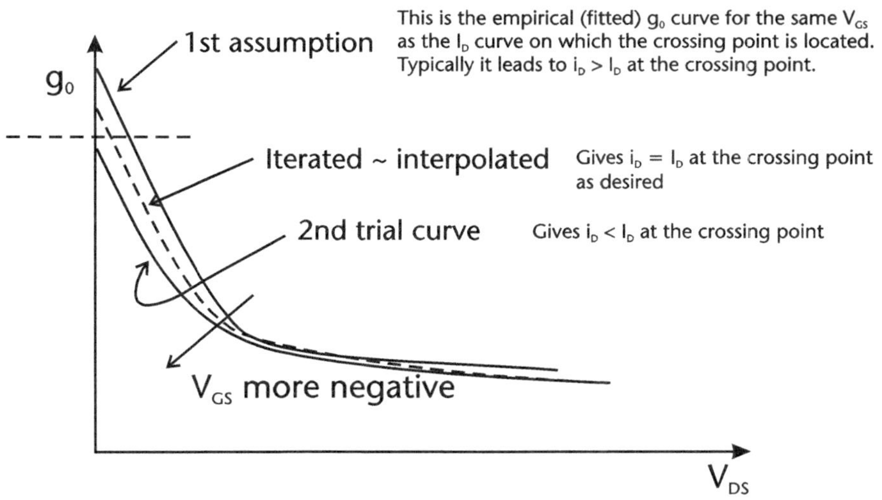

Figure 10.6 Trial-and-error interpolation of the g_0 curves.

The upper frame shows the dc characteristics and pulsed set of curves measured about the point marked + at $V_{DS} = 5$V, just visible on the fifth curve down from the top. In the lower frame, the measured dynamic characteristics and the reconstructed set are compared.

A tougher test is when the bias point is close to one of the extreme corners of the $I(V)$ plane, notably the bottom right corner as in Figure 10.8.

Close inspection reveals differences for curves further away from the bias point, but the agreement is still good enough for many design purposes.

Another example is given in Figure 10.9 for an entirely different type of device.

In the bottom left frame, not only do the reconstructed curves match the pulsed set exactly over the measured range, but they also extrapolate extremely well, always a useful attribute.

The three examples of Figures 10.7, 10.8, and 10.9 are typical of the average, worst, and best agreement from among the wide variety of cases and device types dealt with. All of the results in Chapter 11 were obtained with the reformed model including dynamic $I(V)$ reconstruction as described. Apart from all else, the beauty of this system is that whatever dynamic $i_D(v_{GS}, v_{DS}, V_{GS}, V_{DS})$ function the circuit simulator needs at run time, it can be synthesized in short order within the simulator itself.

For a large-signal model, it is the current function that is required; there is little call for the transconductance to be known separately. Consequently, there has been no incentive to go beyond the foregoing procedure for reconstructing the RF current characteristics. If the transconductance is at times wanted, it can be obtained by differentiating the reconstructed current function.

The technique as described can be viewed as just the first step in a more complete method of undistorting the voltage grid (as outlined below). Given the kind of agreement evident in Figures 10.7, 10.8, and 10.9, which came as something of a surprise, for practical purposes no further refinement was deemed necessary, so the following is untried.

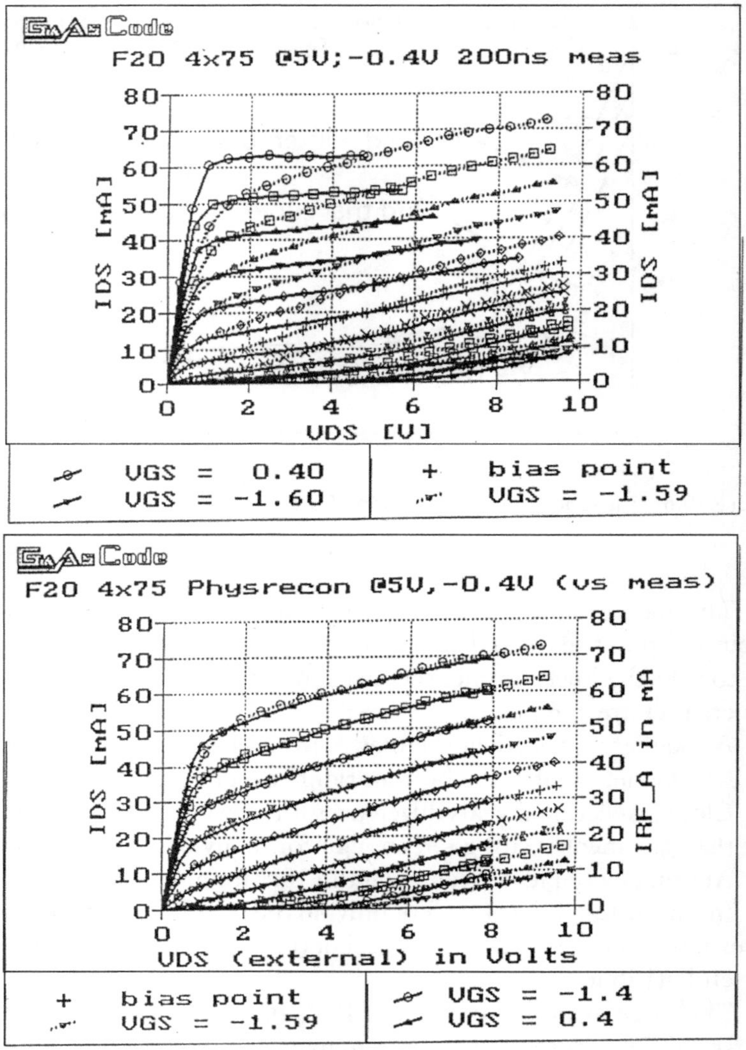

Figure 10.7 Dynamic *I*(*V*) reconstruction example.

At the stage of settling on an interpolated $g_0(v_{GS},v_{DS})$ curve that meets the requirement for reconstructing a dynamic curve, axiomatically we have uncovered how the distorted (V_{GS},V_{DS}) coordinate values need to be transformed to yield the undistorted (v_{GS},v_{DS}) coordinate values. That is to say, by virtue of the premise that the fitted value of g_0 is correct, leading to the imposed condition $g_{0RF} = g_{0dc}$ upon which the method is founded, the transformation $V_{GS} \rightarrow v_{GS}$ and $V_{DS} \rightarrow v_{DS}$ is now known. Thereby, the g_{m0} value that accompanied any particular g_0 value can similarly be transformed onto the RF grid, namely,

$$g_{m0}\left(V_{GS},V_{DS}\right) \rightarrow g_{m0}\left(v_{GS},v_{DS}\right)$$

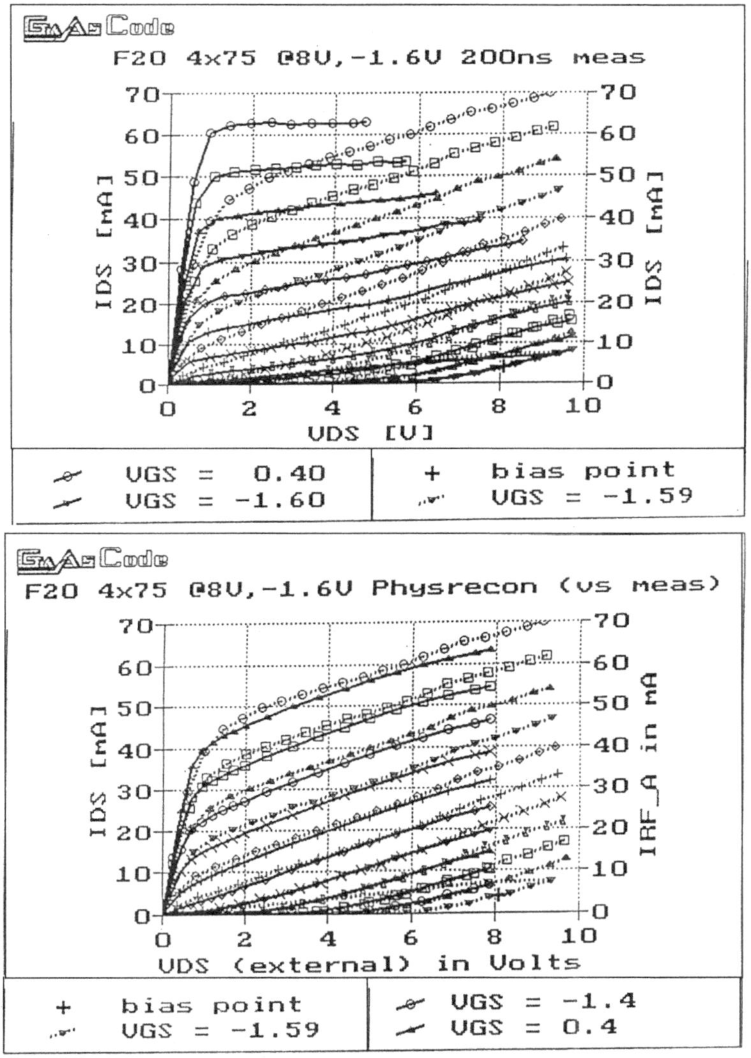

Figure 10.8 Upper frame: dc and pulse-measured curves about the point $V_{GS} = -1.6\text{V}$, $V_{DS} = 8\text{V}$. Lower frame: reconstructed curves (solid lines) compared with the measured set.

By repeating this local transformation for all points along the settled $g_0(v_{GS}, v_{DS})$ interpolated curve, and doing the same for all curves, we thereby reconstruct the RF transconductance function $g_{m0}(v_{GS}, v_{DS})$.

With this added resource, further refinement of the (v_{GS}, v_{DS}) coordinate grid could, if warranted, be achieved by implementing a routine to minimize the error function ϵ, defined as

$$\int_{V_P}^{v_{GSP}} g_{m0}\left(v_{GS}, v_{DSP}\right) dv_{GS} - \int_0^{v_{DSP}} g_0\left(v_{GSP}, v_{DS}\right) dv_{DS} \triangleq \epsilon$$

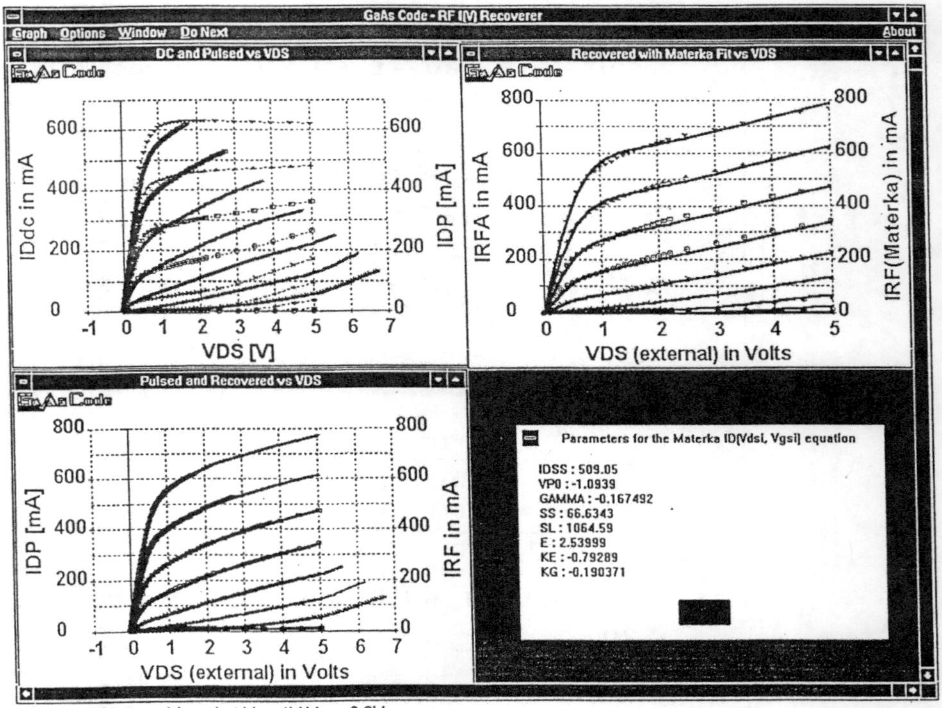

Figure 10.9 Top left frame: dc and pulse-measured curves about the point $V_{GS} = -0.2$V, $V_{DS} =$ 1V. Bottom left: reconstructed curves compared with the pulse-measured set. Top right: fit of the Materka current equation to the reconstructed set (as implemented for a circuit design).

As said earlier, none of it has been tried, such refinement having been unnecessary in the practical circuits worked upon.

Finally, the fact that the procedure worked for all the cases tried is no guarantee that it will work in all instances for all devices. If it fails, reverting to pulsed $I(V)$ measurement is the answer.

10.3 Implications for Slow-Rate Transients

Section 10.2 leaves us with the belief that the size and shape of the depletion region are dependent upon the trapped charge. When the material layers of which a device is composed are lattice-matched, the likelihood is that there are few interface states, leaving the main location for troublesome deep levels as the surface, including passivated surfaces. The extent to which the same holds true for the AlGaN–GaN interface in GaN HEMTs is an open question. Irrespective, if d and X are affected by charge exchange with deep levels anywhere, there are implications for all the elements in an equivalent circuit model.

Using the shorthand τ_{ss} to denote the transient due to charge exchange with a deep level anywhere in the device, τ_{ss} does not just affect $I(V)$; that is a secondary effect. The primary effect is on the space-charge layer dimensions d and X. Consequent upon the primary effect, another secondary effect is on any element that depends upon d and X (i.e., τ, C_{IB}, C_{gdB}, C_N, C_M). Therefore, to model properly the effect of slow transients due to charge exchange with deep levels, it is necessary to account for the slow changes thus imparted to all the elements of the equivalent circuit, not just the current. We can only speculate how this might be done.

Within the framework of the present approach, it would first be necessary to determine the distorted and undistorted values of d and X at all points of the $I(V)$ plane. Next, at each point, the time evolution of how d and X relax would have to be incorporated. By now, however, we are faced with an intractable problem: the transient characteristic is different right across the $I(V)$ plane. Even if we could measure it, the task of embodying such a volume of data in a model is difficult to contemplate.

Still within the framework of the present approach, the first step of finding the distorted and undistorted values of d and X at all points of the $I(V)$ plane could be eliminated by pulsed S-parameter measurements and fitting the model at all such points. That would give the starting values and end values of all the elements, but we would still be left with the intractability of the time change connecting the two not being the same everywhere.

Both the foregoing could be implemented as approximations by assuming the transient to be the same at all points of the $I(V)$ plane, but the overall task would remain unwieldy.

However, due to recent innovations in the field, there would appear to be a solution to the problem: fitting the equivalent circuit model to nonlinear vector network analyzer (NLVNA) measurements [1], essentially a behavioral model embodying transient phenomena. It is always hard to weigh the merits and demerits of a technique without first-hand experience of it, so the particular bother is how to reconcile this approach with the fundamental limitation aired in Section 2.4. Presuming there to be an answer, however, the method centered on the NLVNA would appear to be only manageable way of accounting properly for slow-rate effects.

10.4 Obtaining the Capacitance Functions

From Chapter 15, we have some idea of what to expect for the capacitance functions in the new model, at least in the above-the-knee region. Those results are to serve as checks on the elements freely fitted to practical FET and HEMT data; they have no anterior role in determining what the empirical functions will look like.

It is as well to be reminded that success or failure of a model depends as much upon the efficacy of the fitter-extractor as it does upon the model itself, maybe even more so. Constructing the method and means of fitting is a serious business. It can be costly to skimp it.

Compared with the standard model, the reformed model of Chapter 7 has different elements in different places. As described in Chapter 9, a tried-and-tested software package had long been available commercially for fitting and reverse modeling FETs and HEMTs. Rather than setting out to develop an entirely new fitter-extractor for the reformed model, adapting the tried-and-tested scheme promised at least a first demonstration of the viability of the whole. That purpose was fulfilled to the point where, as so often happens, the temporary survived as the sole implementation.

Two ways can be envisaged of using an existing fitter. The first is to take the output in the form of the standard model elements, all as functions of bias, and make use of the relationships between them and the elements of the reformed model to derive initial estimates for the new element functions. A generic optimizing routine may then be used to match the S-parameters of the reformed model to the measured data. To pursue this line of approach, the first of the required equations is (15.16) from Chapter 15, here renumbered as (10.1):

$$C_{gc} = C_{IB} + g_{m0}\tau_{gm} - C_N \tag{10.1}$$

The following relationships appear later in Chapter 15, Section 15.4.2:

$$C_{gd} = C_{gdB} \tag{10.2}$$

$$C_{ds} = C_M + g_0\tau_{gm} \tag{10.3}$$

The equations also appear in Chapter 15 (15.7) and (15.2), respectively:

$$C_N = \frac{C_{IB}}{1 + \frac{L_G}{X}} \tag{10.4}$$

$$C_{IB} \simeq \frac{g_{m0}L_G}{v_{sat}} \tag{10.5}$$

Finally, (13.9) from Chapter 13:

$$\tau_{gm} = \frac{X}{2v_{sat}} \tag{10.6}$$

To use these equations, g_{m0}, g_0, τ_{gm} can be taken from the fit of the standard model to the basic data resource, which is the multibias S-parameters. Alternatively, the transconductance and output conductance can be obtained by differentiating one of the RF $i_D(v_{GS},v_{DS})$ functions from Section 10.1 or Section 10.2. With C_{gc}, C_{gd} and C_{ds} available from the fit of the standard model:

- C_{gdB} is given immediately by (10.2).
- Equation (10.3) gives $C_M = C_{ds} - g_0\tau_{gm}$.

There are several ways of using (10.1) to find first estimates for C_{IB} and C_N.

Option 1: substitute from (10.5) to give

$$C_N \simeq \frac{g_{m0}L_G}{v_{sat}} + g_{m0}\tau_{gm} - C_{gc}$$

Option 2: substitute from (10.4) and (10.6) to give

$$C_{IB} = \left[\frac{L_G + 2v_{sat}\tau_{gm}}{L_G}\right]\left(C_{gc} - g_{m0}\tau_{gm}\right)$$

Option 3: forget about (10.1) and simply use (10.5) and (10.4).

For those wishing to try the reformed model, the procedure just described offers the quickest route, depending upon the quality of the routine for fitting the standard model and the optimizer. The S-parameters of the reformed model for use by the optimizer can be generated by any basic circuit simulator.

The alternative method is to extend the existing fitter, with a few minor alterations, to derive at run time the new model's elements from those of the commonplace model by network transformation. By some measure, the algebra is voluminous; an overview is given in Chapter 9.

All results involving the reformed model, the element functions of Section 10.7, and the practical circuit comparisons of Chapter 11 were obtained with the rejigged fitter as described in Chapter 9.

10.5 Charge Conservation

In the open literature, much is made of the need to ensure charge conservation, particularly but not exclusively at the gate terminal, it having been proved that failure to do so gives rise to an erroneous dc current during simulation. One argument runs that because a capacitance in the equivalent circuit (say, C_{gc}) is dependent not only upon the voltage across it (V_{GS}) but also another voltage (V_{DS}), after one cycle of both voltages the charge on the capacitor will not be conserved [2]. As a result, various schemes have been proposed for forming a gate-charge function that is conservative [3, 4].

We take a different view: that the imperative to construct a conservative charge function along the foregoing lines is a direct result of approximations or false assumptions that lead to an incorrect representation of how practical devices behave. If the equivalent circuit is configured correctly in the first place, with capacitances to align with each of the charge changes that actually occur, then charge conservation should follow automatically.

In the large, charge Q is stored in the depletion layer underneath the gate; it is affected by both V_{GS} and V_{DS}, so $Q_G = Q = Q(V_{GS}, V_{DS})$. Breaking Q down into its separate components, charge exchange occurs:

- At the depletion edge in region I, being that part of the channel operating below velocity saturation;
- At the depletion edge in region II-III, the part of the channel where electrons are moving at saturated drift velocity;
- At the drain edge of the depletion in region III.

Each of these three regions has a dual character; each responds differently to variations in V_{GS} compared with when V_{DS} varies. Having in view the equivalent circuit, we adopt the terms "branch" and "orthogonal branch" to progress the argument. Straightforwardly, "branch" refers to the part of the network across which a voltage (either V_{GS} or V_{DS}) is applied, whereas "orthogonal branch" signifies the leg of the circuit that is orthogonal to the one where the voltage is applied. With that in mind:

- The first of the foregoing three regions stores branch charge in response to changes in V_{GS} (capacitance C_{IB}) but orthogonal branch charge in response to changes in V_{DS} (transcapacitance C_M).
- The second also stores branch charge in response to changes in V_{GS} and orthogonal branch charge in response to changes in V_{DS}. The single element $g_{m0}\tau$ represents both.
- The third region stores branch charge in response to changes in V_{DS} (capacitance C_{gdB}) but orthogonal branch charge in response to changes in V_{GS} (transcapacitance C_N).

Thus, in the physical device, there are six components of charge storage that together define Q_G. In the equivalent circuit, these six components are represented by five elements as in Figure 10.10. There are two branch capacitances, two orthogonal branch or transcapacitances, and the single $g_{m0}\tau$ element of region II-III, which has the dual character of both capacitance and transcapacitance, neither being true capacitances owing to their origin in pure delay (τ). All are functions of (V_{GS},V_{DS}). In the figure, the inclusion of $g_{m0}\tau$ as a generalized charge element is emblematic to help to picture the problem. In the final equivalent circuit, the function is performed by the gate tapping between the current generators (Figure 7.10).

Apart from the difficulty of foreseeing how a single conservative gate charge could be derived from these five capacitive elements by published methods [3], it should not be necessary as, if properly constituted, they should axiomatically satisfy the conservative dictum. After all, the actual device has no difficulty in conserving the gate charge. Therefore, the need to enforce Q_G conservation as argued in the literature [2] is only secondarily due to the functional dependence of capacitance upon two voltages: it is due primarily to the capacitive elements in the commonplace equivalent circuit not representing correctly how charge is stored in the physical device. For example, C_{gs} (corresponding to our C_{gc}) is a composite of three charge-related elements: C_{IB}, $g_{m0}\tau$, and C_N. The trouble is that, with just C_{gs} in the gate-source branch of the network, all the current associated with it flows in that branch, whereas in truth the C_N component flows in the gate-drain branch. Inescapably, once the gate-source branch embodies a single element C_{gs} (or C_{gc}), there is a built-in error; charge movements are wrongly placed in the model vis-à-vis the physical device.

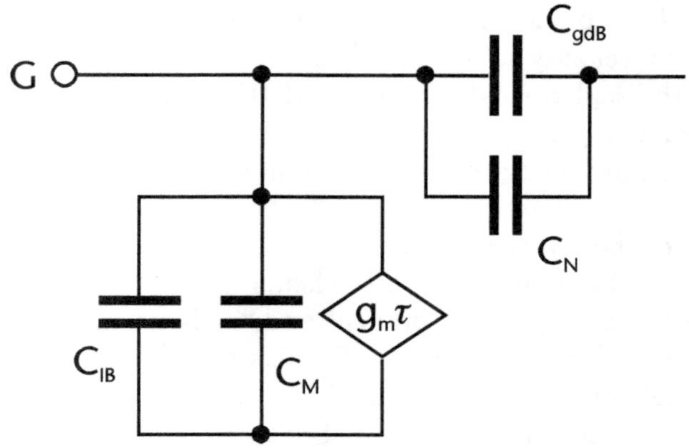

Figure 10.10 Charge storage components in the reformed model.

That said, even with the correct number and placement of capacitances and pure-delay element, failure to conserve terminal gate charge is still possible if the measurements from which they are derived are faulty, if the element functions are fitted badly to the measured data, or if their functional forms are not true to life. Once again, we have to distinguish between errors of principle and lack of attention to detail.

When working at the level of capacitance (or charge partial differentials), there is always the possibility that net charge over a cycle will be other than zero by virtue of numerical approximation and a step size that is not infinitesimally small. In this book, curves were fitted to the capacitances with functions that were smoothly continuous (as in Chapter 9) and which preserved the functional forms of the empirical data extremely well. For all that, in one of the practical circuits dealt with at length, the 500 mW amplifier, initially the simulation failed to converge, a problem that was traced precisely to the appearance of an erroneous dc current. In turn, there had to have been inexactness somewhere in the foregoing string of pitfalls, but exactly where remained undiagnosed. As will be detailed in Section 12.2.3, a simple network expedient allowed the simulation to run without revisiting the entire measurement and model extraction routine, with acceptable results.

10.6 The Defining Case of $V_{DS} = 0V$

We consider the ideal case of a symmetrical device, not to be artificially restrictive but because it serves as the basis for thinking about the practical case of asymmetry.

Let Q be the total positive ionic charge in the depletion region of depth d. Then at $V_{DS} = 0V$, there is a capacitance $\partial Q/\partial V_{GS}$ resulting from changing V_{GS} and $V_{GD}(= -V_{DG})$ equally. In this limit, half of the charge change is accomplished by current flow through the source terminal and half of the charge change is accomplished through the drain (Figure 10.11).

Another way of expressing it is the sameness principle: that the charge change induced by δV_{GD} is the same as that induced by δV_{GS} (= δV_{GD}). However, the charge change due to δV_{GS} is ($C_{IB} - C_N$), whereas that due to δV_{GD} is ($C_{gdB} - C_M$). Thus, we expect

$$\left(C_{IB} - C_N\right) = \left(C_{gdB} - C_M\right) \tag{10.7}$$

We might call this the voltage sameness principle.

Rearranging:

$$\left(C_{IB} + C_M\right) = \left(C_{gdB} + C_N\right) \tag{10.8}$$

We might refer to this as the spatial sameness principle on the grounds that the left term represents the total charge change in the gate-source region (no matter how it came about) and, similarly, the right side represents the total charge change in the gate-drain part of the device. The boundary between gate-source and gate-drain regions is of course ill-defined in a distributed system but allows us to identify with the gate-source and gate-drain branches of the equivalent circuit.

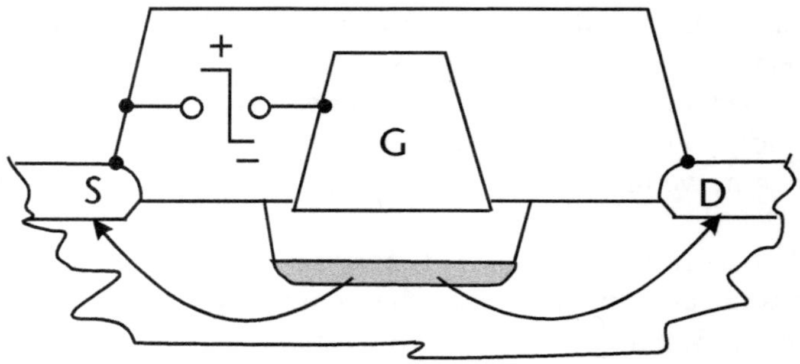

Figure 10.11 Device stimulated with a small pulse with drain and source shorted (so V_{DS} = 0V).

Apart from constituting a key test of whether a model has been well constructed at all stages of measurement and fitting, (10.8) complies implicitly with the requirement of symmetrical charging in the V_{DS} = 0V limit. One of the enduring difficulties with the standard model has been that even studiously measured and extracted C_{gc} and C_{gd} functions often do not satisfy this most basic of physical expectations, and that in turn calls into question whether the standard model is capable of representing what happens in a device in this limit. It may be a consequence of trying to satisfy charge sameness in the V_{DS} = 0V limit that the standard model forces the fitted C_{gd} to look like C_{gdB} rather than the (suspected) correct form yielded by (C_{gdB} − C_M).

The big test is how well the reformed model survives the compliance test of (10.8). Taking typical practical data from Section 10.7, here reproduced as Figures 10.12 through 10.15, we fare only reasonably. It should be stressed, however, that these results came about by applying the measurement and fitting routines straightforwardly with no extra constraints (and there are geometric capacitance contributions not explicitly accounted for). Judging by the ultimate agreement between theory and practice attained for real circuits the compliance nevertheless is good enough for practical purposes.

10.7 Practical Example of Reformed Model Elements

In this section, we look at typical forms or functions for the elements obtained by freely fitting the new model topology to multibias S-parameters. Unlike the plots of Section 3.2.4 where the data points were interconnected by straight lines, in the results to follow curves have been fitted in accordance with the methods of Chapter 9 to create continuous functions suitable for use by the model in circuit simulators. In all cases the drain-source voltage of the horizontal axis is the internal voltage (that is to say, the potential at the drain D' of the intrinsic device).

Figure 10.16 shows the freely fitted transconductance. Although this element does not feature in the nonlinear model, it is relevant in assessing whether the capacitive elements obey the trends elicited in Chapter 15. Here is another example of an HFET exhibiting not only compression near open channel but transconductance rollover too, reminiscent of Figure 4.7. Because the reformulated fitter (Chapter 9) produces elements as functions of the internal voltages, we can take it that they are the internal functions as needed for the model.

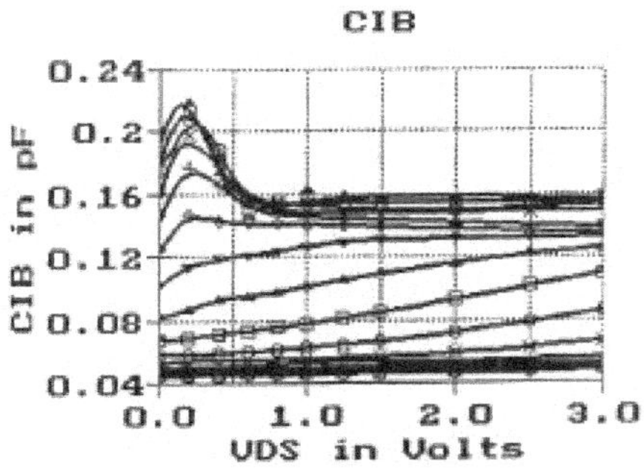

Figure 10.12 Example gate-channel branch capacitance.

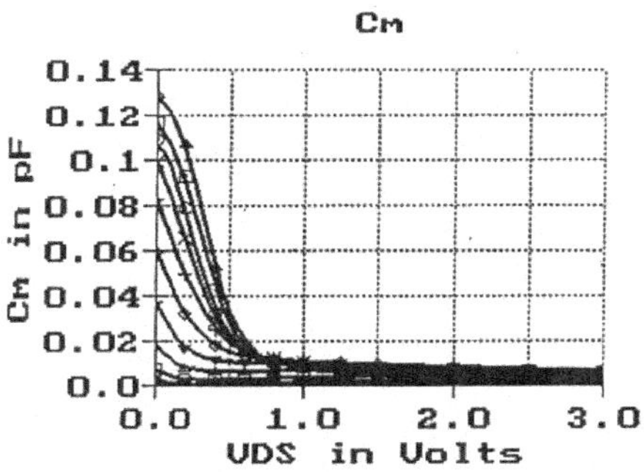

Figure 10.13 Example gate-channel transcapacitance.

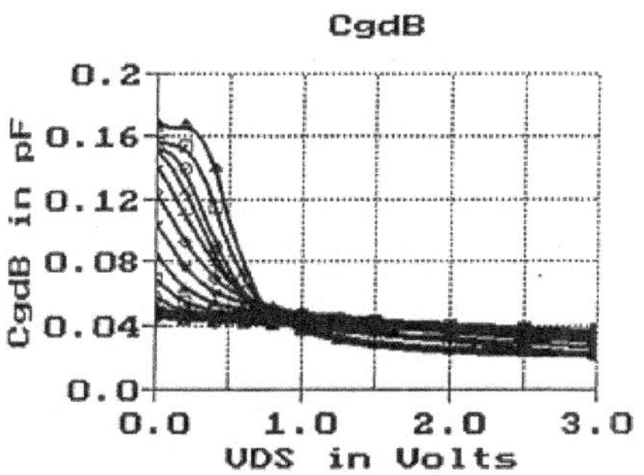

Figure 10.14 Gate-drain branch capacitance.

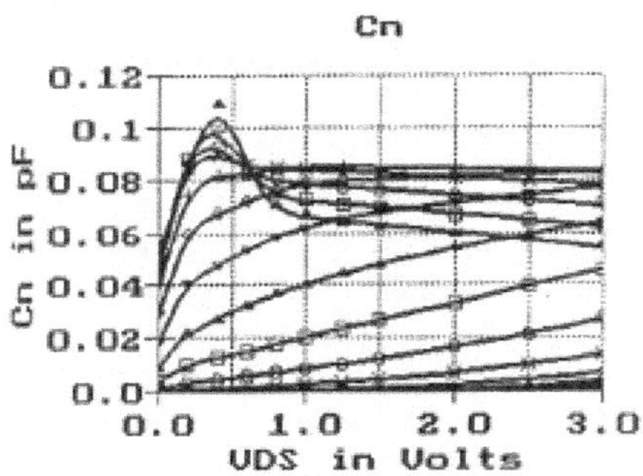

Figure 10.15 Transcapacitance in the gate-drain branch.

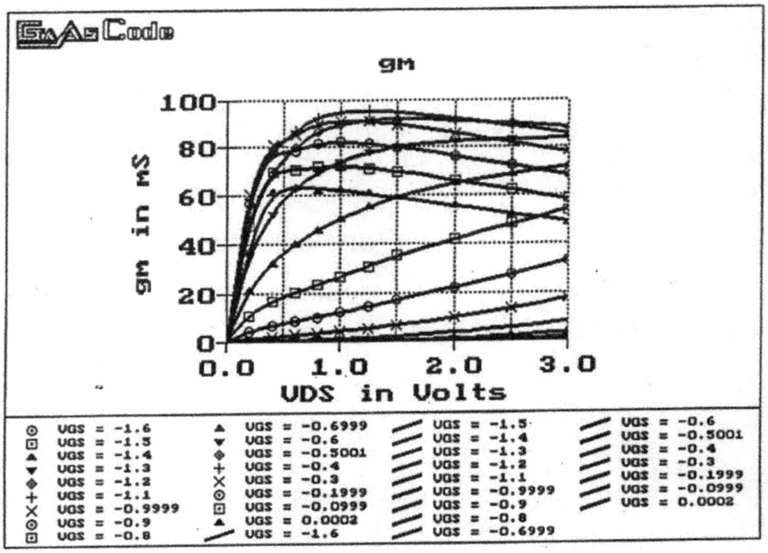

Fit to g_{m0s} as a function of internal V_{DS}

Figure 10.16 g_{m0} as a function of internal voltage $V_{D'S'}$.

The gate-channel branch capacitance is shown in Figure 10.17 with, in Figure 10.18, its companion transconductance in the gate-drain branch C_N.

Via (10.4) and (10.5), in the saturation region, both should resemble g_{m0}, an expectation that holds better for C_N than for C_{IB}. Nevertheless, both comply reasonably in that the general properties are correct.

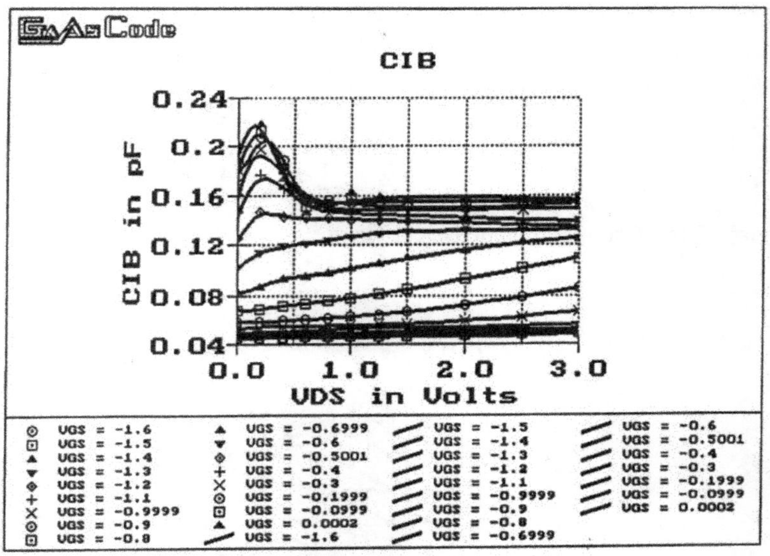

Fit to C_{IB} as a function of internal V_{DS}

Figure 10.17 C_{IB} as a function of internal voltage $V_{D'S'}$.

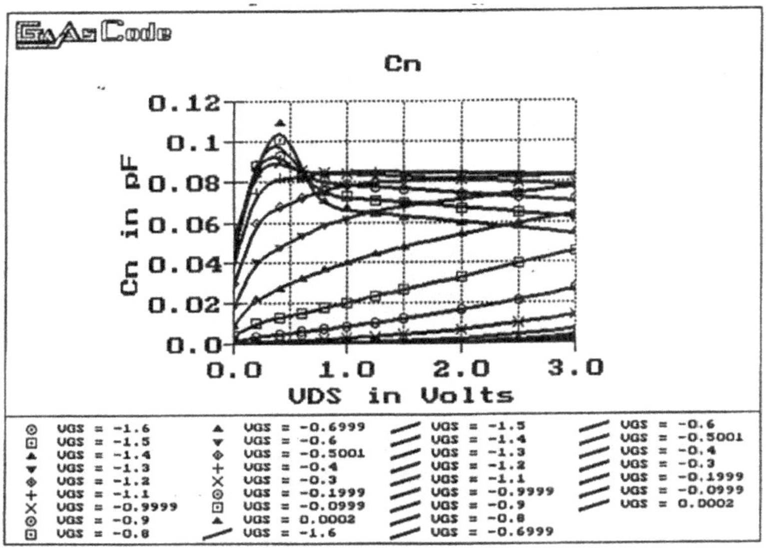

Fit to as a function of internal V_{DS}

Figure 10.18 C_N as a function of internal voltage $V_{D'S'}$.

An unexplained feature is the peak in both capacitances below $V_{D'S'} \simeq 0.5$V; it is not an artefact of the fitted equations, as the data points affirm. As the analyses of Chapter 15 are restricted to the above-the-knee region of operation, we have no yardstick by which to judge whether such a peak is fundamental to all devices or merely a curiosity of this particular device. Neither is sufficient experimental data available from a wide enough variety of device types to be able to tell. From what is available, the pointers are that it is not universal.

Figure 10.19 shows the fitted gate-drain branch capacitance which, as remarked elsewhere, conforms nicely to (10.9) ((6.15) of [5]). Its companion transcapacitance in the gate-channel branch, C_M, is shown in Figure 10.20.

$$C_{gdB} = \frac{2\epsilon Z_G}{\left(1 + \frac{2X}{L_G}\right)} \tag{10.9}$$

These two capacitive elements are supposed to be alike according to:

$$C_M \simeq -\alpha C_{gdB} \quad \alpha \approx 0.5 \tag{10.10}$$

Beyond the general likeness, of particular interest is where (10.11) leads us:

$$C_{gd} = C_{gdB} - g_0\tau_{gm} - C_M \tag{10.11}$$

At the origin, τ_{gm} is zero, leaving just

$$C_{gd} = C_{gdB} - C_M$$

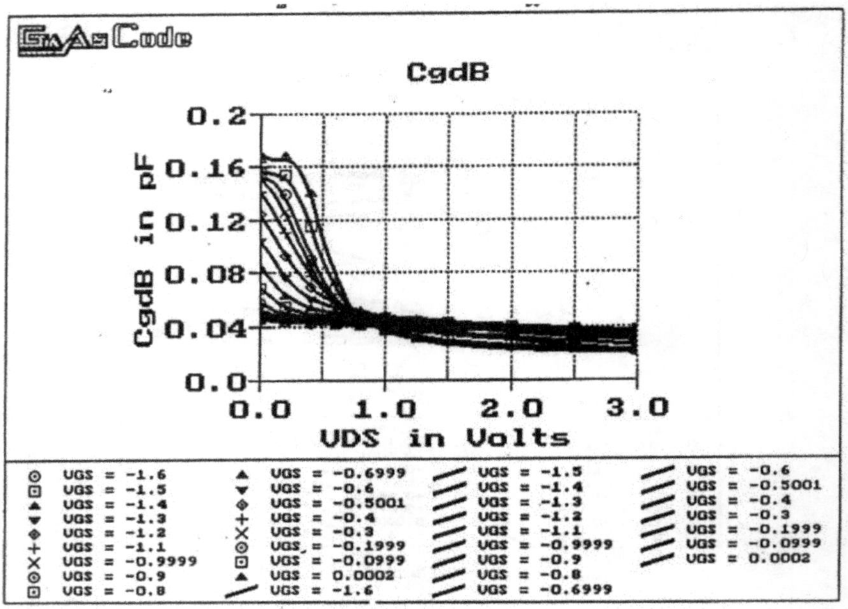

Fit to C_{gdB} as a function of internal V_{DS}

Figure 10.19 C_{gdB} as a function of internal voltage $V_{D'S'}$.

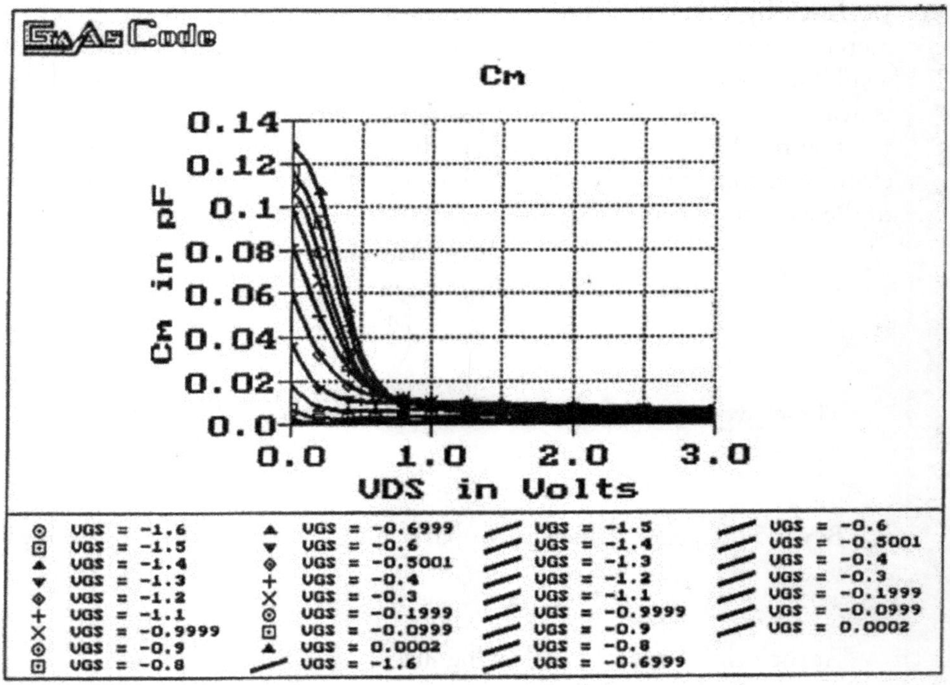

Fit to C_m as a function of internal V_{DS}

Figure 10.20 C_M as a function of internal voltage $V_{D'S'}$.

Enacting this sum eliminates the characteristic upsweep of both capacitances as $V_{D'S'} \rightarrow 0\text{V}$ so the net result, which is the internal element C_{gd} of the standard model, takes on the appearance of Figure 4.10, which is what the macro-cell simulator indicates is the right form for the internal element.

According to (10.12),

$$C_M = \frac{g_0 L_G}{v_{\text{sat}}} \tag{10.12}$$

a further suggestion is that C_M should resemble g_0, which is plotted in Figure 10.21. The general shape is a fair approximation, although the floor values of C_M are too

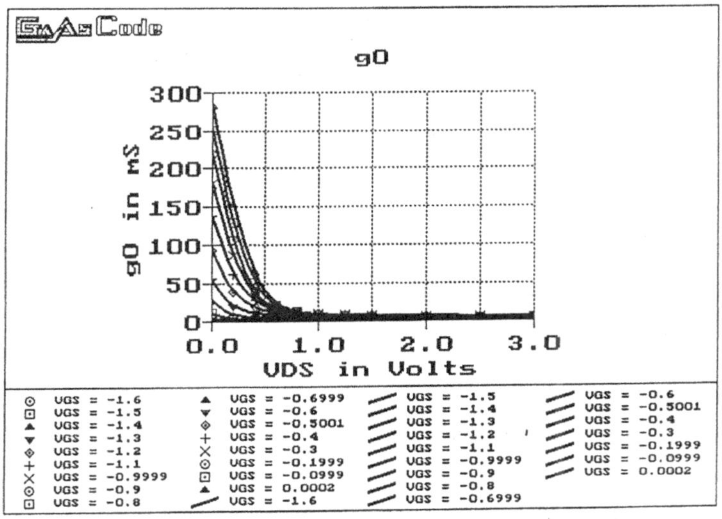

Fit to $g_{0\alpha}$ as a function of internal V_{DS}

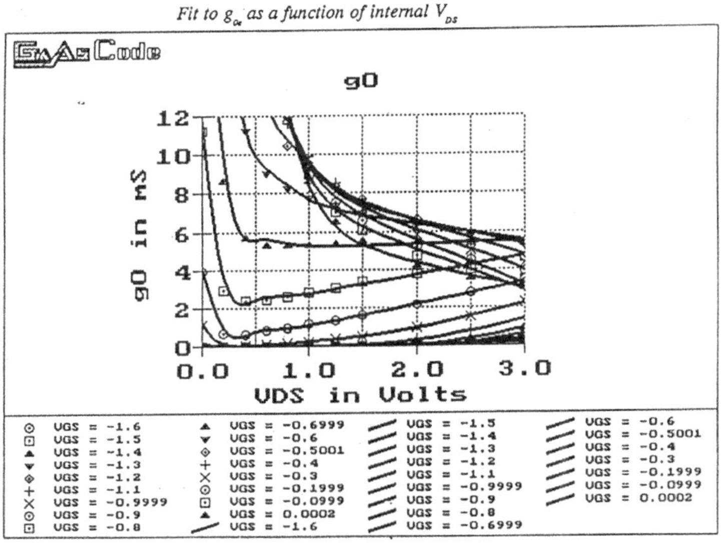

Fit to $g_{0\alpha}$ as a function of internal V_{DS} (detail)

Figure 10.21 g_0 as a function of internal voltage $V_{D'S'}$.

high by a factor of around 2. That said, there may be one or more contributions to the practical value not accounted for by the simple theory.

As fitted, the all-important delay, τ, turns out as in Figure 10.22. Note that in the large-signal nonlinear context, the delay is given the symbol τ, whereas in the small-signal linear case, it is referred to as τ_{gm}. Both symbols refer to the same delay function; it is where the function is applied that determines the symbol. Close inspection of the ordering of τ with respect to V_{GS} (rather, $V_{G'S'}$), and the spacing thereof suggests the device has a nice uniformly doped channel, most likely grown therefore by molecular beam epitaxy (MBE) or metal organic chemical vapor deposition (MOCVD). A lesser possibility is that it is a PHEMT. At first sight, the steeply rising fitted curves emerging from the horizontal axis at approximately $V_{D'S'} \simeq 2.4$V, 1.7V, and 0.7V look erroneous, but they are not. Especially for the first two, there are sparse data points so, if anything, it is the data points that should be questioned.

Figures 10.23 to 10.27 interchange $V_{D'S'}$ and $V_{G'S'}$ as independent variable and parameter with the intention of illustrating not only how well (or badly) the functions of Chapter 9 fit the data but also how well they extrapolate. A sure way of telling how well a function extrapolates is to take a full-range set of data, truncate it, fit the chosen function to the remainder, and compare the extrapolation with the data suppressed. Not enough data were available to apply this moralizing test here.

One comparison stands out: that between Figures 10.24 and 10.27, which the theoretical result of (10.12) connects. By shape or form, the agreement is uncanny.

Bypassing all the steps we have gone through to look back at the very beginning, it will be recalled that the raw data that led to these results were small-signal S-parameters and that the method of extraction ignores nonlinear terms of the type $C\delta V$, as discussed at length in Section 4.2.1. We can use the plots above to see whether the results are indeed consistent with that omission.

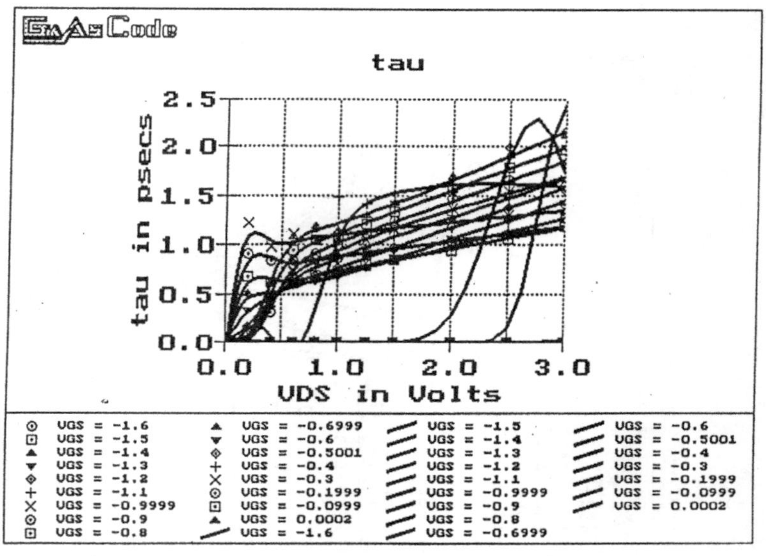

Fit to τ as a function of internal V_{DS}

Figure 10.22 τ as a function of internal voltage $V_{D'S'}$.

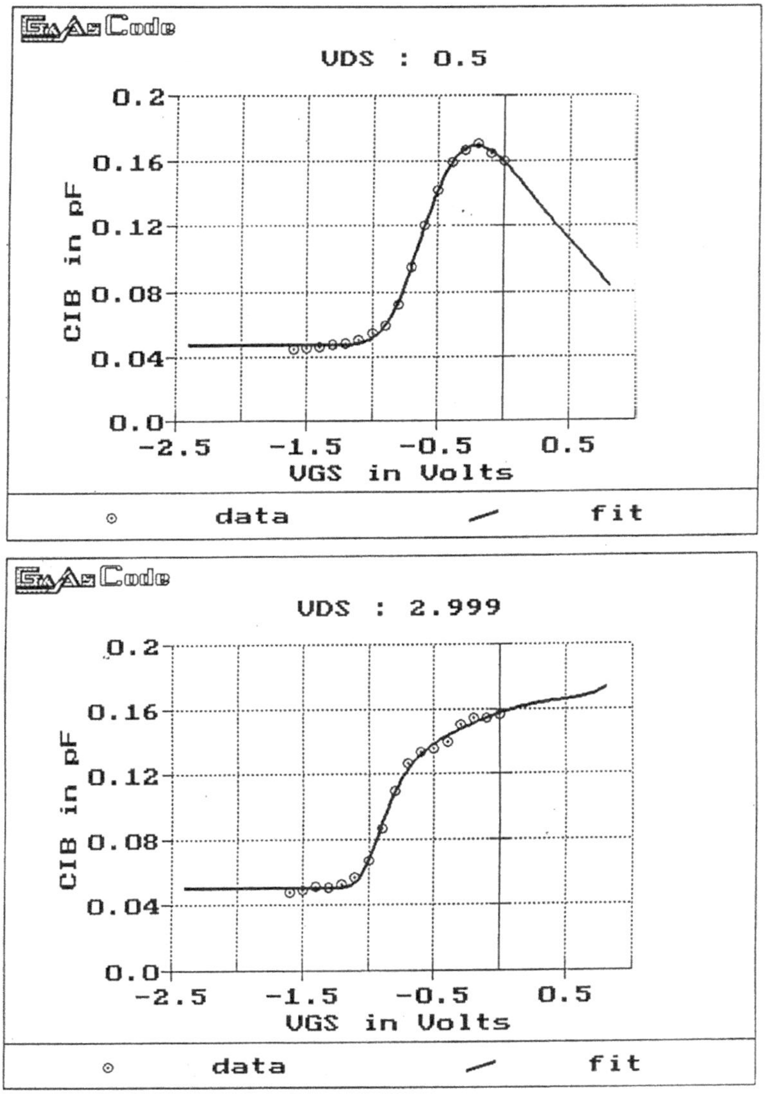

Figure 10.23 C_{IB} as a function of internal voltage $V_{G'S'}$ for internal voltage $V_{D'S'}$ = 0.5V and 3V.

As a case in point, let us examine $C_{IB}(V_{GS},V_{DS})$ in Figure 10.17. Over most of the V_{DS} range, C_{IB} is only a weak function of V_{DS}, so we expect terms $V_{DS}\delta C_{IB}$ to be consistent with the assumption that they would be small.

Now consider $C_{IB}(V_{GS})$ in Figure 10.23, focusing on the steepest section of the curve running through C_{IB} = 0.1 pF. A little simple algebra then shows the ratio of the neglected nonlinear element term to the retained linear element term to be

$$\frac{V\delta C}{C\delta V} \simeq 1.5$$

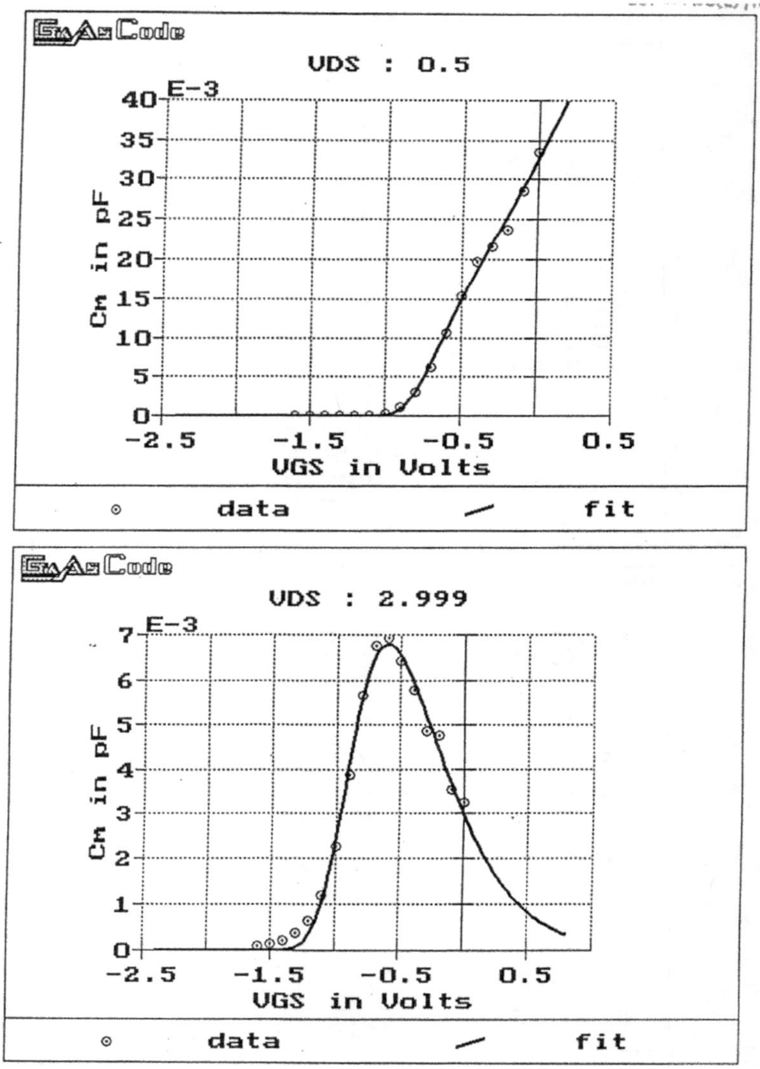

Fits to C_m as functions of internal V_{GS} at V_{DS} = 0.5V & 3V

Figure 10.24 C_M as a function of internal voltage $V_{G'S'}$ for internal voltage $V_{D'S'}$ = 0.5V and 3V.

In other words, the empirical element function $C_{IB}(V_{GS})$ that we have ended up with has non-negligible $V\delta C$-type terms. That does not mean our result is invalid, however, because even if the extraction procedure had taken these terms into account, physics tells us the resultant function necessarily would still possess regions of rapid change. It merely means that we have to view the result with some reservation.

C_{IB} is not alone in such considerations: the same reservation applies (among others) to C_M (Figure 10.24). We could pretty well pick any of the elements and the same would apply.

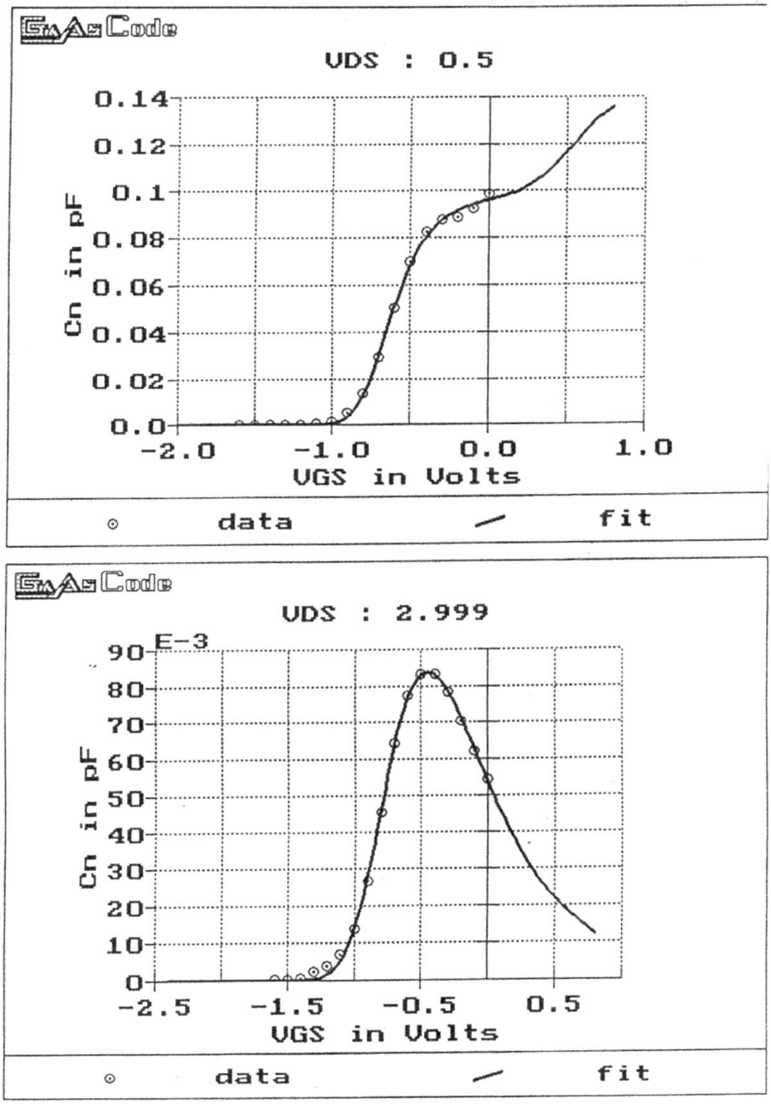

Fits to C_n as functions of internal V_{GS} at V_{DS} = 0.5V & 3V

Figure 10.25 C_N as a function of internal voltage $V_{G'S'}$ for internal voltage $V_{D'S'}$ = 0.5V and 3V.

Yet there is a counterargument to challenge the reservation, which is particularly well illustrated by C_M.

Referring back to (10.12), attention has already been drawn to how well this relationship stands up in Figures 10.24 and 10.27. However, (10.12) emerged from the physics of device operation making no assumption anywhere involving linear approximations. Therefore, it should be valid in the complete large-signal, nonlinear sense. That being so, it must mean the $V\delta C$ criterion is somehow lacking.

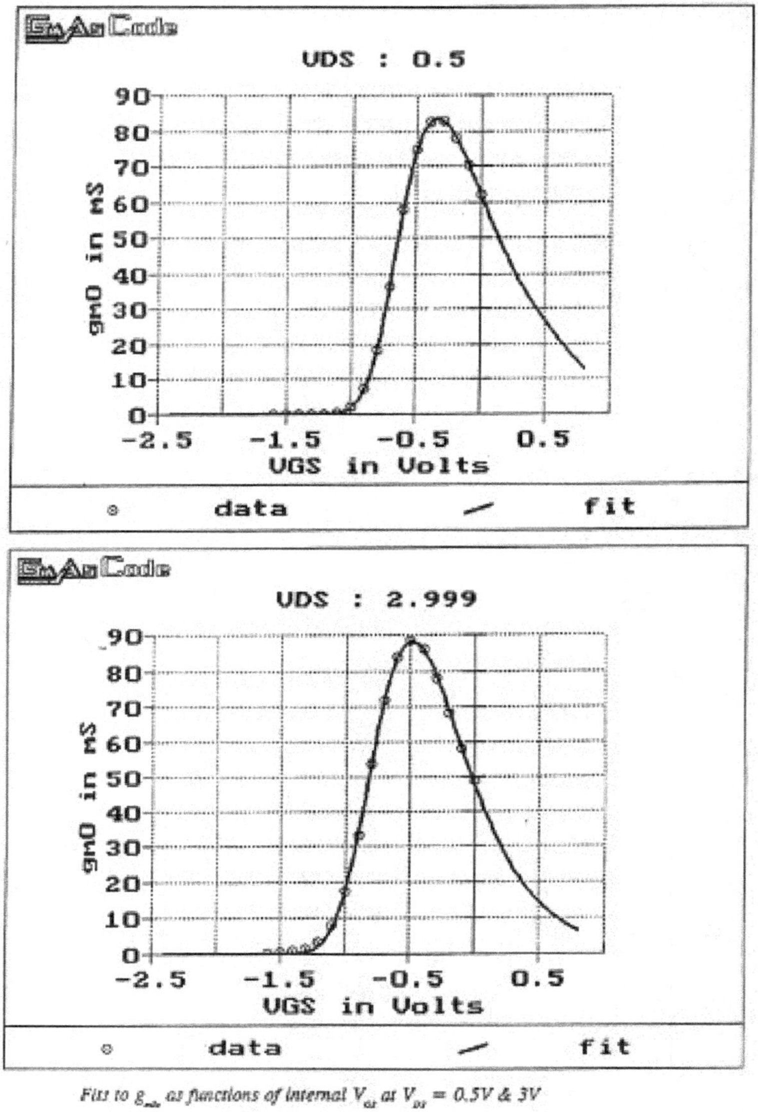

Figure 10.26 g_{m0} as a function of internal voltage $V_{G'S'}$ for internal voltage $V_{D'S'}$ = 0.5V and 3V.

Herein, then, is one more unresolved dichotomy. Which is correct? Should a rigorous extraction procedure account for $V\delta C$-type terms as the ideas of Section 4.2.1 suggest, or does ignoring them lead to element functions that are close enough to the true nonlinear functions anyway, as resorting to the physics of Chapter 15 suggests?

For practical purposes, we have no option but to take the second view. If that is incorrect, it is difficult to fathom from the first principles the consequences for the utility of the model, bearing in mind that the problem affects only parts of the (V_{GS}, V_{DS}) plane. As ever, the answer is to try it.

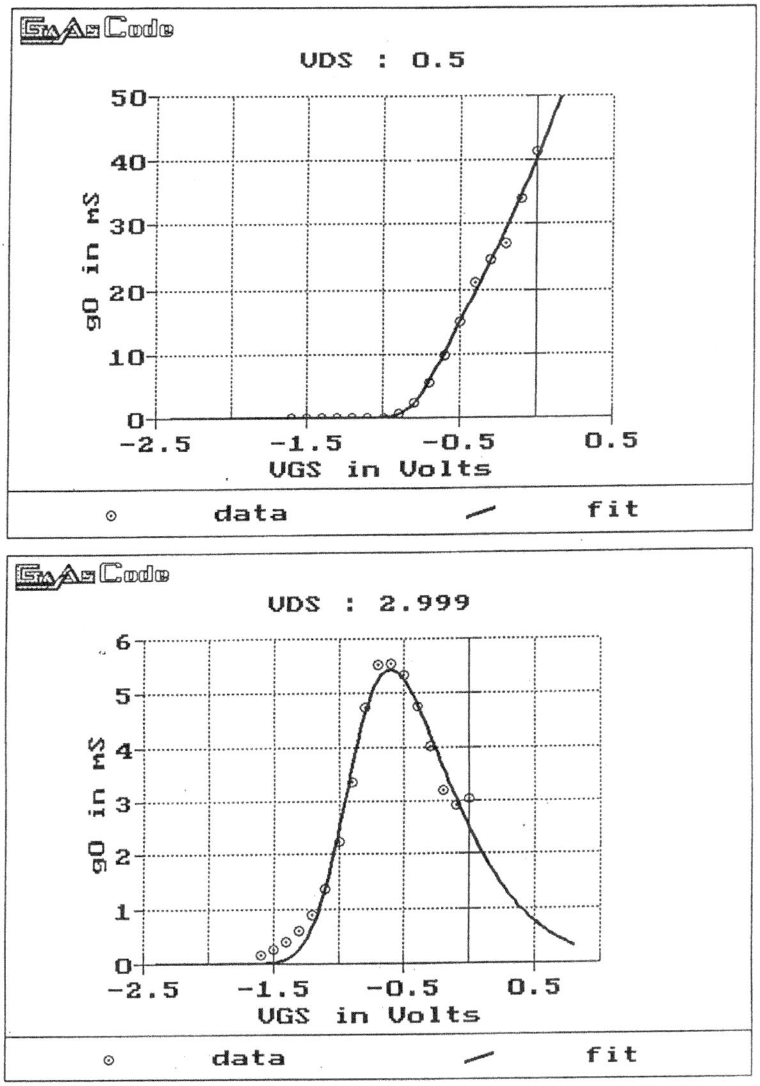

Fits to g_{o} as functions of internal V_{GS} at V_{DS} = 0.5V & 3V

Figure 10.27 g_0 as a function of internal voltage $V_{G'S'}$ for internal voltage $V_{D'S'}$ = 0.5V and 3V.

References

[1] Pedro, J. C., et al., *Nonlinear Circuit Simulation and Modeling*, Cambridge, U.K.: Cambridge University Press, 2018, pp. 282–285.

[2] Tajima, Y., "Progress in RF Device Modeling: from MESFETs to GaN PHEMTs," *Advanced Device Modeling Symposium*, National Nano Device Lab, Taiwan, April 24, 2006.

[3] Pedro, J. C., et al., *Nonlinear Circuit Simulation and Modeling*, Cambridge, U.K.: Cambridge University Press, 2018.

[4] Aaen, P. H., J. A. Plá, and J. Wood, *Modelling and Characterization of RF and Microwave Power FETs*, Cambridge, U.K.: Cambridge University Press, 2007.

[5] Ladbrooke, P. H., *MMIC Design: GaAs FETs and HEMTs*, Norwood, MA: Artech House, 1989.

CHAPTER 11

Practical Results

11.1 Introduction

Before pursuing the ultimate truth of whether it works, it is well to go over some salient facts:

1. A fair test of a model should within reason encompass several different types of circuits, and certainly more than one.
2. In practice, there are many steps involved in determining the electrical properties of a circuit. Similarly, there is a sequence of separate procedures to determine an empirical FEMT model. Inaccuracy can occur in any of these steps.
3. A comparison between theory and practice can therefore be as much a test of the empirical circuit data as of the device model.

The third fact is interesting. We are on familiar ground when there is some disparity, where acceptance or rejection rests on the matter of degree. Sometimes, disagreement is at a level that prompts investigation as to the cause, resulting in improvement in key areas such as process control. However, the real difficulty arises when agreement is near enough exact; it invariably creates suspicion and raises a question: What has gone wrong? It is not an idle concern because agreement to an unfamiliar degree is demonstrated below. Despite being tempered by the fact that the circuit implementation and measurement were done by one organization, with the device model development and construction being done independently by another, a measure of discomfort remains.

Returning to point 1, there are results for a limited mix of circuits, in toto:

- Harmonic generation by a single device in a basic circuit;
- A 500-mW MMIC power amplifier;
- Single FET load-pull amplifiers;
- A frequency doubler implemented as an MMIC.

Even tests limited in scope took many man-months of effort by different organizations to complete. In addition, there is an appraisal, based on simulation alone with no practical data to support it, of the memory or slow-rate effect conundrum. Although it answers no particular practical problem, it generates insight into what is generally involved.

We are left with point 2. Microwave measurements of the kind necessary for circuit and FET characterization require network analysers, power supplies, fixturing, test benches, tuners, calibration pieces and so on together with the knowledge and experience of how to use them. It is a discipline for specialists. De-embedding alone can make or break a project.

In addition to well de-embedded RF data and pulse-measured current-voltage characteristics, constructing a circuit model of a FET or HEMT requires a procedure to fit the elements of the model to the measurements. Further, the resulting element functions (being functions of the device terminal voltages) must, in turn, be put into a form suitable for a circuit simulator, including interpolation and extrapolation.

In between the circuit operatives and the device model constructors, and often independent of both, are foundry models for the passive elements around which a circuit is designed. If everything else is exact and the passive element models are deficient, the whole exercise falls apart at the circuit simulation stage. It all leads straight back to point 3, which can be restated as: Disparity between the measured characteristics of a nonlinear circuit and a simulation thereof is not necessarily due to a defect in the active device model; it could lie in any one of the many intermediate stages.

Excuses made, it is time to look at some results.

11.2 First Test: Power Compression and Harmonic Generation

Figure 11.1 shows the results of the first test of the reformed model. They are for an ion-implanted GaAs FET cell with nominal gate dimensions of 0.5 μm × 300 μm operating at 4 GHz in a basic circuit with mainly resistive loads but consisting also of minor (but unknown) reactance. The model was extracted from multibias S-parameters as described in Chapter 9 and was then coded as a user model for the time-domain simulator outlined in Chapter 12 running on a personal computer.

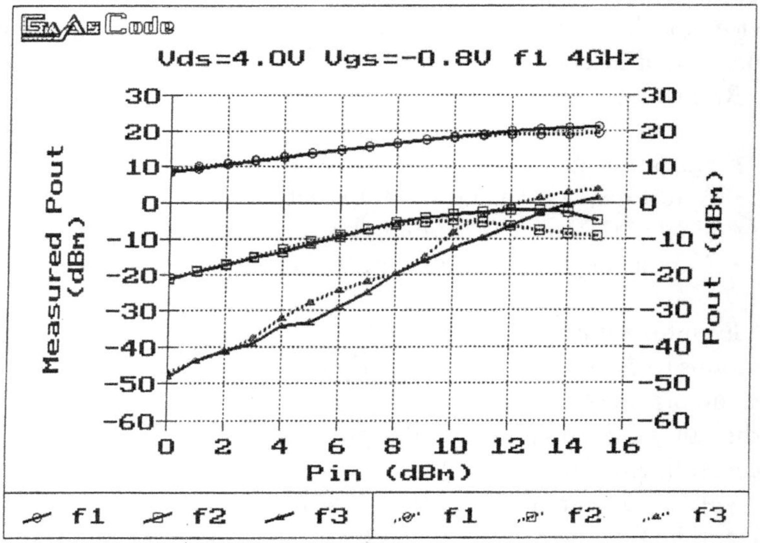

Figure 11.1 Simple first test of the reformed model. The full lines are measurements, and the dotted lines are time-domain simulator results.

Rather than just looking at gain and output power as criteria, the combination of power compression and harmonic generation was adopted as a more exacting first indication of workability.

Magnifying Figure 11.1 shows the agreement at f_1 and f_2 to be excellent up until 8 dBm input power, after which the prediction for the second harmonic starts to fall well below that measured. Satisfyingly, the model, including the RF $I(V)$ reconstruction algorithm, ran faultlessly.

Because this was a first assessment prior to deploying the technique for circuits having a practical application, no experimental effort was put into discovering the cause of the second harmonic disparity. Instead, that was left as a matter of speculation, with one possibility being that little attention was paid to generating an impulse response function for the load.

At 8 dBm input, the device under test just reaches cutoff on the negative v_{GS} swing, but, by then, the waveform should exhibit compression similar to Figure 11.23. The horizontal symmetry about a vertical line is indicative of even harmonic generation, notably second harmonic. To that extent, the result is consistent with what is known about the device, so the f_2 disparity must be due to errors in the RF drain current in the subthreshold region or the load impedance that the current sees (we are here dismissing measurement error as a possibility). The device is of the same type from the same foundry as in Figure 10.7, from which it is evident that the RF $I(V)$ recovery scheme replicates the characteristics at low drain currents rather well. The discrepancy is thus unexplained.

11.3 A 38 GHz Frequency Doubler

Figure 11.2 relates to a frequency-doubler MMIC implemented in a nominal 40 GHz process. The key fact here is that because the frequency of interest (38 GHz) is close

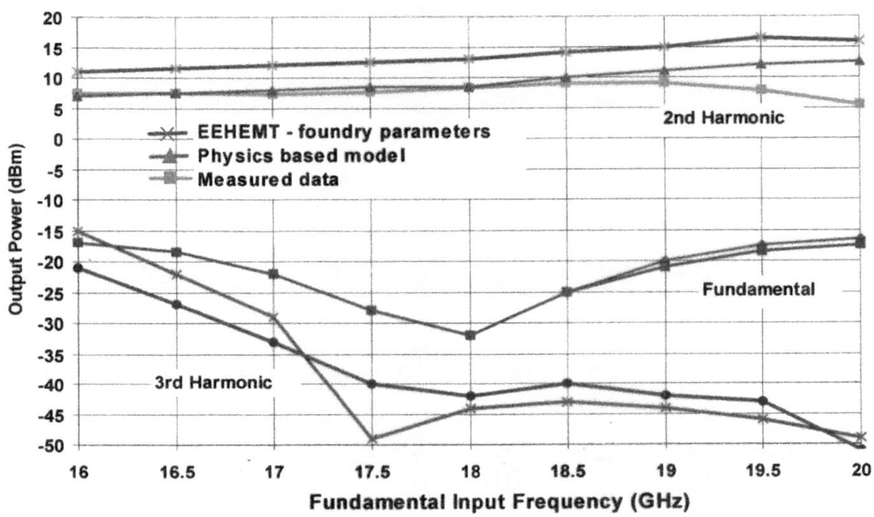

Figure 11.2 Comparison of modelled 38 GHz doubler performance for a –3 dBm input power.

to the nominal cutoff frequency of the devices (40 GHz), it should amount to a stiff test, not just of a model but much else as in Section 11.1.

The circuit was designed using an EEHEMT model by an organization independent from the foundry that manufactured it. All measurements on the finished MMIC were made by the foundry customer. Only when that work was completed were dc and S-parameters data made available for the reformed model of Part II to be constructed and the simulation rerun for comparison.

The reformed model (labeled as "Physics-based model" on the figure) tracks the measured response for the second harmonic nicely up to 37 GHz. At the design frequency of 38 GHz, it is in error by +1.5 dB (a factor of 1.41 in absolute terms), whereas the EEHEMT model is in error by +6 dB (just less than 4 absolute). Significantly, the second harmonic prediction at the upper end of the band is higher than the measured response, the inverse of what happened in Figure 11.1, thin evidence perhaps but evidence nonetheless for there being no systematic error in the reformed device model.

According to the premise that at frequencies low relative to the cutoff frequency it is the conductance properties of the devices that dominate and only at relatively high frequencies do the reactive properties become important, Figure 11.2 is revealing. First, at the fundamental frequency, the EEHEMT and reformed models behave identically (up to 18.5 GHz at least), implying that the conductance characteristics in both are the same. At the second harmonic, however, the +6 dB error of the EEHEMT model can be interpreted as arising from inaccuracy in the way the capacitive elements are handled in that model. In contrast, the fact that the reformed model tracks the second harmonic well suggests the charge storage elements are incorporated correctly.

A performance parameter of interest for this circuit, which a harmonic balance simulator cannot provide, is the start-up time. Accordingly, the reformed device model and the circuit were coded for the time-domain simulator (Chapter 12), which accepts S-parameter blocks for the passive parts of a circuit and translates them into impulse response functions. Figure 11.3 shows the results.

As in the first case, the simulation ran without fault.

11.4 Two-Stage and Three-Stage 500 mW MMIC

Speculation at one time abounded concerning the popularity of simple FEMT models, notably whether they are a prerequisite for circuit simulators (or at least some circuit simulators) to run successfully on circuits containing several devices. Herewith, then, a fair test of the not-so-simple reformed model, resulting in, on the negative side, failure of the simulator to run for the 3-stage circuit, and on the positive side, a complete diagnosis of why, coupled with the reason that the demonstrator circuit did not fully meet specification.

Figure 11.4 shows the layout of a 3-stage amplifier designed and implemented by a consortium as part of the Personal Communicator Network (PCN) project, an initiative of the DTI Link program [1] in the United Kingdom. Equating PCN with handsets for cellular with mobile telephony places the operational frequency in the low gigahertz range.

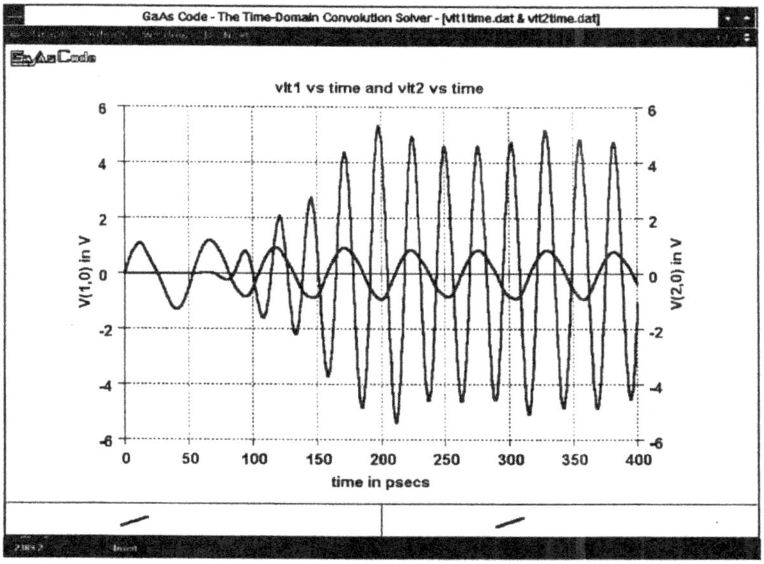

Figure 11.3 Simulation of a frequency doubler showing the input signal and the output signal, including the start-up delay.

The basis for the design was the low-dc power consumption process outlined in Chapter 1, Figures 1.1 and 1.2. A modified Materka-Statz model, as fully detailed in Section 2.10, provided the FET cell parameters needed for the design.

Given the application at a frequency about one-eighth of the nominal 20 GHz for the process and the undemanding specification, the dominant FET property determining overall circuit behavior was expected to be the current-voltage characteristics (for power) and their first derivative (for gain), with the capacitive elements

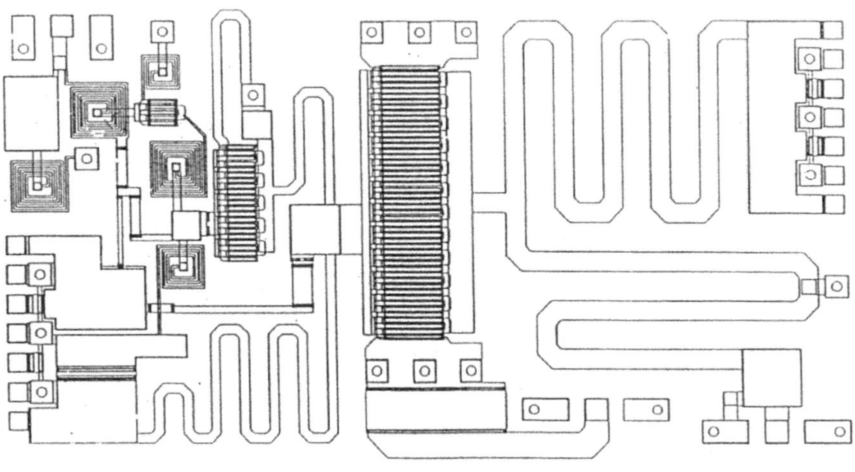

Figure 11.4 A 3-stage 500 mW amplifier.

PCN GaAs PA - 500 mW, 3 stage

Target		Achieved
30	Gain dB	24.5
+27	$CP1_{dB}$ dBm	+26
>40%	Efficiency	40%
	Power In	360mA from 3V

Commentary

Three stage single ended class AB type, with a 48 x 300 μm output device. The PA demands a well controlled GaAs process exhibiting a tightly specified I_{DSS} and G_m.

Figure 11.5 Target specification for the 3-stage circuit compared with the practical figures.

playing only a minor role. As the exercise was exploratory, the specification was limited to power-added efficiency, gain, and power, as in Figure 11.5.

With an uncomplicated specification for the circuit, the modified Materka-Statz model of Section 2.10 was considered workable, particularly as its somewhat rudimentary nature should allow the circuit simulator to run unimpeded, as proved to be the case.

Details expanding upon the practical performance figures in the right column of Figure 11.5 are given in Figure 11.6. Bearing in mind that the low-dc power consumption process was new, there remained much to be learned about which of the many processing steps might need better control to keep the active devices within specification (the passive components were from a long-standing process and so were well characterized). Against such a background, the practical result was adjudged satisfactory as a first effort. Nevertheless, further investigation to trace the origin of the discrepancy was warranted.

Application of the reformed topology model required:

- Fitting the model to the multibias S-parameters of the FET cell as issued by the foundry, also referred to as the characterization data;
- Scaling the model to the gate widths of the devices in the amplifier, being approximately 1.3 mm, 3 mm, and 14.4 mm, respectively.

An early finding was that, partway through the input power sweep the simulator (Libra) failed to converge, thereby raising an unpalatable question: is there something fundamentally difficult, if not intractable, for circuit simulators in other than simple models?

To reduce the problem to more manageable proportions, attention was switched to the 2-stage version of the amplifier (similar to the front end of Figure 11.4 but with FETs of different gate width).

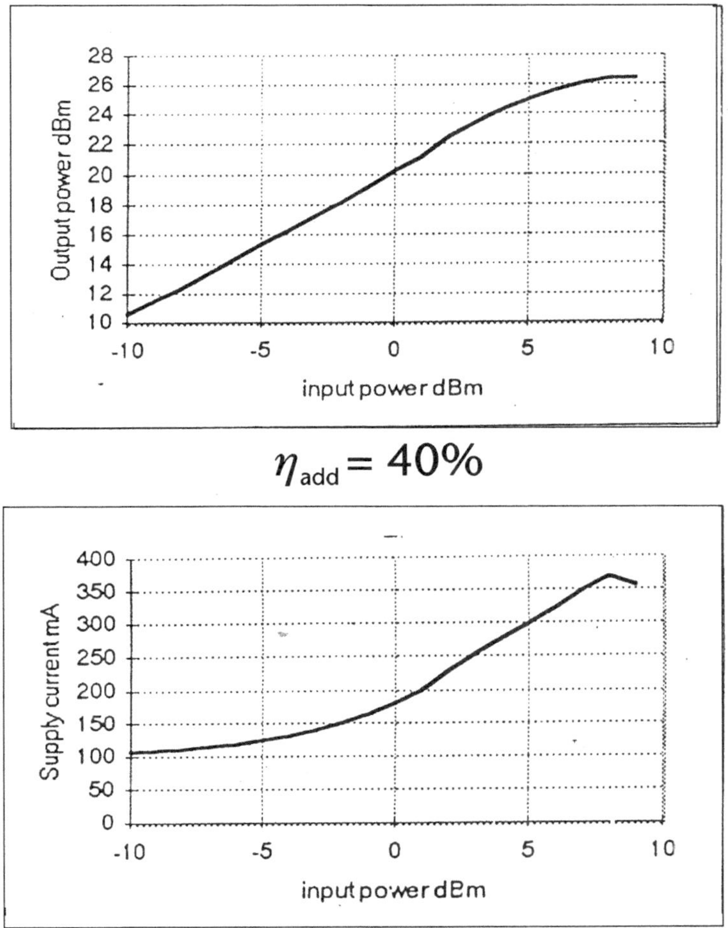

Figure 11.6 Power amplifier gain compression and bias current working from a 3 V supply.

Figure 11.7 shows three curves that were obtained in the following sequence:

1. The lowermost curve, which is the measured response of the finished MMIC;
2. The heavier curve of the upper two, which ends at $P_{IN} = 5.5$ dBm, which is the simulation result with the reformed model fitted to the foundry-issued device data;
3. The finer curve of the upper two, being the first curve shifted upwards to allow for an error in the fabrication process (as detailed below).

Focusing to begin with on the shifted version of the measured curve, when test FETs off the processed wafers were measured they were found to have a lower transconductance than the characterization FETs around which the circuits were designed. That is to say, the practical circuits contained FETs with lower transconductance than planned and, hence, lower gain.

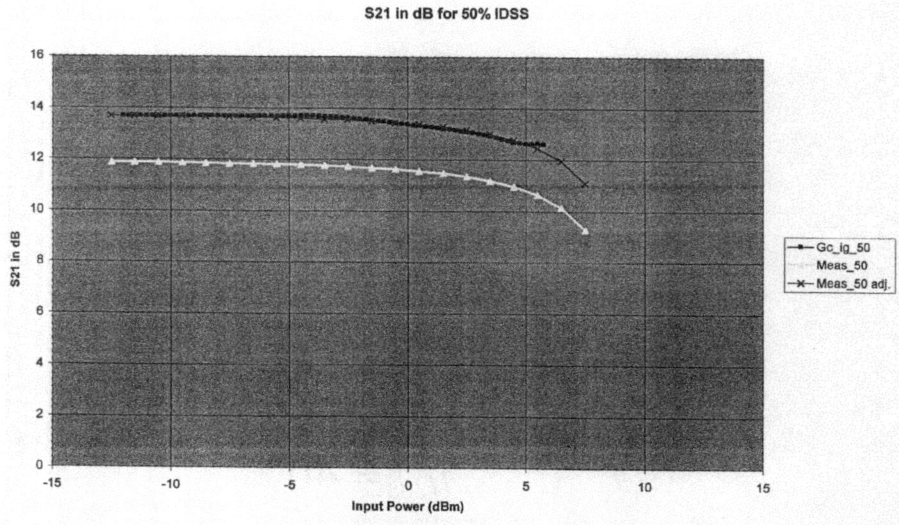

Figure 11.7 Measured gain of the 2-stage power amplifier (lowermost curve); measured gain shifted to allow for higher transconductance of the characterization FETs (feint curve with crosses); prediction using the new topology model fitted to the FET characterization data (heavy curve with dotted points).

Accounting for the transconductance deficit proves straightforward. With an ion-implanted channel, a low-current FET unavoidably has a falling channel doping profile; the device is constructed on the profile tail, confirmed by the transconductance delay function being bunched, as in Figures 13.30 and 13.31. It takes only a slight overetching of the gate recess to result in lower I_{DSS} and lower g_{m0}. With this in mind, the measured response adjusted for the transconductance difference now agrees well with the simulation based on the characterization FET data as issued.

The failure of the simulator to converge above P_{IN} = 5.5 dBm was ultimately traced to a sequence of events as follows:

1. The manifestation of the underlying problem, whatever that happened to be, was that the FETs would suddenly rebias, creating instability.
2. Rebiasing occurred when FET2 saturated (as $v_{DS} \rightarrow$ 1V or so).
3. Suppressing RF-induced dc gate current, and hence its effect on the circuit, allowed the simulator to run to much higher input power levels but altered the shape of $|s_{21}|$.
4. Having determined the cure as in point 3, concern that the voltage dependence of the pure delay element $\tau(v_{GS},v_{DS})$ might be a continuing factor was dispelled (early trials of the model in Libra had proved taxing until expedients to accommodate $\tau(v_{GS},v_{DS})$ were found as in Section 12.2.3).
5. With τ fixed, the simulator would run anyway.

The gate current was found to arise from a marginal failure of the model to conserve gate charge, as discussed in Section 10.5. The current is erroneous and an artefact of capacitive model formulation and a discretized method of solution combined. A full reading of Section 10.5 is necessary to understand the claim of marginal as this was the exceptional case; for all other circuits investigated, the model ran without a hitch.

Suppression of the erroneous dc gate current was accomplished by rearranging the element connections in the gate circuit and introducing an extra node to interpose a highpass filter between the gate resistor and the capacitive elements, the voltage on the node being ignored by the simulator. With this expedient, the simulator would run full-range (Figure 11.8). More details on the arrangement are found in Section 12.2.3.

As so often happens, solving one problem raised another. For input power levels greater than 0 dBm, the freshly simulated behavior stays flat while the practical response falls away. After deliberation along the following lines, no satisfactory resolution was reached.

1. If the behavior predicted with the gate current suppressed is correct, it must have been sheer coincidence that the response with the gate current unsuppressed matched the measurement so well up the point where Libra no longer converged to a solution.
2. Part of the discrepancy could be put down to intercomponent interaction and coupling, which are never fully accounted for by separate component models in isolation.
3. More likely, a legitimate component of the RF-induced gate current, formative to circuit behavior, is generated by parametric pumping of nonlinear capacitance, but is now inadvertently suppressed along with the erroneous component.

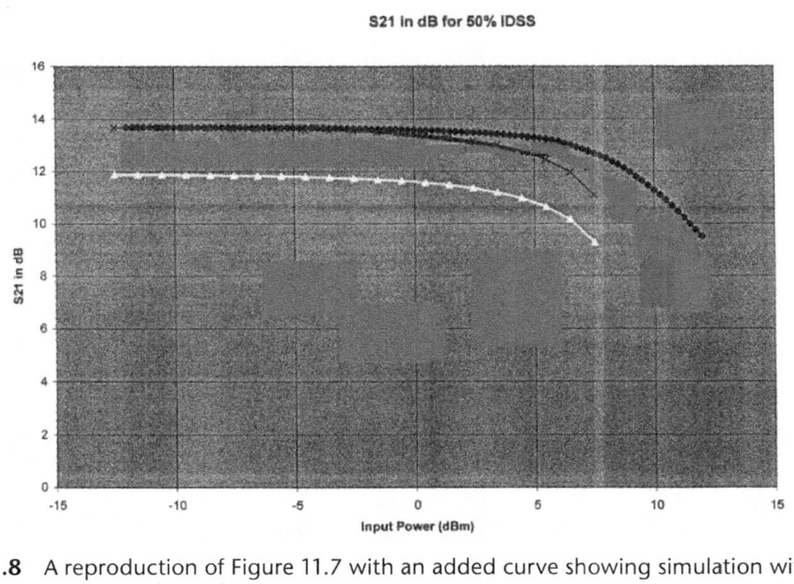

Figure 11.8 A reproduction of Figure 11.7 with an added curve showing simulation with the RF-induced dc gate current filtered out.

Unfortunately, no opportunity arose to go around the loop again and measure test devices off the MMIC wafers, thereby to try afresh to extract model elements guaranteeing gate charge conservation as in Section 10.5. Importantly though, here was evidence to suggest the need for the best possible accuracy and discipline in constructing the new topology model. Put simply, it does not possess the forgiving character of models where the capacitive elements are elementary and simple, like the example in Section 2.10 that was the resource for the amplifier design in the first place. That said, it is not clear whether the reformed model is any more or less forgiving than other models using a capacitance (rather than charge) formulation, the Angelov model for instance.

11.5 Harmonic Load Pull

Arguably, one of the most exacting tests to which a model can be subjected is phase tracking (there is more on the topic at the end of this section). An opportunity to see how the reformed model would fare came at the initiative of, and under the guidance of, Peter Blakey at Motorola SPS, Phoenix, Arizona. Such a trial guaranteed impartiality because the source materials, the circuits themselves, the device data, the load types, admittances, and so on, were all from an organization with which we had no relationship.

The work entailed arm's length simulation of single-device circuits tuned to different general classes of amplifier operation, the results of which were returned to the parent organization and compared with the measured properties of each circuit.

Figures 11.9 to 11.14 are for an average result, the word being subjective.

Figures 11.15 to 11.17 show the corresponding comparisons for the subjective best of the outcomes. The other figures showing the full results for the second load state are Figures 1.10, 1.11, and 1.12.

Of these, the parts of the device characteristics responsible for power compression and gain roll-off are well known, leaving the gate current and phase to be interpreted.

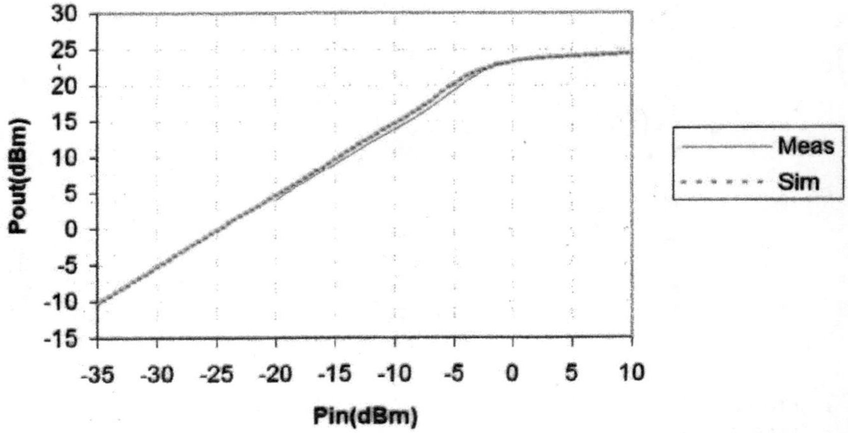

Figure 11.9 Power output and compression for the first load state.

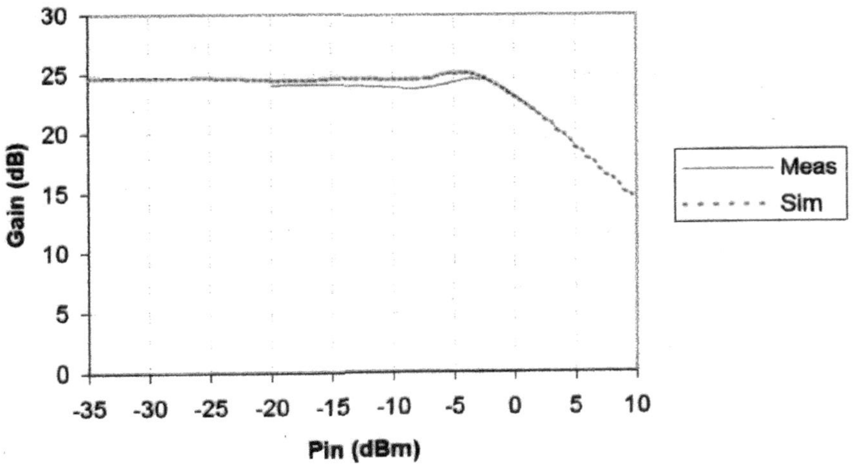

Figure 11.10 Gain for the first load state.

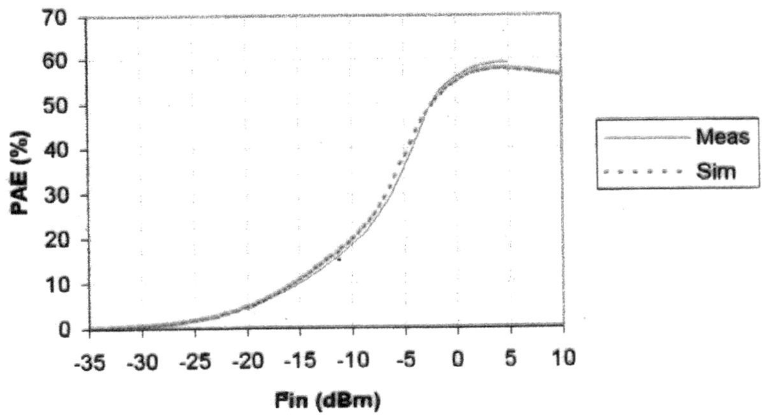

Figure 11.11 Power-added efficiency for the first load state.

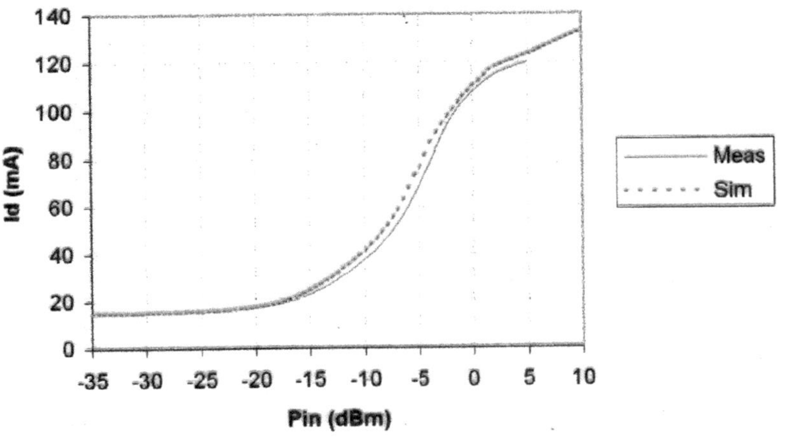

Figure 11.12 Drain supply current for the first load state.

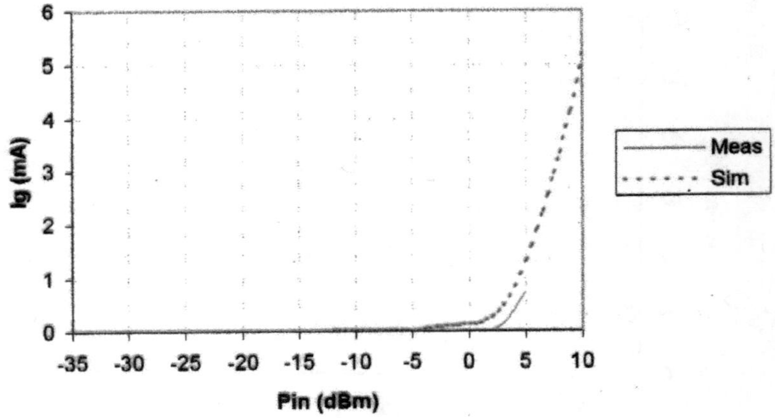

Figure 11.13 Gate supply current for the first load state.

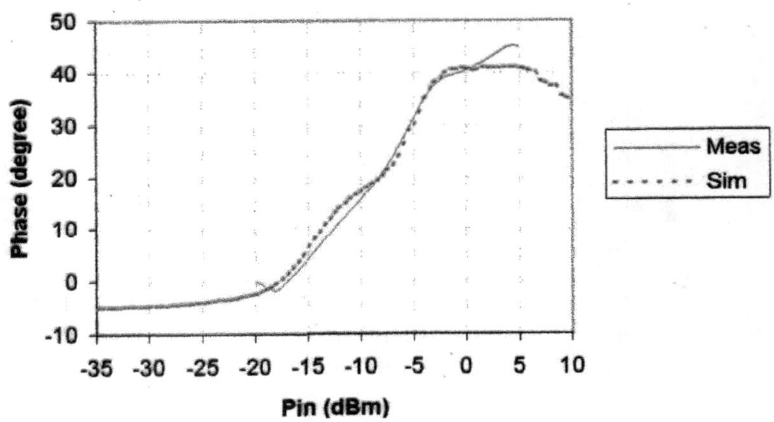

Figure 11.14 Phase shift at the fundamental frequency for the first load state.

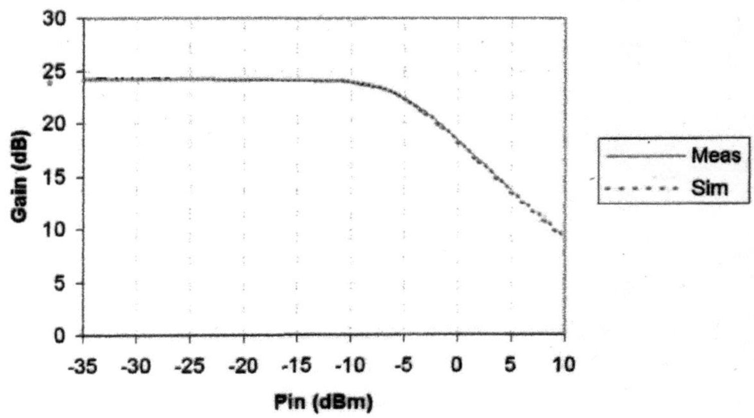

Figure 11.15 Gain for the second load state.

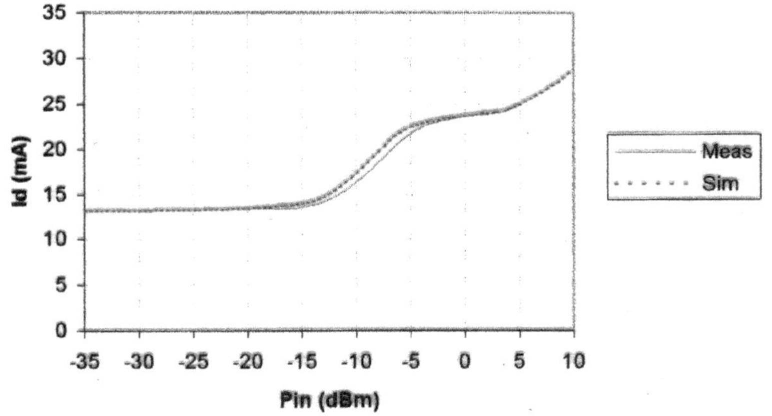

Figure 11.16 Drain supply current for the second load state.

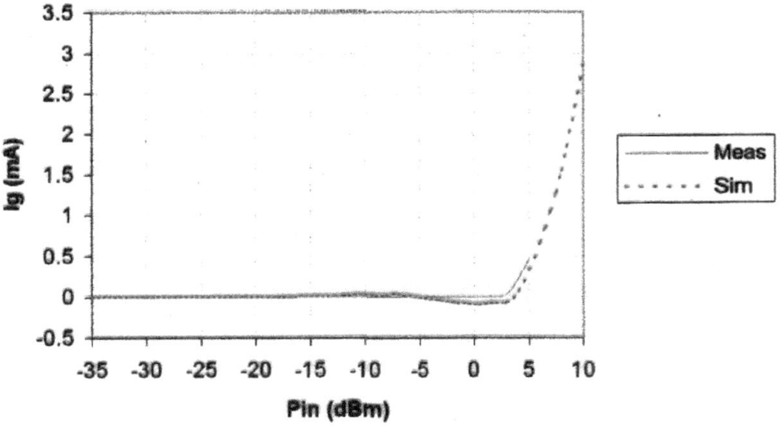

Figure 11.17 Gate supply current for the second load state.

The first showing of the gate current is thought to be determined by parametric pumping of nonlinear capacitance. That being so, C_{IB}, C_N, C_{gdB}, C_M all need to be correct, probably below the knee especially.

Similarly, the first showing in the insertion phase response likely originates from conductance nonlinearity. Waveform distortion affects the phase at the fundamental frequency when the load has a reactive component. At higher input power levels, interaction of the increasingly nonlinear drain current with the complex load impedance leads to continued phase shift by AM-PM conversion.

These broad interpretations do not answer for the role, if any, played by the capacitive elements in the model including charge storage by pure delay and by the delay in the large-signal drain current attributable to τ itself. Without a detailed series of trials where these elements are systematically changed, it is difficult to tell. The degree of agreement evident in the phase comparisons of Figures 11.14 and 11.12 suggests that all aspects of practical device behavior must have been correctly replicated by the reformed model.

Agreement of this order would take the most unlikely of coincidences if there were inaccuracies on both sides. Quite apart from what it may have to say about the ethos of this book, the standard of practical characterization and measurement was what can fairly be described as exemplary.

The relationship between phase tracking and harmonic generation is worthy of comment given the general concern to know whether a model predicts harmonics correctly. At its most basic, the claim herein is that if the phase is predicted correctly, it follows that the harmonics are predicted similarly.

FETs and HEMTs are devices wherein the output current (the drain current) depends not only upon the gate-source voltage but also upon the drain-source voltage. When drain current flows into a complex load Z_L consisting of reactance as well as resistance, the drain-source voltage so developed feeds back and affects the driving force with which we started: the drain current. Hence, $i_D(t)$ is in part driven by $v_{GS}(t)$ and in part by a phase- and amplitude-modified version of itself via Z_L. Remembering that $i_D(t)$ is nonlinearly dependent upon $v_{DS}(t)$, both $i_D(t)$ and $v_{DS}(t)$ will be nonlinear with respect to $v_{GS}(t)$ with different zero crossings, meaning that the fundamental frequency component of both will be shifted in phase with respect to $v_{GS}(t)$ by an amount that is amplitude-dependent. Thus, feedback in part determines the current $i_D(v_{GS},v_{DS})$ via AM-PM conversion involving the load impedance. Waveform distortion, equivalently the fundamental and harmonic components, determines the phase. Therefore, if the phase of the fundamental is predicted correctly, it is implicit that the harmonics are predicted correctly too.

11.6 Memory Effect: Basic Illustration

In Chapter 6, we saw how the instantaneous $I(V)$ characteristics evolve with time to settle finally at the dc characteristics. Measured examples of these transients revealed behavior varying from simple exponential-like changes to complex and nonmonotonic time dependence. As far as it went, it told us nothing about the consequences of such slow-rate effects for a circuit, beyond hinting that waveform distortion, harmonic generation, and phase shift would result.

Via a simple simulation, the purpose of this section is to take a first look at what to expect. There are no practical data with which to compare, partly because there are so many unknowns contributing to the result that a theory versus practice comparison would enlighten us little. The principal unknowns are those complex transients, as will be discussed later. Meanwhile, the problem is to understand the sequence of events and their consequences when a single, simple form of transient is assumed, and that is difficult enough.

So as not to obscure the primary issues with complicating but secondary responses as might arise from a practically meaningful circuit, the simple arrangement of Figure 11.18 was programmed as a user model into the time-domain simulator described in Chapter 12. Only the circuit loads are simple: the reformed device model was constructed for a commercial foundry FET [2], including the RF recovery scheme described in Section 10.2. What is not known, indeed, cannot be known without direct measurement, is the nature of the slow-rate transients, which,

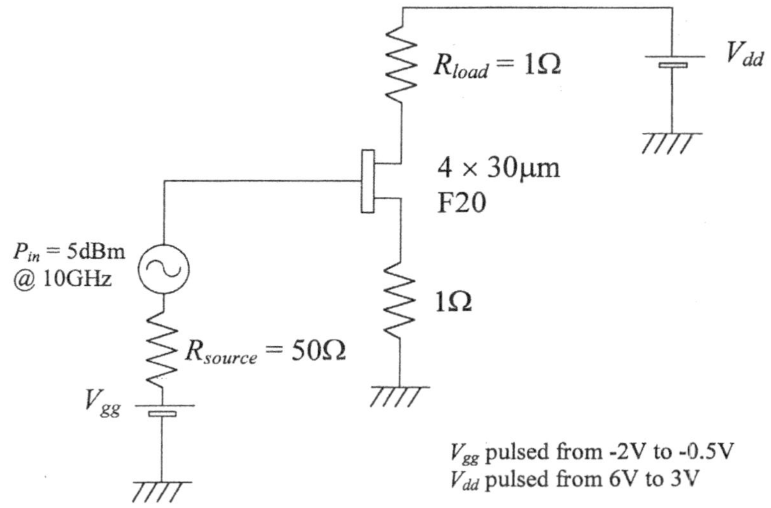

Figure 11.18 Simple circuit for illustrating the consequences of slow-rate memory effects.

point-by-point, connect the instantaneous (or dynamic or RF) $I(V)$ characteristic to the dc set. Therefore, to make progress, assumptions have to be made.

We now assume that the same exponential time dependence governs the way that every point on each dynamic curve evolves. That is to say, the whole set of curves evolves smoothly as time goes on until the final state, the dc state, is reached. Depending upon their location on the $I(V)$ plane, the transition made by some points will be as in Figure 6.6 and others as in Figure 6.7, all as dictated by whether the current needs to fall from a higher value to a lower value or rise from a lower to higher value. Irrespective, the time constant is the same for all, regardless of the rise-or-fall sense of the transition; 100 μs has been taken to be a reasonable value, judging by Figure 6.10.

One step at a time, we now consider the function of the circuit. Both V_{gg} and V_{dd} consist of a dc component and a pulse applied at time $t = 0$, as follows:

$V_{gg} = -2\text{V}$ pulsed to -0.5V;
$V_{dd} = 6\text{V}$ pulsed to 3V.

There is nothing special about these particular values.

As the FET is cut off to begin with, the steady bias at which the device sets up is

$$V_{DS} = 6\text{V},\ V_{GS} = -2\text{V}$$

for which the RF set of characteristics is as shown in Figure 11.19, together with the dc set.

Thus, when the pulses are applied synchronously, the FET's response will be determined by this RF set. From here, we can deduce what is likely to happen as time progresses through the transient.

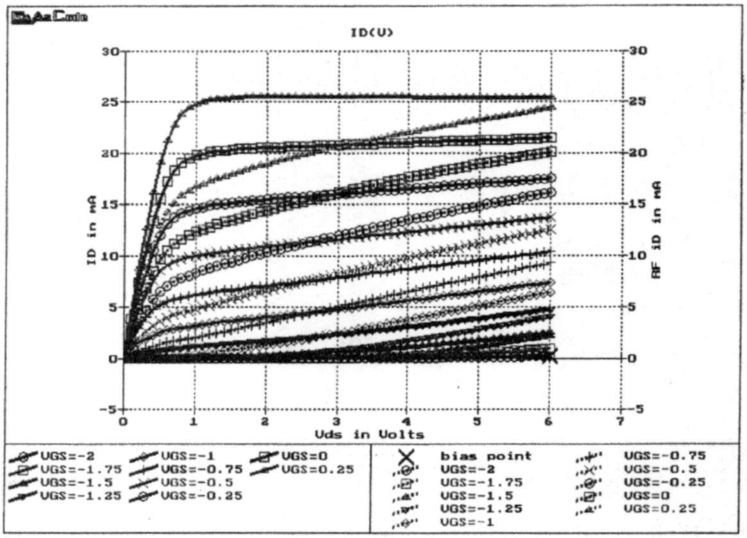

Figure 11.19 RF characteristic for the original bias point (marked X) plotted together with the dc curves.

Ignoring for a moment the presence of RF signals, and ignoring also the small voltage drops across the external drain and source resistors, in response to the pulses the drain current rises immediately to the value on the RF curves at the point $v_{DS} =$ 3V, $v_{GS} = -0.5$V (note the use of symbols appropriate to instantaneous large-signal quantities). Careful inspection of the figure shows this current to be around 8 mA. Eventually, when the transient is over, the drain current will settle at the dc value where $V_{DS} = 3$V, $V_{GS} = -0.5$V, which is approximately 11 mA. In between these starting and end points, the evolution of the drain current consists of a series of points on the exponential transient. In more precise terms, the drain current excursion is formed from a sequence of points on the smoothly evolving sets of current-voltage characteristics, the points being where $v_{DS} = 3$V and $v_{GS} = -0.5$V. Readmitting the RF signal, the drain current can be interpreted as the mean current drawn from the supply V_{dd}, as shown in Figure 11.20.

We now have to consider what happens to the RF voltages and currents, to which end the statement above is key: the determining factor is the time-dependent sets of $I(V)$ characteristics, with the emphasis on sets, plural. As we have previously used the symbol τ_{ss} in the context of slow-rate effects (in Section 10.3), symbolically, we can write for the slowly evolving characteristics

$$i_D = i_D\left(v_{GS}, v_{DS}, t, \tau_{ss}\right)$$

where, in the present example, we have set $\tau_{ss} = 100$ µs.

Without repeating figures endlessly, the best way to envisage what this means is to look again at the upper frames in Figures 10.7 and 10.8. Figure 10.8 is like the dynamic set we started with in the present problem and Figure 10.7 is similar to the set at the end of the transient. Now imagine the one set gradually heading towards the other set through a sequence of intermediate stages where, at each

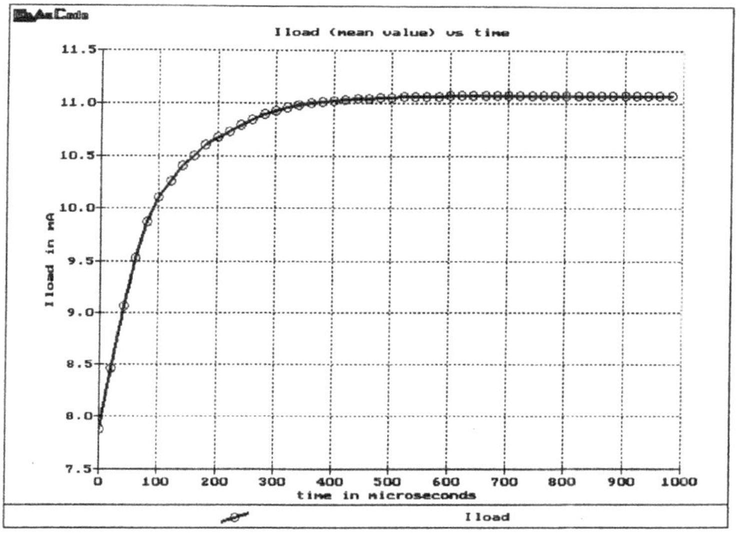

Figure 11.20 Transient in the load current.

stage, there exists a set of $i_D = i_D(v_{GS},v_{DS},t,\tau_{ss})$ characteristics resembling an interpolation between the beginning and end sets, in other words, a whole continuum of evolving sets of curves. At any given time t during the transient, the RF signals are working up and down, or along (depending upon how we care to look at it), one of these intermediate sets of characteristics. The simulator has to know what these evolving characteristics are, so the RF $I(V)$ reconstructor, which is built into the user model, has to be configured to work out a new set of $i_D(v_{GS},v_{DS},t,\tau_{ss})$ at each time step in the transient.

Figure 11.21 shows example waveforms at various times into the transient. They are revealing but not startling; that comes when, based on the insight provided by this example, we try to foresee what is in store for truly practical cases. These particular wave forms exhibit distortion due to overdriving, in itself hardly a surprise, with no discernible phase shift, as there is no reactive load with which the current nonlinearity can interact.

From here, we can begin to see what will happen with a real device. Measurements recorded in Chapter 6 showed the transient is not a simple exponential like Figures 6.6 and 6.7 but generally has a compound form. Also, the nature of the transient varies depending upon position on the current-voltage characteristics, symbolically expressed as

$$i_D = i_D\left(v_{GS},v_{DS},t,\tau_{ss}\left(v_{GS},v_{DS}\right)\right)$$

Instead of evolving smoothly where each set of $I(V)$ curves looks exactly like its predecessors as if they were straightforward interpolations, as time goes on the curves will now develop humps and dips partway along. In turn, the load current will not be smooth or smoothly distorted but will also exhibit irregularities as the transient progresses. No generalities are possible to foresee the consequences for practical circuits.

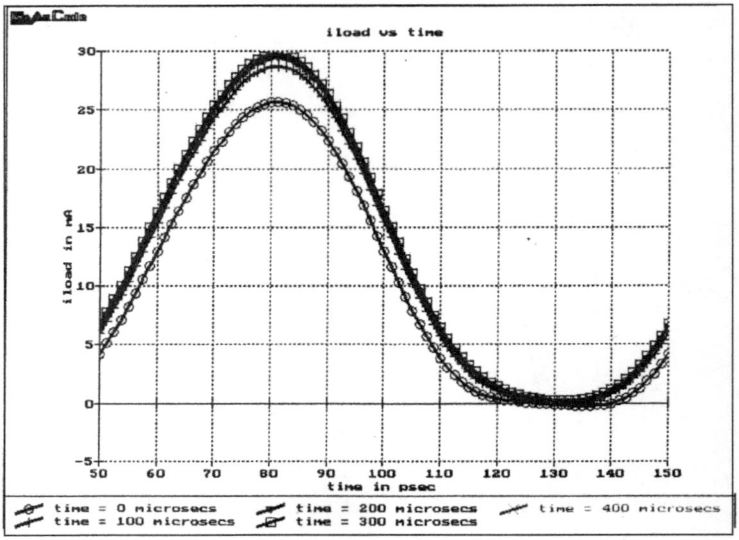

Figure 11.21 Load current at select instants during the transient. Time on the x axis is relative to the start of the sample, while time in the key is relative to the start of the transient.

At this point, it is time to take stock and concede that we now have too big a problem to handle, with implications for a sizeable array of measurements, data storage, incorporation into the model, simulation overhead, and so on. If we must deal with the intricacies of multiple forms of a slow-rate transient, the methods of [3] may be the only way, but at least we first have an understanding of the nature and depth of the problem.

References

[1] DTI LINK Project PC012, "Evaluation of Semiconductor Technologies for the Radio Functions of Personal Communications Network Applications," U.K Department of Trade and Industry, 1992–1995.

[2] Plessey Research (Caswell) Ltd., F20 process.

[3] Pedro, J. C., et al., *Nonlinear Circuit Simulation and Modeling: Fundamentals for Microwave Design*, Cambridge, U.K.: Cambridge University Press, 2018, p. 283ff.

CHAPTER 12

Circuit Simulators

12.1 Introduction

This chapter recounts the experience of getting the reformed model into circuit simulators in order to apply it to real-world circuits, as in Chapter 11. It is therefore a user's view, not that of a simulator specialist.

As the work was done in the late 1990s, some of it may be out of date; simulators, especially harmonic balance simulators, may have progressed by leaps and bounds since then. Nevertheless, some of what was done may be of relevance to latter-day efforts to populate circuit simulators with user models.

One of the chief challenges arose from having to develop approximations for voltage-dependent delay (Section 12.2.2), given that the harmonic balance simulator made no provision. Judging by the results of Chapter 11, these expedients proved to be workable, if not ideal. The key finding is to base the approximation on an expansion of $\tau(v_{GS},v_{DS})$ about its value at the bias point. Occasional convergence failure had to be diagnosed and circumvented and, for some circuits, run time extended to many hours on a mid-range workstation of the day.

A time-domain simulator licensed from a third party proved to be a more receptive platform for the reformulated model. Although limited to circuits having only a few active devices, it ran quickly and faultlessly, thereby providing valuable pointers as to how the principles of the book might best be exploited in the future (Section 12.4).

12.2 Implementation in a Harmonic Balance Simulator

12.2.1 Particularizing the Model

Figure 12.1 shows the model in the form coded for the harmonic balance simulator. Compared with Figure 7.10, there is the addition of conduction currents i_{DGC} and i_{DGD}, represented by diode symbols but actually functions of (v_{GS},v_{DS}), typified by Figure 3.7. The only other change of note is the replacement of nonlinear capacitances by current generator equivalents. Note that, as said in connection with Figure 7.10, C_{ds} needs to be added as a geometric-only capacitance (the space-charge part that appears in the standard model is already accounted for by this model, as in Section 15.4.2).

The voltage driving current through C_{IB} and C_N is taken to be v_{36}, rather than v_{37}. Sections 14.1 and 14.5 deal with the contrasts between the physical device,

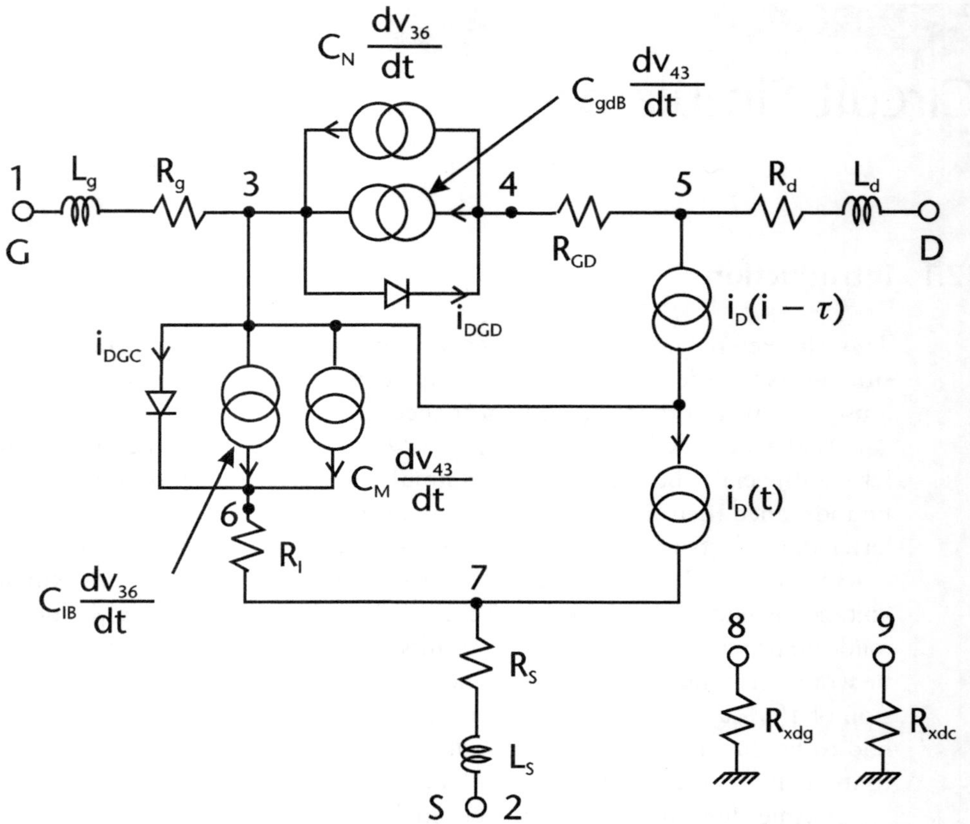

Figure 12.1 The reformed nonlinear model coded as a user model for a harmonic balance simulator.

wherein the capacitive and resistive properties of region I are distributed, and the circuit model, which, of necessity, represents those properties by lumped elements. In the physical device, the voltage driving the distributed capacitance is itself distributed along the channel, whereas in the circuit model the driving voltage has to be at a point. Whichever point is chosen cannot be correct in all circumstances. On balance, we take the more correct of the two possibilities to be v_{36}, which is equivalent to the potential between the gate metal and the junction between region I and region II in Figure 13.22.

Any element with a lowercase subscript is treated as linear (constant); any quantity with one or more uppercase subscripts is taken to be nonlinear. In accordance with the time-honored adage "there is an exception to every good rule," here the exception to constancy is the source resistance. As much as we might like to think it should be constant (or near enough constant), now and again in the odd device it may not be, as, for example, in the PHEMT examined in Section 4.2.2. It is a complication we can do without, so we assume that R_s has a fixed value.

We do not place any nodes between the linear *R* and *L* components. It is more robust to implement a single *LR* series element, rather than *R* and *L* separately, as *L* has infinite admittance at dc.

A model written for the harmonic balance simulator, Libra, must include an expression for the current at each nonlinear node together with an expression for the partial differential of each current with respect to every relevant voltage. The nonlinear nodes in Figure 12.1 are 3, 4, 5, 6, 7, 8, and 9.

The algebra is straightforward but voluminous, so we confine ourselves here to equations for the forming of differentials.

Nodes 8 and 9 are auxiliary nodes whose purpose is to provide a means to calculate the differentials. "Calculation" is the wrong word; a better word is "sense," because the differentials are de facto generated by the simulator as the numerical solution progresses. What we need to do is retrieve their value. Seeing how a subcircuit can be used for the retrieval begins with the basic expression

$$i = C\frac{dv}{dt}$$

Next, we use a resistor to measure i via

$$V = iR$$

Providing we prescribe values for C and R, by measuring V, we find

$$\frac{dv}{dt} = \frac{V}{CR}$$

All that is needed, therefore, are simple subcircuits involving small capacitances and small resistances, the resistances already being shown connected to nodes 8 and 9 in Figure 12.1. Introducing incidental charges q_8 and q_9, we can set up a system of equations to evaluate the differentials of the key voltages v_{36} and v_{43}.

$$q_8 = C_{xdg}\left(v_4 - v_3\right) = C_{xdg}v_{43}$$

$$V_8 = -\frac{dq_8}{dt}R_{xdg} = -C_{xdg}R_{xdg}\frac{dv_{43}}{dt}$$

$$q_9 = C_{xgc}\left(v_3 - v_6\right) = C_{xgc}v_{36}$$

$$V_9 = -\frac{dq_9}{dt}R_{xgc} = -C_{xgc}R_{xgc}\frac{dv_{36}}{dt}$$

$$i_8 = \frac{V_8}{R_{xdg}} = -\frac{dq_8}{dt} \qquad i_9 = \frac{V_9}{R_{xgc}} = -\frac{dq_9}{dt}$$

Thus, for example, wherever the time differential of the voltage v_{36} appears, it can be replaced by

$$\frac{dv_{36}}{dt} = -\frac{V_9}{C_{xgc}R_{xgc}}$$

so the current attributable to C_N is

$$C_N\frac{dv_{36}}{dt} = -C_N\frac{V_9}{C_{xgc}R_{xgc}}$$

and so on right throughout the 50-odd equations that are needed to meet the simulator's requirements. Similarly for dv_{43}/dt.

The only requirement is that the values chosen for C_x and R_x should be compatible with the general magnitudes of the variables with which the simulator is dealing. $C_xR_x \sim 1$ ps is reasonable as an order of magnitude.

In circuit simulation, the model executes in three phases. First, the generators $i_D(t-\tau)$ and $i_D(t)$ are replaced by the dc function $I_D(V_{GS},V_{DS})$ and the simulator run to find the dc solution, the bias condition. Second, the RF $I(V)$ recovery routine built into the model (Section 10.2) constructs the set of dynamic characteristics appropriate to the now-found bias point. Third, simulation is resumed to find the RF solution. It is only in the third phase that the full equivalent circuit model comes into play.

12.2.2 Accommodating τ

Incorporating τ correctly in the user model, or at least as close to correct as possible, is fundamental if the merits of the model are to be fairly tested, then realized. The way that harmonic balance works does not make it easy.

To review briefly why it is important to include the delay properly, it is not just that the drain current is delayed relative to the source current, it is that the tapping branch between these two generators contributes branch charge storage $g_{m0}\tau$ in response to v_{GS} changes, and orthogonal branch charge storage $g_0\tau$ in response v_{DG} to changes as dealt with in Sections 15.2.4 and 15.3.4, respectively. Bearing in mind that the delay is a function of voltage,

$$\tau = \tau\left(v_{GS}, v_{DG}\right)$$

it becomes apparent that these are major contributions to the total nonlinear capacitance.

From a device engineer's perspective, it is not easy to discern why $\tau(v_{GS},v_{DS})$ is not provided for by a harmonic balance simulator. The best guess is that it is the Fourier transform difficulty discussed in Section 12.2.3. Irrespective, it is a reality that has to be worked around. We now outline the various expedients tried.

As a first trial, consider the Taylor's expansion of the current regarded as a function of time:

$$i_D(t-\tau) = i_D(t) - \tau\frac{di_D}{dt} + \frac{\tau^2}{2!}\frac{d^2 i_D}{dt^2} - \cdots \tag{12.1}$$

Very crudely, we can approximate

$$i_D(t-\tau) \simeq i_D(t) - \tau\frac{di_D}{dt} \tag{12.2}$$

Now

$$i_D(t) = i_D\left\{v_{GS}(t), v_{DG}(t)\right\}$$

Differentiating:

$$\frac{di_D}{dt} = \frac{\partial i_D}{\partial v_{GS}}\frac{dv_{GS}}{dt} + \frac{\partial i_D}{\partial v_{DG}}\frac{dv_{DG}}{dt}$$

Substituting in (12.2),

$$i_D(t-\tau) \simeq i_D(t) - \tau\left\{\frac{\partial i_D}{\partial v_{GS}}\frac{dv_{GS}}{dt} + \frac{\partial i_D}{\partial v_{DG}}\frac{dv_{DG}}{dt}\right\} \tag{12.3}$$

Identifying v_{GS} with v_{36} and v_{DG} with v_{43}, we can write, from the equations of Section 12.2.1,

$$\frac{dv_{GS}}{dt} = -\frac{V_9}{\tau_x} \qquad \frac{dv_{DG}}{dt} = -\frac{V_8}{\tau_x}$$

where $\tau_x = C_x R_x$.

Substituting into (12.3):

$$i_D(t-\tau) \simeq i_D(t) + \frac{\tau}{\tau_x}\left\{g_{m0}V_9 + g_0 V_8\right\} \tag{12.4}$$

This equation is not a viable answer as it stands because it still explicitly depends upon some value of τ without having defined what that value actually might be. All that can be said is that, in writing (12.1), τ was treated as independent of time, which is untrue given that $\tau = \tau(v_{GS}, v_{DG})$ and v_{GS}, v_{DG} are both functions of time.

The logical fixed value of τ that would render (12.4) viable is the value at the bias point, which we shall call τ_0. That being so, an improved approximation should follow from Taylor's expansion specifically about τ_0:

$$\begin{aligned} i_D(t-\tau) &= i_D(t-\tau_0-\delta\tau) \\ &= i_D(t-\tau_0) - \delta\tau\frac{di_D(t-\tau_0)}{dt} + \frac{(\delta\tau)^2}{2}\frac{d^2 i_D(t-\tau_0)}{dt^2} - \cdots \end{aligned} \tag{12.5}$$

Retaining the first two terms only:

$$i_D(t-\tau) \simeq i_D(t-\tau_0) - \delta\tau\frac{di_D(t-\tau_0)}{dt} \tag{12.6}$$

where

$$\frac{di_D(t-\tau_0)}{dt} = \left[\frac{\partial i_D}{\partial v_{GS}}\right]\frac{dv_{GS}(t-\tau_0)}{dt} + \left[\frac{\partial i_D}{\partial v_{DG}}\right]\frac{dv_{DG}(t-\tau_0)}{dt} \tag{12.7}$$

and the barred terms signify qualified conditions; thus:

$$\left[\frac{\partial i_D}{\partial v_{GS}}\right]_{v_{GS}(t-\tau_0), v_{DG}(t-\tau_0)}$$

$$\left[\frac{\partial i_D}{\partial v_{DG}}\right]_{v_{GS}(t-\tau_0), v_{DG}(t-\tau_0)}$$

We now need extra subcircuits (like those at bottom right in Figure 12.1) to create nodal voltages proportional to the differentials of the delayed voltages. Separate nodes are required for each of the two first-order derivatives. Similarly, two further nodes are required for each of the higher-order derivatives in (12.5) if we choose to include them.

An alternative requiring half as many new subcircuits is to sense $i_D(t - \tau_0)$ directly, which can be done by inserting a small resistor in series with the upper current generator in Figure 12.1 and labeling the sensing node "node 10." Then, by the same principles as in Section 12.2.1, with extra subcircuit nodes 11, 12, 13, ..., nodal voltages can be generated:

$$V_{10} = \frac{1}{R} i_D\left(t - \tau_0\right)$$

$$V_{11} = \frac{1}{R}\frac{di_D\left(t - \tau_0\right)}{dt}$$

$$V_{12} = \frac{1}{R}\frac{d^2 i_D\left(t - \tau_0\right)}{dt^2} \ldots$$

In operable form, (12.5) then becomes

$$i_D(t - \tau) = \left(\frac{1}{R}\right)\left(V_{10} - \delta\tau V_{11} + \frac{\delta\tau^2}{2} V_{12} - \frac{\delta\tau^3}{3!} V_{13} + \ldots\right) \quad (12.8)$$

$\delta\tau$ is obtained as the difference between $\tau = \tau(v_{GS}, v_{DG})$ at the present operating point and $\tau_0 = \tau(V_{GSB}, V_{DGB})$ at the bias point.

This construct for handling the delay element proved workable in the nonlinear circuits described in Chapter 11 and has the advantage that it could be implemented in other simulators that do not allow delayed voltages.

An alternative to the Taylor expansion is to consider a simple harmonic:

$$i_D(t) = \mathrm{a}\sin(\omega t) + \mathrm{b}\cos(\omega t) \quad (12.9)$$

so

$$\frac{di_D}{dt} = \omega\left(\mathrm{a}\cos\omega t - \mathrm{b}\sin\omega t\right) \quad (12.10)$$

The delayed current is

$$i_D(t - \tau) = \mathrm{a}\sin(\omega t - \omega\tau) + \mathrm{b}\cos(\omega t - \omega\tau)$$

After expanding then collecting terms:

$$i_D(t - \tau) = (\cos\omega\tau) i_D - \left(\frac{\sin\omega\tau}{\omega}\right)\frac{di_D}{dt} \quad (12.11)$$

As $i_D(t)$ is known, a' and b are known so (12.10) can be used with (12.11). Another option is to make the substitution for di_D/dt as in (12.4) to give finally

$$i_D(t - \tau) = (\cos\omega\tau) i_D(t) + \left(\frac{\sin\omega\tau}{\omega\tau_x}\right)\left\{g_{m0}V_9 + g_0 V_8\right\} \quad (12.12)$$

We started with the artifice (12.9), but that is not to say the method is restricted to drain currents conforming to (12.9) (it would defeat the entire ethos of nonlinear analysis anyway). It simply requires ω to be the fundamental frequency of $i_D(t)$, whatever its form.

As a closing comment, the various approaches of this section had little bearing on convergence of the two-stage MMIC of Section 11.4.

12.2.3 Run Time and Convergence

Typical run times varied from many tens of minutes to many hours on a mid-range workstation of the day (late 1990s) depending on the circuit (Sections 11.3 to 11.5). Forbearance proved to be an asset insofar as results were eventually forthcoming.

For some circuits, getting the simulator to run at all was the main problem. Time had to be committed to gaining a basic understanding of how harmonic balance simulators work to find the detailed cause of failure, even though, in general terms, the voltage-dependent delay had long been held up as the prime contender. The following account is an outline, the purpose being to convey an idea of what had to be faced and may still have to be faced with this type of circuit simulator.

In order to balance the signal and harmonics produced by the nonlinear device with those resulting from the constraints imposed by the linear part of the overall circuit, the waveform generated by the time-domain part of the simulator containing the active device has to be transposed to the frequency domain by Fourier transformation. The simulator converts

$$\frac{\partial i(t)}{\partial v(t)} \text{ to } \frac{\partial I(\omega)}{\partial V(\omega)}$$

via Fourier transformation for use in an iterative scheme, often Newton-Raphson, to balance the harmonic components on both sides of the interconnection points between the nonlinear and linear parts of the network: the nonlinear nodes. It may be shown that the kth harmonic component of the drain current depends upon the mth harmonic of the driving (e.g., gate) voltage as

$$\frac{\Delta I_{Dk}}{\Delta V_m} = \frac{1}{T}\int_{-\frac{T}{2}}^{+\frac{T}{2}} \frac{\partial i_D}{\partial v} e^{-j(m-k)\omega t}\, dt$$

Up until this point, there is nothing untoward; the trouble starts when $i_D(t)$ is a delayed version of a current $i(t)$ so the transform becomes:

$$\frac{\Delta I_{Dk}}{\Delta V_m} = \frac{1}{T}\int_{-\frac{T}{2}}^{+\frac{T}{2}} \frac{\partial i_D(t-\tau)}{\partial v(t)} e^{-j(m-k)\omega t}\, dt$$

Now

$$\frac{\partial i(t-\tau)}{\partial v(t)} = \frac{\partial i(t-\tau)}{\partial v(t-\tau)}\frac{\partial v(t-\tau)}{\partial v(t)} \qquad \text{where } \tau = \tau(v)$$

$$= g_{m|\tau \text{ ago}} \frac{\partial v(t-\tau)}{\partial v(t)}$$

Also, it may be shown that

$$\frac{\partial v(t-\tau)}{\partial v(t)} = f(\tau,v)_{\text{now}} \frac{\left(\frac{\partial v}{\partial t}\right)_{\tau \text{ago}}}{\left(\frac{\partial v}{\partial t}\right)_{\text{now}}}$$

If $v(t)$ is periodic, then the term

$$\frac{\left(\frac{\partial v}{\partial t}\right)_{\tau \text{ago}}}{\left(\frac{\partial v}{\partial t}\right)_{\text{now}}}$$

will pass through $\pm\infty$ every cycle, and the simulator will object.

One way to work around this problem is to consider $v(t)$ to be sinusoidal, so

$$\left(\frac{\partial v}{\partial t}\right)_{\tau \text{ago}} = e^{-j\omega\tau}\left(\frac{\partial v}{\partial t}\right)_{\text{now}}$$

Consolidating:

$$\frac{\Delta I_{Dk}}{\Delta V_m} = \frac{e^{-j\omega\tau}}{T}\int_{-\frac{T}{2}}^{+\frac{T}{2}} g_{m|\tau \text{ ago}} f(\tau,v)_{\text{now}}\, e^{-j(m-k)\omega t}\, dt$$

In trying to solve one problem, we have created another. The term $e^{-j\omega t}$ is dependent upon the harmonic but is modulated by τ being a function of v, so we cannot specify this term to the simulator.

Once more, a compromise has to be sought. One option is to substitute the offending term with $\tau(v)$ replaced by τ_0, the value of the delay variable at the bias point; thus:

$$\frac{\Delta I_{Dk}}{\Delta V_m} = \frac{e^{-j\omega\tau_0}}{T}\int_{-\frac{T}{2}}^{+\frac{T}{2}} g_{m|\tau \text{ ago}} f(\tau,v)_{\text{now}}\, e^{-j(m-k)\omega t}\, dt$$

A second option is to expand $e^{-j\omega t}$:

$$e^{-j\omega\tau} \approx (1 - j\omega\tau)$$

We now end up with two parts for the Fourier integral, the second involving $\tau(v)$ as a linear multiplier which is more manageable. Should it prove unworkable, $\tau(v)$ could be replaced by τ_0 as a further level of approximation.

Compromises based roughly on the foregoing produced outcomes as summarized in Table 12.1 for trials arbitrarily designated A, B, and C.

In summary, the impression remains that a harmonic balance simulator is a difficult host for the reformed model. Given the account of this section and the easy assimilation of the model by the time-domain simulator discussed in Section

Table 12.1 Circuit-by-Circuit Harmonic Balance Convergence Trials

Derivative Approximation	*38 GHz Doubler*	*Harmonic Load Pull*	*Two-Stage Power MMIC*
A	Worked	Failed	Failed
B	Not tried	Worked	Failed
C	Worked	Worked	Worked

12.3, it is not the time-domain part of the harmonic balance simulator, which is the source of difficulty but the Fourier transform.

In Section 11.4, we dealt with an entirely different type of mechanism that led to convergence failure for a two-stage MMIC: the buildup of an erroneous dc gate current arising from marginal nonconservation of gate charge. Although this was the only instance, it highlights an imperative: all measurements and fitting with extraction of the model must be done carefully and systematically. It cannot be too highly stressed that second-best risks ending up with capacitive elements that will not satisfy the requirement of gate charge conservation. Section 10.5 has more.

Returning to the two-stage amplifier, a way had to be found to suppress the erroneous dc gate current (Figure 12.2).

In the model as implemented in Libra, the four capacitive elements C_{IB}, C_N, C_{gdB}, C_M are replaced by the four current generators. When these four, together with the charge storage properties of the tapping branch, fail to conserve gate charge an erroneous dc gate current appears at the node marked *. By separating node 3 from node * and introducing a highpass filter between the two, the erroneous component of dc current is prevented from passing out of the gate and into the

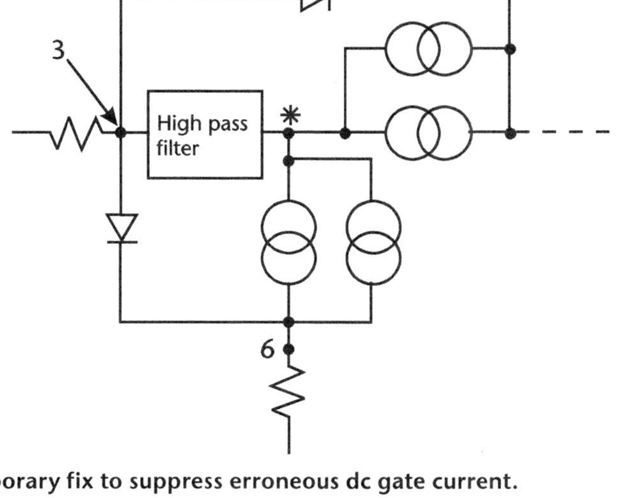

Figure 12.2 Temporary fix to suppress erroneous dc gate current.

circuit at large while still allowing RF components of current to pass unhindered as before. Importantly:

- The signal voltage driving the appropriate capacitance elements is still that between nodes 3 and 6 in Figure 12.2.
- The voltage on node * is ignored in the simulation.
- Any genuine component of dc gate current, as might arise from parametric pumping of nonlinear capacitive elements, is also suppressed, a distinctly undesirable outcome. Consequently, at best, the artifice of Figure 12.2 has merit only when no such legitimate dc gate current exists.

12.3 Experience with a Time-Domain Simulator

As outlined in Chapter 1, collaborative programs gave access to facilities and expertise, which proved invaluable to the present effort, among them an association with University College Dublin through a project called EDGE, the acronym for Enhanced Design of GaAs in Europe. The time-domain simulator, which emerged as an indispensable tool in testing and proving the reformed model, was the product of many years of research and development by the Department of Electronic Engineering run by Professor Tom Brazil at University College.

EDGE was a program of research and development conducted under the auspices of the European Commission in the mid-to-late 1990s [1] with the objective of fostering exchange between European innovators in all aspects of microwave circuit design. Through this program, the time-domain simulator was enhanced to cater for the reformulated FEMT model in full, with no approximations or compromises.

A user interface was grafted onto the simulator engine to bring the package into line with the commercial standards of the time, and a license was acquired for limited commercial exploitation under the name Evolver. Importantly, the simulator was configured to work with, not replace, proven commercial circuit simulators. It catered for circuits containing several devices with the following attributes:

- Two-time-frame solution accounting for dispersion transients affecting RF behavior;
- Convolution-based with passive network characteristics imported as N-port S-parameters with up to 99 ports allowed;
- Excitation by single tone signal including power and frequency sweeps, two CW sources, two-tone power sweep, pulsed input, pulsed bias and signal, and arbitrary waveform input (CDMA and GSM are examples);
- Output: voltage and current waveforms, power-added efficiency, insertion phase, power in mixing products, AM-to-PM conversion, and dc supply current;
- Fast execution, measured in minutes, not hours.

The passive N-port S-parameters had to be generated in Touchstone format by any commercial circuit simulator or electromagnetic simulator. Thereby, complex microwave structures could be accommodated including intercomponent coupling, parasitic interactions between bond wires and packages, and dispersion and frequency-dependent losses in passive elements. A special algorithm assuring physical realizability (or causality) automatically converted the N-port S-parameters to impulse response functions.

Above all, the EDGE project provided a host for the reformed model, as in Figure 12.1 (it should be restated that the diode symbols are not literal, but are symbolic for the more complicated dependence as per Figure 3.7), without approximation for the voltage-dependent delay function $\tau(v_{GS},v_{DS})$. Subsequent application and tests against practice included:

- Exploitation of highly nonlinear HEMT characteristics to generate millimeter-wave harmonics over wide bandwidths;
- AM-to-PM conversion as a function of RF input power and harmonic loading;
- Self-biasing studies of circuits due to RF-induced dc gate current.

Among the many applications foreseen for this kind of model-simulator combination were:

- Generation of behavioral models of devices and subcircuits;
- Prediction of circuit reproducibility and manufacturability by linking the RF and microwave properties of the active devices to technology via methods based on Part III of this book.

12.4 Simulation Prospects

Speculatively (on the grounds that it is untried), a two-step sequence involving time-domain simulation with subsequent conversion to X-parameters [2, 3] appears most promising. The basis is as follows.

In Section 12.3, it was recorded that the reformulated model could be coded in full, without omissions or compromise, for a time-domain simulator that would solve a single-device circuit in typically a few minutes, depending upon the circuit complexity.

By deconstructing a multistage circuit into single stages, each could be simulated in the time domain and the results recast as X-parameters. The key attributes of X-parameters are that they not only retain the nonlinear characteristics of the original circuit, but also individual blocks can be cascaded to predict the dc and RF behavior of the overall circuit.

An extensive literature exists on X-parameters. Among the most instructive in the above sense are two web-based videos by Jack Sifri [4, 5], the second of which deals with arbitrary (i.e., not 50Ω) loads.

It is necessary to qualify the contention. First, not all nonlinear circuits can be broken down into small enough separate stages. That said, experience with the time-domain simulator was restricted to circuits containing a single active device, so the limit on number of devices while retaining workability was not determined.

Concerning X-parameters, there is the question of whether the bounds on the terminating impedances on each stage are tight enough to satisfy the black box memory reservation raised in Section 2.4.

On the positive side, circuit simulation along these lines would allow the key physics-compliant features of the reformed model to be kept, notably:

- A capacitance (not nodal charge) formulation;
- Charge storage by pure delay, which is voltage-dependent;
- Charge flows in the correct branches;
- RF current-voltage characteristics that are bias-point dependent.

Total simulation time promises to be reasonable.

Finally, because the first phase is equivalent circuit-based, the asset of scalability remains.

None of the foregoing answers for what could be achieved through directed simulator development by a major CAD provider were that considered to be a worthy use of resources. The facility to design nonlinear circuits intended to operate near the frequency limits of the active devices is a clear area of interest, as well as high-specification circuits operating at a fraction of f_T, which can prove problematic to design by long-standing methods.

How to extend the capability to slow-rate transient or memory effects remains the outstanding unknown. The complex forms of transient observed in practical devices, with multiple decay rates that vary right across the current-voltage plane, present a formidable problem.

References

[1] EDGE, funded under FP4-ESPRIT 4, March 15, 1996–May 14, 1998, record number 31557, CORDIS_project_21404_en.pdf.

[2] Sifri, J., "X-Parameters Aid MMIC Design," *Microwaves & RF*, July 15, 2010.

[3] Pedro, J. C., et al., *Nonlinear Circuit Simulation and Modeling*, Cambridge, U.K.: Cambridge University Press, 2018.

[4] Sifri, J., "Generating X-Parameter Models from Circuits," Part 1 of 4, https://www.youtube.com/watch?v=JdX_VRC9nL8, 01 Aug. 2014.

[5] Sifri, J., "Generating Load-Dependent X-Parameter Models," Part 2 of 4, https://www.youtube.com/watch?v=qFd6m1JLLZw, 01 Aug. 2014.

Part III

CHAPTER 13

Fundamentals of FET Operation

An epigram from one of the great literates, George Bernard Shaw, reads as: "The reasonable man adapts himself to the world: the unreasonable one persists in trying to adapt the world to himself. Therefore all progress depends on the unreasonable man."

The relevance of this saying to the present endeavor is that, to some, to involve physics in a circuit-oriented discipline risks the tag of being "without reason." Having borne witness to the successes and failures of largely nonphysical approaches of the kind outlined in Chapter 2, however, perhaps there was not much to be lost by being unreasonable and involving physics to see if further progress could be made.

13.1 Introduction

Part III of this book is about device physics, not for its own sake but as a toolkit for Part II. It is not about rigor, but practicability. As will become apparent, device physics by itself in the form of a purely physics-based model has proved unworkable as a practical aid to circuit analysis and design. If it is to have a role, it must perforce be as a feed of ideas and principles for a circuit model.

Rigor is not a prerequisite to serve in this capacity. Indeed, its complexity is a hindrance, but then the problem becomes one of knowing which of the finer principles can be thrown away and which must be kept. Inevitably, the compromises hit upon are not unique, so different workers may favor different approximations, and all may be just as workable. What is described in Part III should therefore be viewed as the efforts of just one group of unreasonable men (the use of gender is justified, as all were male).

Despite the abandonment of rigor, the concepts may retain a certain mystique for the circuit engineer, so to follow them may be far from easy, daunting even. Recognizing that as a possibility is the reason for physics being relegated to third place in the book. Hopefully, the remainder will be intelligible and useful to the circuit fraternity without having to struggle with the intricacies of device physics.

Throughout Part III, as in *MMIC Design: GaAs FETs and HEMTs* [1], all analytical developments and derivations are undertaken assuming that the channel doping N is uniform. For deducing the essential forms and trends needed for reformulating the equivalent circuit model, the assumption works well enough. All the equations are readily extended to the more general case of $N(y)$ but at the risk of losing analytical tractability. Similarly, structural features can be included like recess depth and shape, together with physical features like surface states D_{ss}, both of which in the first instance affect the space-charge layer extension X (and hence much else). To do so, however, is clumsy for analytical use.

The purpose of this chapter then is to examine those aspects of device operation that either are necessary or will help us in constructing a physics-compliant equivalent circuit model of HFETs suitable for nonlinear use. First, we need general rules for laying out the topology of the circuit, that is, deciding the nature of the elements required in each branch of the network. Second, we need specific rules for what these elements should look like, upon which voltages they depend, what their functional forms might be, and whether there are relationships between different elements that, later, can be used as a guide to tell whether the element functions extracted from measurements are in line with the expected form.

As a genus, compound semiconductor FETs can either be a fascinating lifetime's work or intractable objects, depending upon whether your interests are physics or designing and manufacturing circuits using them. In sharp contrast to the early days of semiconductor devices from the first transistor to Gunn and IMPATT diodes and their derivatives, tackling the physics is not something on which many companies are willing to spend money nowadays. For many years, device physics has been the province of academia, with a foreseeable result: while the outcomes are disciplined and correct in an academic sense, they are verging on the obscure to the practicing engineer; they are not practicable.

To come up with a workable physics of FETs requires compromise and, where necessary, a willingness to forsake learned principles. To be able to say something practically useful at all, much of the time the need is to identify the first-order physical mechanisms that really matter then discard the rest, what might be called second-order effects. In the first instance, experiment is the only way to learn which is which, although subsequent confirmation for either retaining or discarding some beloved idea can be had by resorting to a more rigorous theory. An inevitable consequence is that the purist can only wince, so we have to look to the practicing engineer for justification.

So we deal with a workable device physics. It is the foundation for reforming the equivalent circuit model to render it valid for nonlinear design through being compliant with the mechanisms of device operation. Among the effects retained, the purist will find no mention of diffusion or velocity overshoot; practice has it that we can do quite well without them. At the other end of the scale is a 2-D electric field originating in the drain end of the depletion region; without it, there is no hope of a physical understanding, let alone model, of how FETs, HFETs, and HEMTs behave the way they do. In between, there are awkward questions not fully resolved, for instance the role of the dielectric relaxation time.

Dielectric relaxation time theory depends upon the electron mobility, which, in turn, involves the momentum relaxation time, and that leads headlong into a conundrum: the momentum relaxation time is of the same magnitude as the dielectric relaxation time [2]. So, in ascribing charge movement in a low-field region to dielectric relaxation, we are involving a physical process based on a mechanism much like itself. What to do? Discarding labels, we merely assume that charge can rearrange itself on a picosecond timescale, regardless of what we call it. Here is just one example of the hidden truth that, no matter what we do, there is always approximation involved somewhere; it is simply a matter of degree. Perhaps, then, the acceptability of an approximation can be judged by whether it leads to practically useful outcomes.

Stretching the rationale to its limit, sometimes approximations that are inconsistent with one another have individual application. For example:

- A 2-D electric field terminating nonuniformly on the gate leads naturally to the depletion edge *d* being almost flat (Section 13.4).
- While retaining the flat *d* approximation, assuming uniform termination of the 2-D electric field on the gate allows expressions to be deduced for the gate-channel and gate-drain voltages, which turn out to be most valuable (Sections 13.5 and 13.6).

Perforce, all the developments in this chapter apply to the velocity-saturated regime of operation only. Whereas these methods yield thoroughly useful results for the v_{sat} regime, they are inadequate to deal with the region below the knee in V_{DS}. That is the province of the macro-cell solver (Chapter 16). Neither is accurate enough to cope with the nuances of full nonlinear design, but both are indispensable to evolving an empirically based equivalent circuit model that is.

A fair part of this chapter comes from [1] and so leans heavily on that work. Here we give more background to some of the equations, derivations, and results.

13.2 Electron Depletion and Transport

All the FET variants of concern here are metal semiconductor FETs (MESFETs). Structurally, they differ mainly in the composition of the material layers of which they are composed. In the round:

- The term FETs is used generically but originally referred to bulk devices, FETs constructed on a heavily doped N-type layer of compound semiconductor of thickness typically one-fifth of the gate length or less. The N-type layer itself is grown on a semi-insulating buffer layer grown on a semi-insulating substrate.
- HFET refers to a FET fabricated on material layers of different composition forming a heterojunction somewhere. A GaAs FET example is where the semi-insulating buffer layer is replaced by an undoped AlGaAs layer, which has a wider bandgap. A wider bandgap buffer helps to reduce deconfinement of energetic electrons from the channel, thereby improving predictability and reproducibility of the device characteristics. In this book at least, HFET is also used as shorthand for any of the three categories of MESFET and their variants.
- HEMTs involve doping the wider bandgap material rather than the bulk layer, which is now undoped. The doped layer can be either on top of or underneath the undoped layer. Either way, mobile electrons transfer from the highly doped layer to the undoped layer forming a dense but very thin layer of electrons, so thin it has come to be referred to as a 2-D electron gas (2-DEG). In turn, that has given rise to the alternative tag of TEGFET. To add to the richness of the subject, some workers prefer MODFET to reflect the modulation doped nature of the layers. Whatever we choose to call it, the general idea is to create

a layer of mobile electrons isolated from the positive ions that gave rise to them, thereby reducing ionized impurity scattering and leading to high electron mobility. Variations around these basic themes are possible.

Despite the structural differences, when it comes to operating the various subtypes under large-signal conditions, they all behave in much the same way. Taking a microscopic view and looking down from the gate to what is below, it is the net positive space charge that matters more than the detail of its makeup. In the bulk FET, the net positive space charge is that contained within the depletion region. Because the depletion region varies in depth between the source and the drain, so too does the net positive charge. In the HEMT, the block of positive charge is fixed in size and shape, but the mobile charge of electrons occupies a layer where there are no positive ions, so the mobile charge is a mass of negative space charge. Now it is the carrier density that varies with position between source and drain (Figure 8.4 of [1]). Adding the two together yields a net positive charge with a similar spatial dependence to that in the bulk FET.

What overrides the material differences is the sameness of how electrons are transported in high electric fields irrespective of their origin. The essential features of a device consequently are:

1. A layer, immediately underneath the Schottky gate but removed a distance d from it, consisting of mobile carriers (electrons) of sheet density n_s cm^{-2}. For the purposes of analysis, n_s can be regarded as a channel of electrons having volume density n cm^{-3} and thickness $(W - d)$, where d is variable and W is fixed.
2. A layer of positive ionic charge, usually positioned between the gate and the mobile charge but sometimes below it (as in the inverted HEMT). Either way, the positive ionic charge gives rise to an electric field that terminates on the gate.
3. An extension of the layer of positive ionic charge into the gate-drain region where there is no gate metal immediately above. Lacking a close-by metallic surface on which to terminate, the electric field arising from the positive charge in the extended region curls through the device structure eventually to terminate on the gate. Thus, the field is 2-D and is absolutely key: it drives electrons down the channel (thereby giving rise to the drain current), it affects the depth d in 1, it is a major factor contributing to the gate-channel (or gate-source) and gate-drain voltages, and, finally, it has a profound effect on electron transport (in simple terms, the velocity field or $v(E)$ characteristic).

Figure 13.1 shows a cross-section of this minimal structure together with the various voltages that we will need for analyses to follow. For simplicity, the diagram omits other features, such as the gate recess and surface charge, which have a bearing on device behavior.

Several comments in connection with Figure 13.1 are in order. First, the voltages V_{SG} and $V_{S'G}$ differ by the voltage drop across the source-gate region $I_D R_S$. Similarly, V_{DG} and $V_{D'G}$ differ by $I_D R_D$ (these ohmic drops are not shown to avoid cluttering the figure). Second, the arrows indicating a potential difference point in

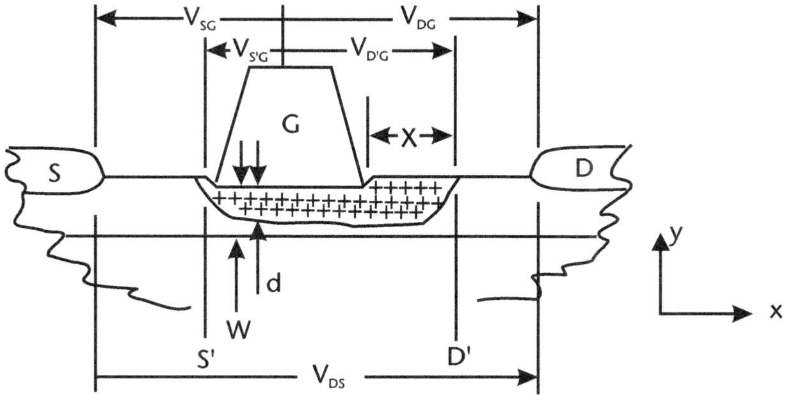

Figure 13.1 Minimal FET structure in aid of analysis.
(Note that $V_{GS'}(= -V_{S'G})$ is used interchangeably with V_{GC}, the gate-channel voltage.)

the direction of the leading subscript. For instance, if we were to refer to V_{GS} instead of V_{SG}, the direction of the arrow at the top left in the diagram would point the other way, toward the gate. Third, reversing the order of the subscripts is equivalent to inverting the polarity of the voltage, so if $V_{GS} = -1\text{V}$, $V_{SG} = +1\text{V}$.

As noted beneath the figure caption, the voltage $V_{GS'}$ is used synonymously with V_{GC}, the gate-channel voltage. There is no special reason. When writing it out in full, the gate-channel voltage is less cumbersome than the gate-virtual source voltage. Of the two, $V_{GS'}$ is the most precise in defining the points between which the potential exists.

In addition to the space charge of positive ions, or depletion for short, the second crucial piece of information we need is how the electrons in 1 move in response to the electric field arising from 2 and 3. Expressing the result in the first-order form of a velocity-field characteristic, all first-hand empirical and theoretical evidence points to the conclusion of Figure 13.2 (Figure 6.A.7 of [1]).

Having given the conclusion, it is appropriate to explain the rationale (the remainder of this section can be skipped those who are satisfied with just the conclusion; a more detailed account may be found in Appendix A to Chapter 6 of [1]).

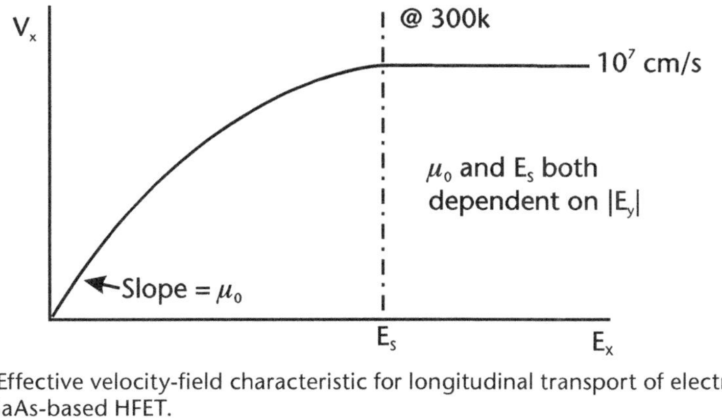

Figure 13.2 Effective velocity-field characteristic for longitudinal transport of electrons in the channel of a GaAs-based HFET.

The rationale begins with the history of how people thought about nonsteady-state transport of electrons injected into a region of high electric field in a semiconductor (not just GaAs). For FETs, the excitement started with an article by J. G. Ruch in 1972 [3]. Using Monte Carlo methods, Ruch analyzed the flight of electrons injected into an N-type semiconductor and found that the velocity initially overshot any and all values in the steady-state $v(E)$ characteristic. If practically proven, the phenomenon promised higher transition frequencies, f_T, for FETs fabricated on selected semiconductor materials, GaAs being one of them. For some, however, the optimism was offset by the suspicion, not all speculative but evidence-based, that y-directed electric field transverse to the direction of motion of electrons may have a quenching effect on electron transport in the longitudinal direction or x-direction [4]. For that to be so, the transverse field had to impart energy to the electrons so they become hotter more quickly than if the field were absent, and hotter electrons attain the energy sooner at which the drift velocity saturates. The idea arose in connection with planar Gunn devices, much like a MESFET but without a gate, constructed on nonuniform N-layers. A static field arises naturally from transverse doping gradients: in the equilibrium state, the negatively charged electrons become displaced from the positive ions (which are the ionized dopant atoms) until a field is created that restrains them everywhere. Translated to the FET, Figure 13.3 is a sketch showing where static electric fields can arise.

Doubt was cast on the transverse field proposition by the observation that any static electric field fundamentally cannot impart energy to free carriers. In the case of Gunn devices, the question thus became whether the field arising from a transverse doping gradient is truly a static field or whether a traveling domain creates nonequilibrium transverse fields (it is the latter). In any event, Brennan and Hess [5] showed that, whereas a static transverse electric field cannot by itself give rise to electron heating, when a longitudinal field is also present, the conclusion is different.

For two reasons, it is beside the point whether the electrostatic fields in Figure 13.3 have any effect on HFETs even when a longitudinal field is present. First, the application of a drain-source voltage inescapably gives rise to y-directed field E_y as in 3 above. Second, the characteristics of the device are dominated by the high-field region at the drain end of the gate, by which point the electrons have gained enough energy to be traveling at their saturated drift velocity. It follows that, even if velocity overshoot were to occur immediately upon injection of electrons from the source into the channel layer, it would have no effect at the terminals.

Domination of the microwave characteristics by the high-field region at the drain end of the gate accounts for why all first-hand experimental evidence tells us that the velocity that matters is $v_{sat} = 10^7$ cms^{-1}, at least for GaAs and its tertiary derivative materials. Within the bounds of estimative error, reverse modeling[1] [6] of numerous FETs, HFETs, HEMTs, and PHEMTs from different fabrication houses has always led to the same number. Figure 1.7 is a typical result.

The term "first-hand" refers to the direct experience of the group behind this book. Second-hand reports, of which there are few, claim velocities around 2×10^7 cms^{-1} [7, 8]. Taken at face value, these accounts create uncertainty as to what the true value, the value that matters from a practical point of view, might be. Detailed

[1] Data generated by the FETLink fitting and reverse modeling tool.

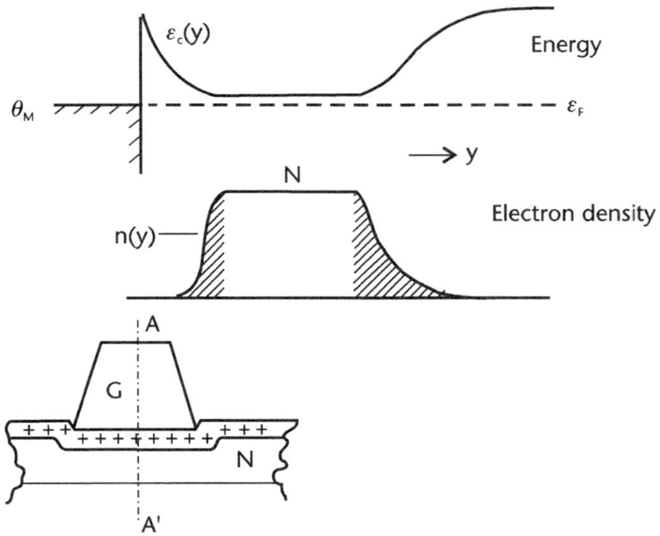

Figure 13.3 Showing (shaded) the channel electrons which experience y-directed fields in equilibrium.

accounts of the experiments that led to these conclusions, supported by at least the semi-quantitative transport theory, would help to justify the claim. Meanwhile, for Si, GaAs, and its derivatives, we stick with Figure 13.2.

For the wider bandgap materials GaN and SiC, a practicable $v(E)$ characteristic as might apply to the electron transport in HFET channels in the presence of transverse fields can be established similarly, starting with the 1-D velocity-field curve. For SiC, [9] contained plots of $v(E)$ from studies covering different crystal polymorphs, temperatures, and methods of acquisition (experimental versus theoretical). In light of such a variation, generalities are not possible and each must be dealt with on its own merits.

Turning to GaN, an especially informative article is due to Shul and Barker [10], who studied the high-field transport properties of bulk GaN and AlGaN-GaN heterostructures. Their article is of interest for its practicality: it does not abandon consideration of factors affecting practical HFETs in favor of measuring the $v(E)$ characteristic under sanitized conditions. HFETs are messy structures compared with idealized 1-D samples so, as we have endeavored to show for GaAs-based materials, what happens in real devices can be quite different from transport in 1-D. They were refreshingly forthright about surface breakdown, the need for passivation, and unwanted carrier injection at the contacts, all problems that had to be solved for the ideal-case measurements.

For bulk 1-D samples, a peak electron velocity of 2.5×10^7 cms^{-1} was obtained. For AlGaN-GaN heterostructures, the first set of results gave 1×10^7 cms^{-1}, a familiar number. The peak velocity found for a second sample set with different ohmic contacts was just over 3×10^7 cms^{-1}. Shul and Barker made the point that 1×10^7 cms^{-1} coincided with the value deduced from practical HEMTs.

The foregoing GaN results have an air of credibility about them for the variations and disparities to which they own up. Early semiconductor device engineering is seldom definitive. For the present, as we are above all interested in electron motion in practical HFETs, perhaps the best way of resolving the matter is reverse modeling. For the answers to be dependable, however, the procedure must be carefully done at all stages from measuring S-parameters through fitting the standard model and allowing for geometric capacitances. The gate length and width must also be known with a high degree of certainty.

To conclude, the velocity-field characteristic for GaN can be expected to look like Figure 13.2 with ranges as follows:

$$1 \times 10^7 \text{ cms}^{-1} < v_{sat} < 3 \times 10^7 \text{ cms}^{-1}$$

$$150 \text{ kVcm}^{-1} < E_S < 180 \text{ kVcm}^{-1}$$

13.3 The Space-Charge Layer Extension *X*

Decades ago, it was clear from computer simulations that the depletion layer extends into the gate-drain space. Less easy to tell is how much importance was accorded to it as a determinant of the FET's electrical properties, both dc and RF.

Our particular understanding of how FETs and HEMTs operate grew out of experiments to unravel the mechanisms limiting RF output power. It revealed the overriding importance of the space-charge-layer extension between the gate and the drain, a dimension that we call X (Figure 13.1). Not only does X play a key role in determining self-regulating impact ionization in the gate-drain region, it gives rise to a 2-D electric field that affects the whole device, coupled with a distinctive time τ to charge and discharge it.

In contrast to other compact physics approaches [11] where the dominant mechanism is assumed to be electron transport in the channel governed by the x-directed field, the key factor is identified as a 2-D electric field which emanates from the space-charge-layer extension X (not from a departure of space-charge neutrality in the conducting channel).

Before delving into a step-by-step account of how the field emanating from X affects device behavior, it is well to consider what it affects and how it arises.

Something has to absorb the potential difference that exists under operating conditions between the gate and the drain. A simplified view gives the general idea. If we imagine looking down in the y direction into the FET structure from anywhere along the gate, we would see the channel layer depleted to a depth d, as in Figure 13.1. Between the gate and the far edge of d are positive ions that give rise to a triangular-shaped field, the peak of which is at our point of observation: the gate. Next, imagine rotating a vertical slice of this block of charge at the drain end of the gate so that it extends along x towards the drain. We now have a means of supporting the gate-drain voltage, just as the charge with which we started underneath the gate supports the gate-channel (or gate-source) voltage. The larger the gate-drain voltage, the further the block of charge will extend towards the drain. Just as with the y-directed electric field under the gate, the x-directed field in the gate-drain region is triangular (very roughly) with the peak at the gate metal, the corner. (As an aside, one manifestation of the intense electric field at the gate corner is light emission accompanying soft breakdown [12].)

Progressing from this rather rudimentary view of the depletion extension to what it affects or brings about, the list is:

- The gate-drain capacitance of the standard model;
- The gate-drain branch capacitance in the revised model;
- The gate-source branch transcapacitance in the revised model;
- The output conductance;
- Electron transport in the channel;
- The subthreshold region;
- The transconductance delay τ_{gm}.

Apart from the transconductance delay, all the foregoing derive from the 2-D electric field to which the space-charge layer extension gives rise. Each is considered at greater length in the rest of the book. For now, however, we attempt the step-by-step account of how the field emanating from X affects device behavior.

Being a vector, the field is written as $\mathbf{E}_X(x,y)$, where the subscript X serves to remind that this field emanates from X. Underneath the gate, $\mathbf{E}_X(x,y)$ determines the channel opening through which electrons are propelled in the x direction at a velocity governed by the velocity-field characteristic of Figure 13.2. The x-directed component of electron velocity, $v_x(E_x,E_y)$, depends not just upon the x component of $\mathbf{E}_X(x,y)$ but on the y component too (as in Section 13.2).

$\mathbf{E}_X(x,y)$ affects the depth of depletion d at all points of x, the position along the gate that we shall refer to as x_2, thereby altering the contour of $d(x_2)$. Section 13.7 is devoted to what comes next. Crucially, at all points of x_2, $\mathbf{E}_X(x,y)$ leads to a decrease in $d(x_2)$ to one degree or another depending upon X, which in turn depends upon $V_{D'G}$ or, in terminal voltage terms, V_{DG} or V_{DS}. Given that the channel current, which is eventually collected by the drain, is governed by the channel opening $(W - d)$, it follows that i_D and $I_D$are dependent not only upon V_{GS} but also upon V_{DS}. That is, the shape or form of $i_D(v_{DS}),I_D(V_{DS})$ is determined by $\mathbf{E}_X(x,y)$. Fast HFET simulators built around this dominant mechanism produce results that match the dc and RF behavior of practical FETs and HEMTs without departure of space-charge neutrality in the conducting channel (although the meaning of that statement for HEMTs is less clear than it is for FETs).

In the conducting channel there is an x-directed component of electric field as is required to propel electrons from the source to the drain. Although in our way of looking at the world this field is a component of $\mathbf{E}_X(x,y)$, in other models it has to exist without there being any X, as for instance in the gradual-channel approximation.

Towards the drain end of the gate the channel field rises rapidly. The electron mobility decreases, also rapidly, as the saturation electron velocity is approached. Corresponding to the rising field, the potential $V_{CH}(x)$ in the channel rises rapidly, which in turn increases the depth of depletion d. In the self-same region, however, the y-directed component of $\mathbf{E}_X(x,y)$, which we shall call $E_{yX}(x_2)$, is largest, and that causes d to decrease or lift. Thus, by chance or design, we have two mechanisms that tend to compensate one another, thereby flattening d. We deal with the phenomenon of flat in Section 13.4.

The 2-D field emanating from X terminates nonuniformly on the gate. If we were to pursue a rigid theoretical treatment, a field singularity at the gate corner would result, which is unhelpful and cannot occur in practice anyway (voltage applied to the drain would incur an avalanche straightaway). To stand a chance of practical relevance we need an analysis, necessarily approximate to be tractable, that implicitly suppresses the singularity.

Figure 13.4 is a solution for the gate charge distribution $Q_G(x_2)$ induced by $E_X(x,y)$as calculated from an approximate Green's function method (Figure 6.22 in [1]). Unsurprisingly, the induced electron distribution is nonuniform, reminiscent of an exponential, rising towards the drain. (Note that the reversed horizontal axis with values increasing from right to left is a consequence of the method of calculation.)

On the vertical axis, $Q_G(x_2)$ is normalized with respect to qNd, where qNd is the sheet charge density in the depletion region immediately under the gate. Thus the minimum value of 1 on the vertical axis corresponds to the negative charge induced on the gate by the positive ionic sheet charge qNd below. On top of that, the curves represent an additional electron charge induced on the gate by the 2-D field emanating from the depletion extension X. For example, 2 on the vertical axis means

$$Q_G(x_2) = qNd$$

that is, the incoming field terminating on the gate metal is as large as the field coming from the depletion charge immediately below. For large-signal operation, the full range of (X/L_G) values shown on the figure typically can be traversed.

Lastly, to make sure that we understand all that Figure 13.4 is telling us, $X/L_G = 1$, for example, means that the total charge induced on the gate by the depletion extension equals that induced by the block of charge in the depletion underneath the gate of length L_G (for $X/L_G = 2$, it is twice as much, and so on) (Figure 13.5).

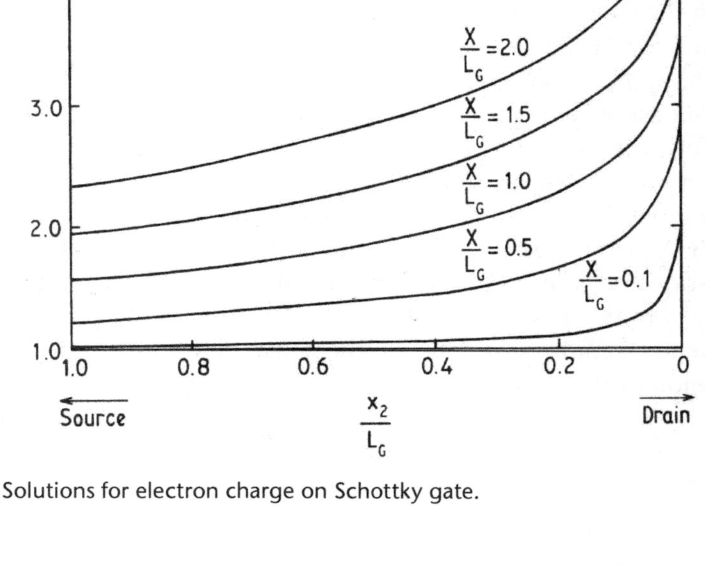

Figure 13.4 Solutions for electron charge on Schottky gate.

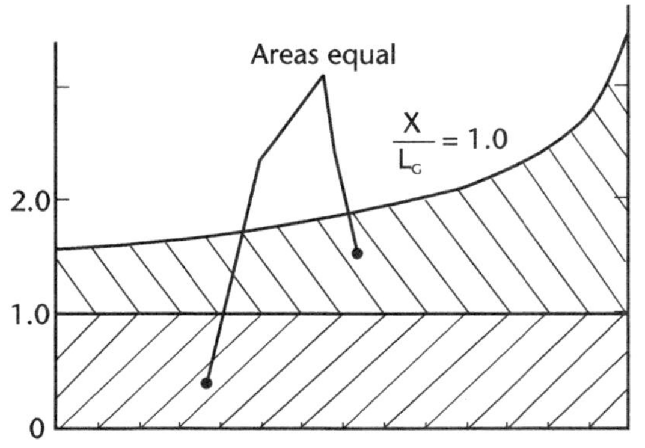

Figure 13.5 Illustrating charge conservation between the gate and the space-charge-layer extension for the case $X/L_G = 1$.

The area hatched to the left represents the total charge induced by the field $E_X(x,y)$ coming from the extension X, and that hatched to the right represents the electron charge on the gate induced by the positive ions immediately underneath.

While being somewhere between plausible and realistic, the results in Figure 13.4 are easier to use in accounting for some aspects of FET behavior than others. Bearing in mind that the objective is to develop relationships for trends, general functional dependencies, shapes, and magnitudes, further compromise may be possible. If the forms of Figure 13.4 lead to analytical impasse, continued approximation can be tried to see if it leads to anything useful. Sometimes it does.

13.4 The Flat *d* Approximation

To predict the various forms of the elements in the revised circuit model, even roughly, means mustering some equations to derive them from the physical processes that they are meant to represent. One concept that makes the job easier is that the depletion edge may be flat or approximately so.

Ignoring for the moment any y-directed field arising from the X region, the potential $V(x)$ in the channel increases positively from source to drain, progressively reverse biasing the Schottky barrier; the depth of depletion d increases correspondingly. A much exaggerated form is shown in Figure 13.6.

Now, if the y-directed electric field E_y is introduced into the depletion layer as in Figure 13.7, it adds to the field that is already present and so, to support the

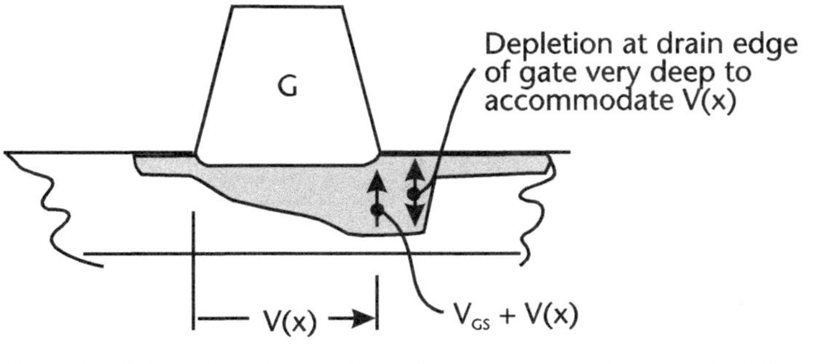

Figure 13.6 The potential gradient in the channel increases the depth of depletion.

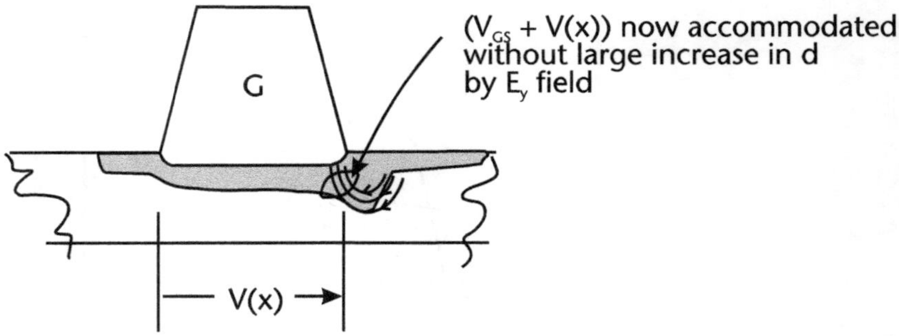

Figure 13.7 Introducing a field that is *y*-directed where it enters the depletion region leads to a decrease in the depth of depletion.

total voltage ($V_{GS} + V(x)$), d can be smaller. In other words, introducing E_y causes the depletion to shrink. Another way of describing the effect is to say that d lifts.

The d-lift principle offers the prospect that there could exist along the gate a distribution $E_y(x)$ such that the d-lift everywhere compensates the d pull-down due to $V(x)$. The result would be a flat d.

The purpose of this section is to give credence to the idea that the form of the charge induced on the gate by the field arising in the space-charge-layer extension naturally tends to flatten the depletion edge under the gate. Such a finding is in sharp contrast to the gradual channel approximation wherein no 2-D E field originating outside the channel is allowed for.

A rectilinear approximation for the electric field allows a tractable analysis of the form that the field must have (Figure 13.8).

Every line of electric force that enters the plane $x = x_2$ from the right turns from the x direction to become a line of y-directed electric force, eventually terminating on the gate. Thus, the total flux terminating on the gate between $x = 0$ and $x = x_2$ equals the total flux entering the channel at $x = x_2$.

In the usual way, flux balance would entail summing the y-directed flux density on the gate over the interval $x = 0$ to $x = x_2$ and setting that equal to the x-directed flux density summed from $y = d$ to some larger value of y where the flux density becomes negligible. The problem starts to look messy.

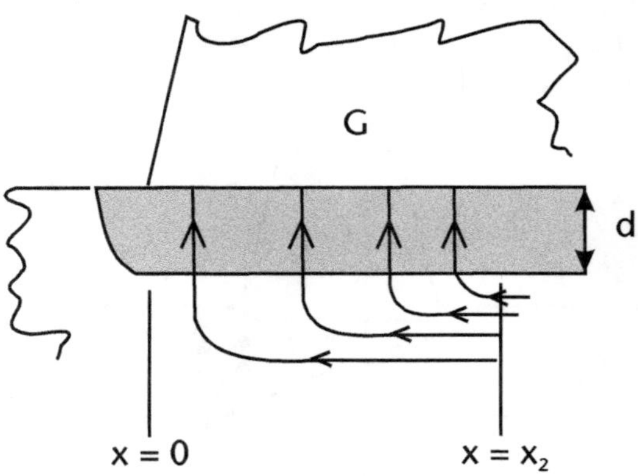

Figure 13.8 Rectilinear field approximation

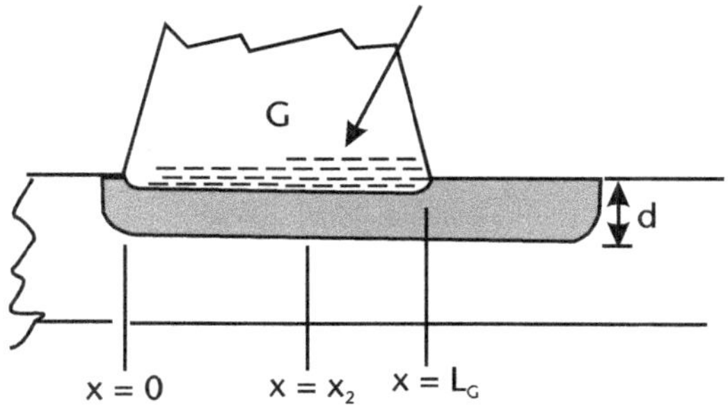

Figure 13.9 Charge per unit area induced on the gate by field originating from the space-charge-layer extension.

Because the objective is to establish trends rather than absolute accuracy, a heuristic alternative is called for. The following development bypasses the requirement for conservation of flux between the x-directed and y-directed fluxes with a contrivance, the implications of which are examined at the end of this section by dealing with the omission retrospectively.

The essence of the reduced argument focuses on a single function, the electron charge induced on the gate $Q_G(x)$ (the units of which are C.cm^{-2}, signifying per unit L_G and per unit Z_G); see Figure 13.9.

At any position $x = x_2$, the voltage developed across the depletion layer of depth d on account of $Q_G(x)$ is:

$$\frac{Q_G(x_2)d}{\epsilon} = E_y(x_2)d$$

and we require this to be equal to $V(x_2)$ where

$$V(x_2) = \int_0^{x_2} E_x(x)\,dx$$

Our task is therefore to find $Q_G(x_2)$ which satisfies this requirement subject also to the boundary conditions $V(0) = 0$ and $V(L_G)$ (the voltage at the drain end of the channel) and, importantly, such that $d \neq d(x)$, that is, d = CONSTANT. Thereby, we will have invoked the flat d condition.

Referring to Figure 13.10, the flux crossing $x = x_2$ from right to left is

$$qNdX - \int_{x_2}^{L_G} Q_G(x)\,dx \tag{13.1}$$

Similarly, the flux to the left of the plane $x = x_2$ is

$$\int_0^{x_2} Q_G(x)\,dx \tag{13.2}$$

At any position x_2, the potential developed across the depletion layer by $Q_G(x_2)$ is

$$\frac{Q_G(x_2)}{\epsilon}d = E_y(x_2)d = V(x_2)$$

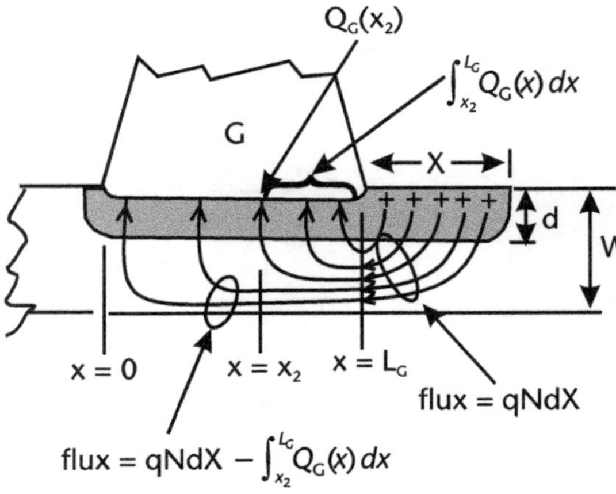

Figure 13.10 Flux and induced electron charge on the gate.

Now (13.2) can be rewritten as

$$Q_G^{-1}(x_2) - Q_G^{-1}(0) = Q_G^{-1}(x_2)$$

where the superscript −1 signifies "the integral of." $Q^{-1}_G(x_2)$ has the dimensions of potential multiplied by ϵ, that is

$$\frac{Q_G^{-1}(x_2)}{\epsilon} = V(x_2) = \frac{Q_G(x_2)d}{\epsilon}$$

The unauthentic step in the argument is the left side of this equation, which sets $\int_0^{x_2} E_y(x)\,dx = V(x_2) = \int_0^{x_2} E_x(x)\,dx$. In other words, it implies that $E_y(x) = E_x(x)$ with the connotation that x-directed field simply turns to become y-directed field. It is an artifice that avoids having to deal with unknowns associated with flux conservation governing the proper relationship between these fields. It is just about passable only for the insight the result brings. A more disciplined account is to be found at the end of the section.

Finally, given the equality of (13.1) and (13.2):

$$Q_G(x_2)d = qNdX - \int_{x_2}^{L_G} Q_G(x)\,dx$$

As a reminder of the overall aim, the embodiment of d in this equation as independent of x is the condition that we set out to satisfy.

Rearranging and dividing through by ϵ, the equation to be solved is

$$E_y(x_2)d + \int_{x_2}^{L_G} E_y(x)\,dx = \frac{qNdX}{\epsilon}$$

which has the solution

$$E_y(x_2) = \frac{V(L_G)}{d}\exp\left(\frac{x_2 - L_G}{d}\right)$$

where $V(L_G) = V(x_2 = L_G)$ is the potential in the channel at the drain end of the gate.

When a HFET first enters the velocity saturated regime, the potential in the channel near the drain end of the gate has a value just sufficient for velocity saturation to occur, that is, $V(L_G) = E_y(L_G)d = E_S d \sim 0.3\text{V}$ for a GaAs device as an order of magnitude (and around 10 times that number for a GaN HEMT).

By way of illustration for a 0.5 μm HFET with $d = 0.1$ μm, the distribution of the y-directed field along the gate that would give rise to a uniform or flat depletion edge is as in Figure 13.11.

The form of $E_y(x)$, equivalently $Q_G(x)/\epsilon$, bears some resemblance to that given in Figure 13.4.

What the development lacks in rigor it makes up for in insight. The contrivance that $E_y(x)$ and $E_x(x)$ are the exact same field, in a vague sense as if in Figure 13.8 the field lines were straightened out, reduces the mathematical problem to 1-D, thereby avoiding embroilment in flux in 2-D. As the considerations at the end of the section make apparent, however, to a degree the contrivance forces the answer that we want: that the general form of the gate charge for a flat depletion edge should resemble that of Figure 13.4. As preconceived as that sounds, we could have tackled the problem in reverse and asked what must be the relationship between $E_y(x)$ and $E_x(x)$ to achieve that form of $Q_G(x)$. Trying to answer the question thus inverted is no easier.

Despite the lack of rigor, the result of Figure 13.11 is representative of reality, or at least what reality must be given singularity suppression at the gate corner in practical devices. That said, it is important to air what the development further overlooks.

First, it is purely electrostatic. It takes no account of whether the E_x field is compatible with the $v(E)$ characteristic and the need for current continuity in the channel, and it says nothing about the consequences of conduction nonlinearity. Second, in the general case, the voltage applied at the drain is well in excess of 0.3V, so there is additional flux emanating from X to terminate on the gate somewhere. Bearing in mind that it took only 0.3 V worth of flux to render the depletion edge flat, if the whole field profile could be lifted uniformly, d would still be flat. The induced charge on the gate that terminates $E_y(x)$ would then resemble the curves in Figure 13.4. Qualitatively, the following sequence of events is envisaged as drain voltage is applied, starting from zero:

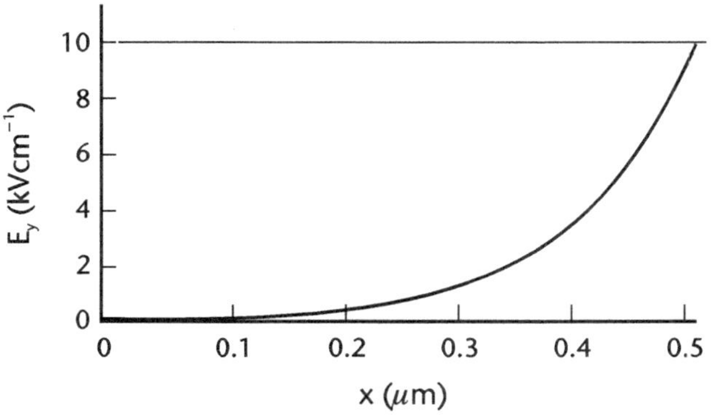

Figure 13.11 Form of the field terminating on the gate that would give rise to a flat edge to the depletion region.

- At first, the drain-source voltage is absorbed as $I_D R_S, I_D R_D$ together with the (initially resistive) voltage drop along the channel. The depletion edge becomes tapered as current starts to flow, being deepest at the drain end of the channel.
- As V_{DS} is gradually increased, a point is reached where the field at the drain end of the channel reaches the critical value E_S where velocity saturation sets in. The x-directed field in the channel and the y-directed field terminating on the gate are both nonuniform, having a natural distribution tending to flatten the depletion edge.
- Further increase in V_{DS} leads to expansion of the space-charge layer into the gate-drain region and added flux. After allowing for a small part of this extra flux as needed to drive current through the gate-source access resistance R_S, the greater part terminates on the gate, thus lifting the whole nonuniform field distribution of Figure 13.11 and preserving the flatness of d.

An impression of how similar to, or how far removed from, reality all this is can be gained by comparison with Figure 13.12, which is a sketch taken from typical computer simulations.

The close spacing of the equipotentials near the drain end of the gate signifies intense electric field broadly consistent with Figure 13.4 and with Figure 13.11. Note that, although the depletion edge under the gate may be flat, the equipotentials are not.

A depletion edge that is flat in region I does not require channel fields in L_I as high as those in region II and III. Bearing in mind that the onset of velocity saturation is not an abrupt phenomenon as would result from a simple two-line $v(E)$ characteristic but occurs gradually, there is no sharp dividing line between regions I and II. Instead, the boundary is indistinct, properly regarded as a transition. Irrespective, the electron velocity over a fair part of region I may approach v_{sat} and, like other approximations, assuming it to be near enough v_{sat} can sometimes be justified by the result to which it leads. We further need to bear in mind that, in practice, the depletion edge is diffuse, not sharply defined.

Instead of attempting further justification of the flat d idea from first principles, the criterion of whether the approximation proves of value in practice has been taken as proof enough. The proposition that not only does flux emanating

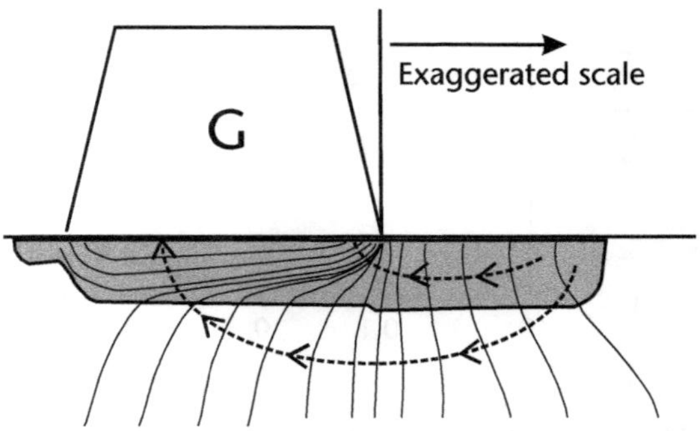

Figure 13.12 General form of equipotentials from computer simulation.

from the space-charge-layer extension lead to a flat edge to the depletion layer, but also that the depth of depletion decreases with increasing V_{DS}, has proven worth.

The flat *d* approximation has many uses, including developing insights into the dependence of $I(V)$ below the knee, the form of $I(V)$ above the knee as determined by *d*-lift, and the forms and magnitudes of various components of capacitance in the equivalent circuit model.

Before leaving the matter of flat *d*, it is important to outline where the foregoing treatment bypassed the imperative of flux conservation between the x and y coordinates.

At any position x in Figure 13.8, imagine there to be a small Gaussian box, with horizontal dimension Δx and vertical dimension Δy, with its top edge coincident with the depletion edge at d. We regard Δy as generally being a function of x. Writing the equation for flux entering and leaving the Gaussian box:

$$D_x(x)\Delta y + Q_G(x)\Delta x = D_x(x + \Delta x)\Delta y$$

After dividing by ϵ

$$E_x(x + \Delta x)\Delta y = E_x(x)\Delta y + E_y(x)\Delta x$$

Put

$$E_x(x + \Delta x) = E_x(x) + \frac{dE_x}{dx}\Delta x$$

Then

$$\frac{dE_x}{dx}\Delta x\Delta y = E_y(x)\Delta x$$

Integrating from $x = 0$ to $x = x_2$

$$\int_0^{x_2} E_y(x)\,dx = \int_0^{x_2}\left\{\frac{dE_x}{dx}\Delta y\right\}dx$$

However,

$$V(x_2) = \int_0^{x_2} E_x(x)\,dx$$

If the right sides of these two equations are to be equal, being the contrived condition referred to earlier, it implies

$$\frac{dE_x}{dx}\Delta y = E_x(x)$$

that is,

$$\frac{dE_x}{E_x} = \frac{dx}{\Delta y}$$

or

$$\ln\{E(x)\} = \int \frac{1}{\Delta y} dx$$

for example,

If Δy = CONSTANT: $E(x) \sim \exp(x)$.

By implication, the artifice of setting $E_y(x) = E_x(x)$ has to a degree forced the exponential form of the result in Figure 13.11.

However,

if $\Delta y \propto x$: $E(x) \sim$ CONSTANT

The point is that, without a full 2-D solution, we have no way of knowing what $\Delta y(x)$ might be. It likely lies somewhere between the two limits, meaning $E(x)$ will have some sort of concave-up form reminiscent of Figure 13.4.

13.5 The Uniform E_{yX} Termination Approximation

Figure 13.13 is a sketch of a FET showing the space-charge-layer extension X. Electric field emanating from the positive ions within X terminate on the gate somewhere, inducing an electron charge $Q_G(x)$.

In the previous section we saw how, when velocity saturation sets in at the drain end of the channel, a nonuniform distribution $Q_G(x)$ could arise which would tend to flatten the depletion edge. Thereafter, as the drain-source voltage is increased, the extra lines of electric force arising from increases in X would mostly end up on the gate.

Taking Figure 13.4 as a guide, the general outcome is one where an offset is added to the original nonuniform distribution so the whole is lifted. The first question is: exactly how does this lift occur? The second question is: how do we represent it in a manageable or workable form to help foresee how practical devices behave?

In connection with the first, the detailed manner in which the extra field terminates on the gate depends upon the operating condition of the FET, whether near pinch-off, open channel, the magnitude of the drain-source voltage (mainly via the ratio X/L_G), and so on. Attempting to pursue the problem on this basis leads

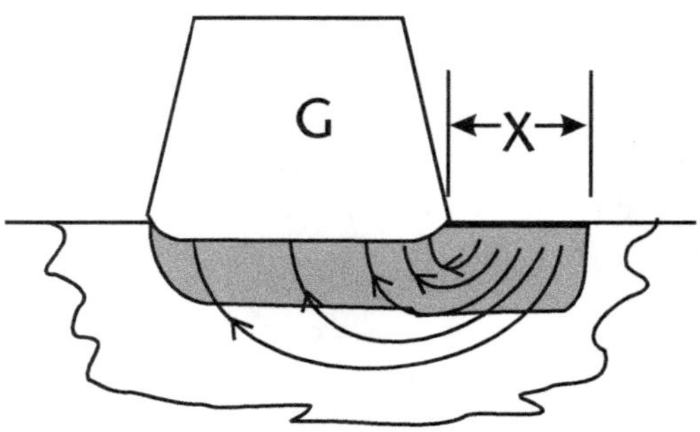

Figure 13.13 General disposition of the field from the depletion layer extension terminating on the gate.

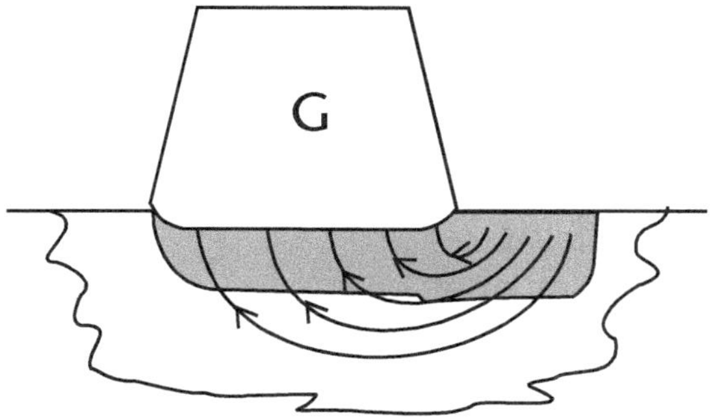

Figure 13.14 Uniform termination of part of the electric field on the gate.

nowhere in the quest for a workable physics. Consequently, a broader view is the only way forward. Yet again, whether any proposed simplification has merit has to be decided by its practical utility. Broadly, does it tell us anything useful?

Figure 13.14 is the compromise. Regardless of operating condition, the extra field is assumed to induce electron charge uniformly on the gate irrespective of the path, or paths, taken by the field between its point of origin (the space-charge-layer extension) and its point of termination (the gate). Paying no heed to the path is at first sight a risky business: it is discarding factors that might have more than a little influence on the FET's characteristics.

Continuing with the catechism, if nevertheless the uniform termination approximation is followed, where does it lead?

- It allows approximations to be set up for the gate-source and gate-drain voltages in terms of the key parameters of the depletion layer: the doping density N, the depth of depletion d, the extension X, and the gate length L_G.
- It accounts for the output conductance g_0 via a mechanism we are calling d-lift whereby the field progressively lifts the depletion edge as the applied drain voltage is increased.
- It enables expressions to be developed for the capacitive elements in the revised equivalent circuit model, notably the gate-drain branch capacitance C_{gdB}, the gate-drain branch transcapacitance C_N, and the gate-source branch transcapacitance C_M.

In all cases, we are looking for trends and functional forms to help to construct the revised equivalent circuit model for nonlinear applications and to provide a yardstick for assessing the plausibility of extracted elements when it comes to the final application of the model in practical circuits.

13.6 Expressions for V_{GC} and $V_{D'G}$

The next step in the development is to derive equations for the gate-channel potential $V_{GC}(= V_{GS'})$ and the gate-drain potential $V_{D'G}$, both as defined in Figure 13.1.

They are needed if we are to formulate ideas for how the elements in the circuit model should look with respect to the voltages across them.

Figure 13.15 shows the general form of the induced charge on the gate and the electric field along a path P running horizontally through the space-charge-layer extension from the gate towards the drain.

The exponential-looking curve labeled $Q_G(x_2)/\epsilon$ represents the field terminating on the gate. $Q_G(x_2)$ is the distribution of electron charge induced on the gate by all the positive ions in the depletion region. The quantity qNd represents the component arising from by the positive ionic charge in the depletion layer immediately below the gate. Between the $Q_G(x_2)$ curve and the level qNd is the component due to the ionic charge in the extension X. The area under the curve marked E_x along path P is the gate-drain voltage $V_{D'G}$.

We now need to simplify the whole picture to arrive at an approximate expression for $V_{D'G}$. On the left in Figure 13.16 is a sketch that repeats the form of Figure 13.15.

With the field along path P now approximated by a triangle, the expression for the gate-drain voltage is

$$V_{D'G} \approx \frac{qNd}{2\epsilon}\left(X + \frac{X^2}{L_G}\right) \tag{13.3}$$

Notwithstanding its apparent rough-and-ready origins, (13.3) works rather well.

We turn next to the gate-channel voltage V_{GC}, or $V_{GS'}$. V_{GC} differs from the gate-source voltage by the potential drop across the source resistance R_S; thus:

$$V_{GC} = V_{GS} - I_D R_S$$

The distinction is important because, in extreme cases like the PHEMT oddity examined at length in Section 4.2.2, R_S can become nonlinear, thereby creating a large and nonlinear difference between V_{GC} and V_{GS}. For now, however, the task is to find an expression for V_{GC}.

To evaluate the potential, it helps to draw the field profile through a vertical slice anywhere between $x = 0$ and $x = L_G$on the right part of Figure 13.16. Ionic charge in the depletion layer underneath the gate gives rise to a linearly increasing electric field rising from the depletion edge at d to the gate. The cross-hatched component

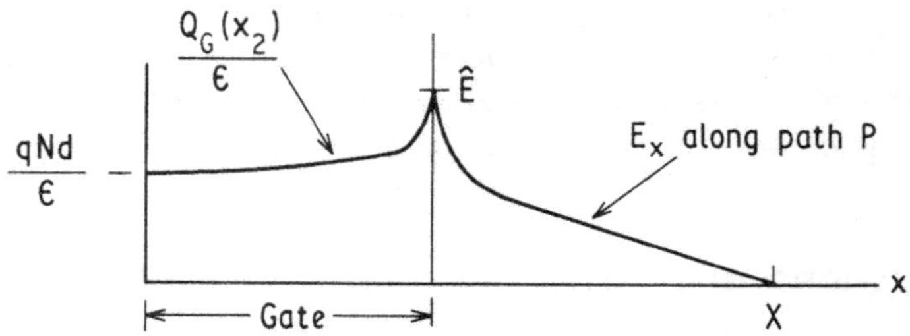

Figure 13.15 General form of gate charge and electric field solutions. (*From:* [1, Figure 6.24]. Reprinted with permission.)

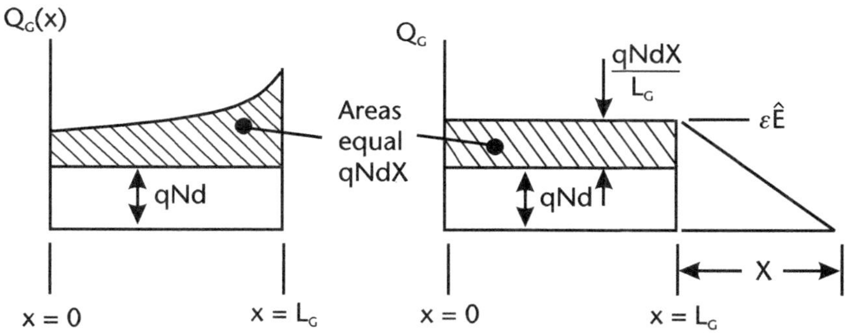

Figure 13.16 True-to-life form of the gate charge with simplified form for deriving an expression for $V_{D'G}$.

in Figure 13.16 represents the field coming in from X, roughly as in Figure 13.14, and distributed uniformly along the gate. This field penetrates the depletion layer and is a constant addition to the other component, which is triangular in y. Drawing the variable part on top of the constant part, we end up with Figure 13.17.

As before, the area under the field profile is the voltage that we are after:

$$V_{GC} \simeq \frac{qNd^2}{2\epsilon}\left(1 + \frac{2X}{L_G}\right) \tag{13.4}$$

(Note that for consistency with the polarity inferred by the subscript notation, V_{GC} should strictly be written as a negative number. We drop the negative sign for ease and take it as implied.)

Inherent in this approximation is the assumption that the incoming y-directed field is all akin to the leftmost three lines in Figure 13.14. Put another way, all the field arising from within the space-charge-layer extension is assumed to contribute a potential across the depletion layer that is the same for all x. For the time being, such a simplification is good enough (it requires rethinking when it comes to evaluating an expression for the transcapacitance C_M).

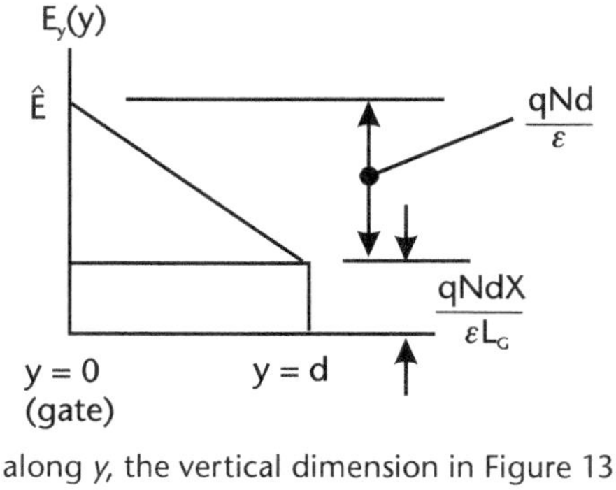

Figure 13.17 Field profile along y, the vertical dimension in Figure 13.16.

Sooner or later, we have to start putting in numbers to check that the emerging equations make sense. From Figure 13.17:

$$\hat{E} \simeq \frac{qNd}{\epsilon}\left(1 + \frac{X}{L_G}\right)$$

Knowing the breakdown electric field strength E_B, a limiting value of X/L_G can be determined for various N and d for HFETs fabricated in different semiconductor materials. For instance:

$$E_B \simeq 640 \text{ kcm}^{-1}, \; N = 3.5 \times 10^{23} \text{ m}^{-3}, \; d = 0.1 \; \mu\text{m}$$

$$\frac{X}{L_G} \leq 0.3$$

If instead $N = 1.5 \times 10^{23}\text{m}^{-3}$, then

$$\frac{X}{L_G} \leq 2$$

Both are compatible with Figure 13.4. Usually, however, a device would not be operated so close to breakdown so smaller values of X/L_G are de rigeur. Note that the peak electric field has a minimum value dictated by the bulk charge (when $X \ll L_G$).

13.7 The *d*-Lift Principle

To understand the general idea of *d*-lift, 1-D suffices. What happens in FETs is then easier to follow.

Figure 13.18 shows a Schottky gate biased negatively with respect to the bulk semiconductor or channel. Equivalently, the channel represented by the plane A–A′ is held at a positive potential V_{CH}, with respect to the gate. We use V_{CH} interchangeably with V_{GC}, depending upon the emphasis required. For the present, V_{CH} handily distinguishes the 1-D case. No current flows, so the depletion edge at d is also at potential V_{CH}.

Now imagine that an electric field, which we shall denote by E_{yX} is introduced as in Figure 13.19. For the time being, we are not concerned where E_{yX} comes from: to digress into its origin risks distracting from the main argument.

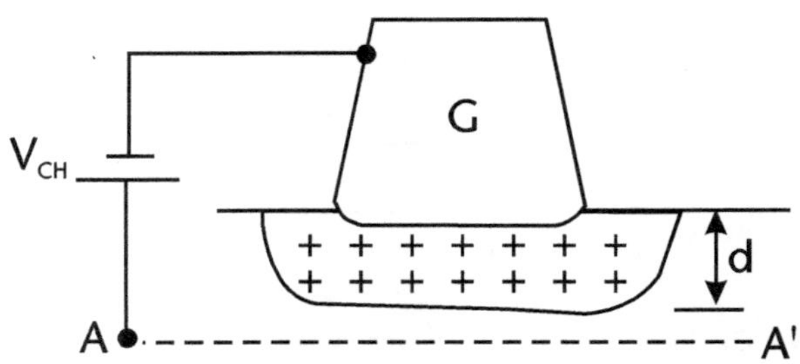

Figure 13.18 Simple Schottky contact with reverse bias.

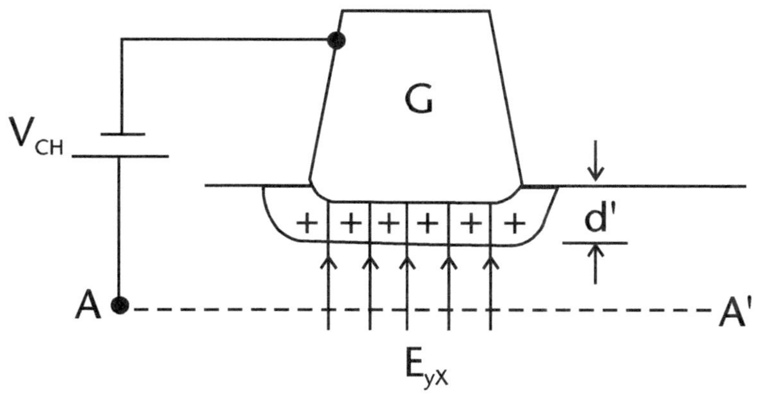

Figure 13.19 Schottky contact with the incoming electric field.

For the potential difference between the gate and the depletion edge to remain the same (it is held at V_{CH}), the depletion edge must lift towards the gate. Figure 13.20 shows the electric field in the two cases.

The introduction of E_{yX} lifts the depletion edge from d to d' such that the shaded areas, being the potential V_{CH}, are the same.

In practical FETs and HEMTs, the lines of electric force that we have called E_{yX} arise from the positive ionic charge in the space-charge-layer (or depletion layer) extension X in the gate-drain region. d-lift has a profound effect on the shape or form of the $I(V)$ characteristics, the associated conductances g_{m0} and g_0, together with other elements that depend secondarily upon them. In [1], the output conductance g_0 was ascribed to the substrate current. We now believe this to be at best a secondary effect, the principal mechanism being d-lift.

It is now possible to interpret qualitatively how the $I(V)$ characteristics are made up. d-lift depends on X increasing, but X does not increase until the v_{sat} regime is

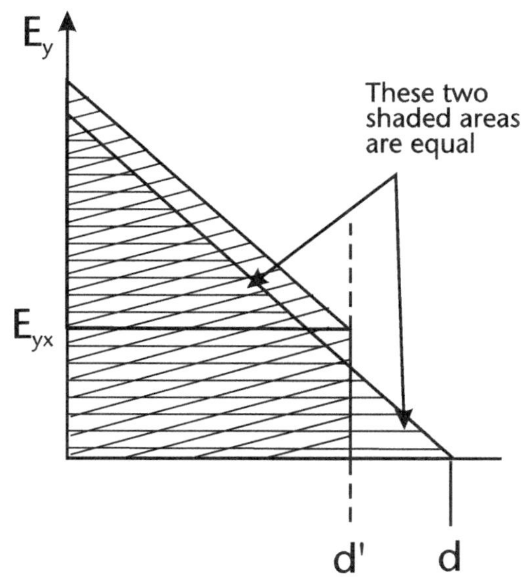

Figure 13.20 Field profiles without and with E_{yX}.

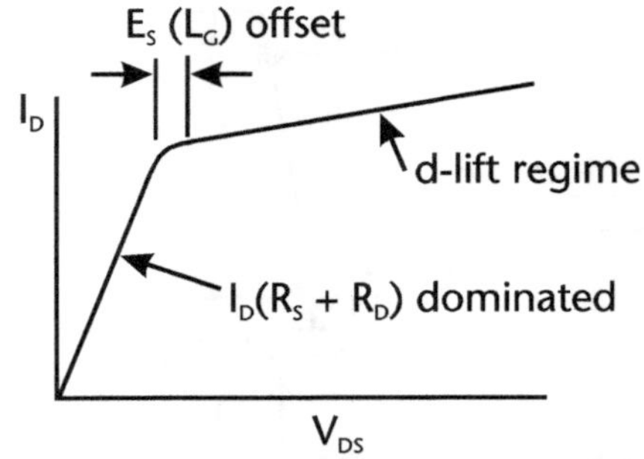

Figure 13.21 Sketch showing the composition of the *I*(*V*) characteristics in terms of individually identifiable processes.

reached. Therefore, *d*-lift does not set in until the *x*-directed field at the drain end of the channel attains the value E_S and, in GaAs devices, the potential drop along the channel $V(L_G) \simeq 0.1$V. Thus, the *I*(*V*) curves are made up as in Figure 13.21.

For later use in working out equations for g_0 and other elements, we now establish a fundamental relationship governing *d*-lift: that between changes in *d* and changes in *X*. By "g_0 and other elements" we are referring to elements attributable to changes in the drain voltage when, importantly, the gate-channel voltage is maintained constant, that is, preserving the condition under which the foregoing principle of *d*-lift was established. As the concern is now with practical 2-D devices, not idealized 1-D variants, we revert to the symbol V_{GC} for the gate-channel potential (rather than continue with V_{CH}).

Referring to Figure 13.17, the field of interest is

$$E_{yX} = \frac{qNdX}{\epsilon L_G}$$

A small increase in applied drain voltage leads to an incremental increase in *X*, δX, which, in turn, means the *y*-directed field changes; thus:

$$\delta E_y = \frac{qNd\delta X}{\epsilon L_G}$$

As a result of this change, $d \rightarrow d' = (d - \delta d)$, as per Figure 13.20. The voltage supported by bulk charge of density N cm^{-3} of depth d is

$$\frac{qNd^2}{2\epsilon}$$

Therefore, the change in voltage when $d \rightarrow d'$ is

$$\frac{qN(d - \delta d)^2}{2\epsilon} - \frac{qNd^2}{2\epsilon} = -\frac{qNd\delta d}{\epsilon}$$

For $\delta V_{GC} = 0$, this must equal $\delta E_y d$, so that

$$\delta d = -d\frac{\delta X}{L_G} \qquad (13.5)$$

The negative sign signifies that an increase in X leads to a decrease in d, a physical outcome not to be forgotten.

Reviewing the foregoing development critically reveals the assumption, mentioned just after (13.4), that all of δE_y contributes to the change in voltage across the shrunken depletion layer. Equivalently, all the incremental changes in the electric field follow paths exactly as in Figure 13.19. Such an assumption ignores the fact that some field lines, like the two rightmost in Figure 13.14, are only partly y-directed and in any case exist only through part of the depletion layer. Consequently, such lines of electric fields exert way less than the full lifting effect assumed. Thus, it should come as no surprise that the full-lift assumption does not hold in some circumstances, notably when comparing the relative magnitudes of the branch capacitance C_{gdB} and the associated transcapacitance C_M. That and other practical evidence suggest in such cases that a better approximation is

$$\delta d = -\alpha d\frac{\delta X}{L_G} \qquad (13.6)$$

where α has a value of less than unity, typically around 0.5.

Equations (13.5) and (13.6) are key relationships to which we shall resort several times in working out functional trends for the equivalent circuit elements. As in Figure 13.21, one of the important influences that d-lift has on FET behavior is that it governs the output conductance g_0.

$$g_0 \overset{\text{def}}{=} \frac{\partial I_D}{\partial V_{D'G}}\bigg| V_{GC}$$

where $| V_{GC}$ signifies that V_{GC} is constant.

Differentiating (13.3),

$$\frac{\partial V_{D'G}}{\partial X} = \frac{qNd}{2\epsilon}\left(1 + \frac{2X}{L_G}\right)$$

therefore

$$\delta X = \frac{2\epsilon}{qNd\left(1 + \frac{2X}{L_G}\right)}\delta V_{D'G}$$

Substituting into (13.6):

$$\delta d = -\frac{2\epsilon\alpha}{qNL_G\left(1 + \frac{2X}{L_G}\right)}\delta V_{D'G}$$

However,

$$\delta I_D = qNv_{\text{sat}} Z_G \delta d = \frac{2\epsilon\alpha v_{\text{sat}} Z_G}{(L_G + 2X)} \delta V_{D'G}$$

So finally,

$$g_0 = \frac{\delta I_D}{\delta V_{D'G}} = \frac{2\epsilon\alpha v_{\text{sat}} Z_G}{(L_G + 2X)}$$

This equation turns out to have unanticipated significance.

- First, it shows that the output resistance $R_{ds} = g_0^{-1}$ is proportional to the space-charge-layer X extension or, equivalently, to the transconductance delay τ_{gm}.
- Second, after deducing an expression for the gate-channel transcapacitance C_M in Section 15.3.3, it will become apparent that

$$g_0 = \frac{C_M v_{\text{sat}}}{L_G}$$

- Third, it is demonstrably an integral part of the elegant symmetry that emerges among the circuit model elements, as discussed in Section 15.4.

13.8 The Delay τ_{gm}

Following the brief introduction in Section 3.3, the time has come to give an account of what τ_{gm} is in physical terms and the function that it performs in an equivalent circuit. From the measurement of this delay function, it is possible to retrieve, sometimes only qualitatively but sometimes quantitatively, data concerning the makeup the device from which it came. One such role was covered at length in [1, p. 184].

Figure 13.22 is a cross-section of a device delineating the three regions where the combination of the way that electrons are transported and spatial position in the structure are different. In region I, the motion of electrons is governed by the nonlinear part of the velocity-field characteristic for $E \le E_S$ in Figure 13.2. At the end of region I, the field has attained the value E_S where the electron velocity saturates. Thereafter, through regions II and III, electrons travel at v_{sat}. Region II is the

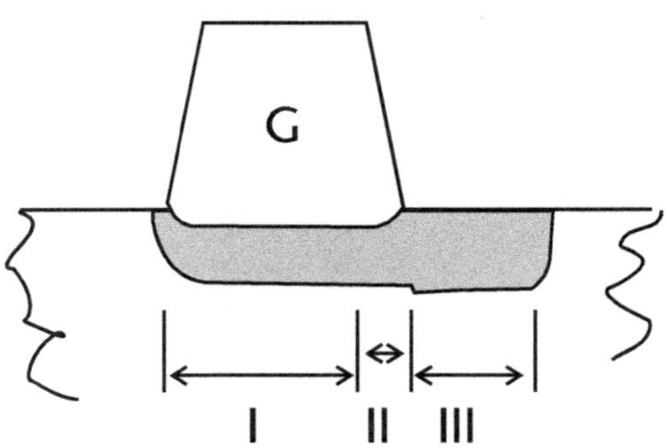

Figure 13.22 The channel divided into three zones.

velocity-saturated part of the channel that is in the shadow of the gate; region III is beyond the drain edge of the gate, in the shadow of the depletion layer extension X.

By one means or another, the task is to account for the general form of τ_{gm}, as in Figure 13.23, in terms of what happens in these three zones.

τ_{gm} arises from regions II and III. First thoughts suggest that it might be the time taken by electrons traveling at v_{sat} to traverse from one end of region II-III to the other. On second thoughts, other considerations, as below, suggest more like half this time is a better figure.

To pursue the initial train of thought, in an operating FET, electrons enter the channel at the left end and, because the mobility is relatively high, they can respond rapidly to any changes in the time-dependent voltages applied to the gate and drain. For simplicity, and so as not to obscure the principle at stake, we temporarily assume the channel current in this section of the channel to be an instantaneous function of the voltages. Now, however, any instantaneous change in the channel current that appears at the region I–II boundary cannot appear at the drain until it has propagated, at velocity v_{sat}, right through regions II and III. So where is this reasonable expectation lacking?

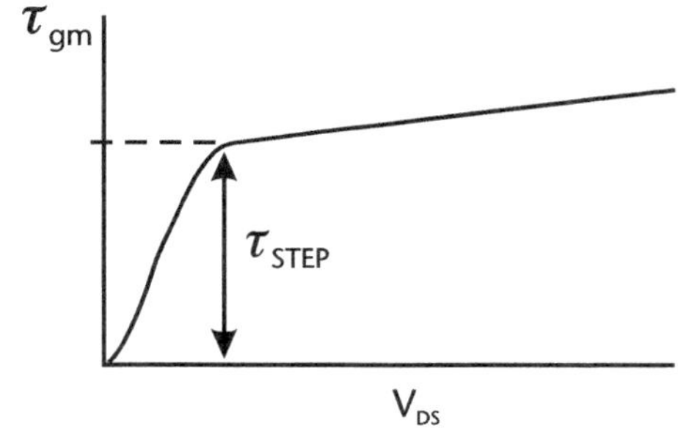

Figure 13.23 General form of τ_{gm}.

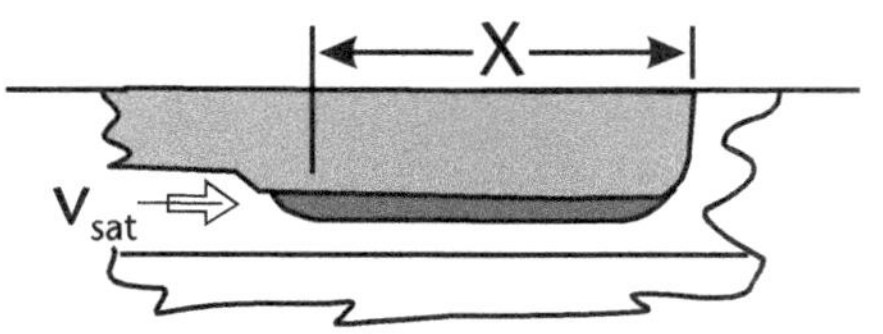

Figure 13.24 Electron propagation into an incremental change in the depletion edge.

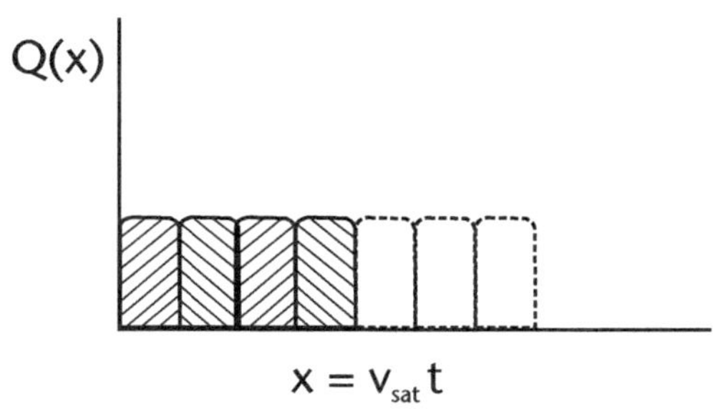

Figure 13.25 Time and space dependence of the charge deposition in region II-III.

Figure 13.24 is an expanded view of the depletion edge showing electrons, as shaded, gradually propagating through regions II–III in response to a step change, say v_{GS}, in the gate voltage. The charging current is $g_{m0}v_{GS}$. Figure 13.25 is a diagram showing how the charge is deposited.

Charge is deposited progressively, akin to having n small capacitors $\delta C = C_{TOT}/n$ and adding an extra small capacitor, fully charged, at each time interval. The difference between a parallel plate capacitor and the region II-III capacitance, $C_{II\text{-}III}$, is that, in the latter case, the charge is fed in from one edge, not straightforwardly onto the plates as in a conventional capacitor. So, by nature, $C_{II\text{-}III}$ is a time-dependent capacitance (Figure 13.26), the composite nature of which was detailed in [1, p. 107].

Unlike a time-dependent capacitance that is systematically pumped by external means in parametric converters, a de facto time-dependent capacitance is arguably not a capacitance at all. If sticking with the concept of capacitance is inescapable, how are we to set down a relationship for a capacitance and its associated charging time? Perhaps the only way is to reduce the problem to a time-average capacitance, a weird concept perhaps, but one that gets as close as we can to the idea of capacitance in a circumstance where charge is stored by pure delay. In due course, charge storage by pure delay will emerge as one of the cornerstones of the reformed equivalent circuit model.

For now, staying with the notion of a time average capacitance, from Figure 13.26, we can write:

$$C_{II-III} = \frac{\epsilon X Z_G}{2d} \tag{13.7}$$

If at the same time we take it from Section 14.4 that

$$C_{II-III} = g_{m0}\tau_{gm} \tag{13.8}$$

and

$$g_{m0} \simeq \frac{\epsilon v_{sat} Z_G}{d}$$

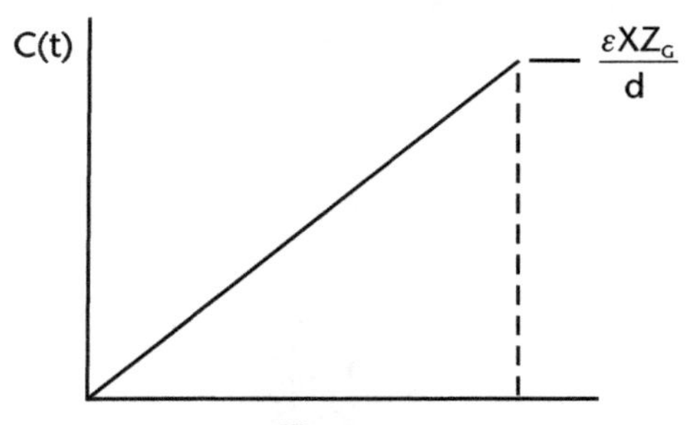

Figure 13.26 Region II–III charging expressed as capacitance.

the result is

$$\tau_{gm} = \frac{X}{2v_{\text{sat}}} \tag{13.9}$$

We have now ended up with the idea that τ_{gm} is but one-half of what we originally thought it should be. Both cannot be correct.

It has proved awkward to resolve these apparently conflicting principles, and a pragmatic view has to suffice.

First, the function for τ_{gm} in the reformed circuit model is to come from measurement. Consequently, if the measurements are without fault, any aspect of circuit behavior that depends upon τ_{gm} should be correct, all other things being equal. Second, if we adopt (13.9) as the best option, if it proves to be a poor choice it is only going to affect the value of X as interpreted from the measured variable, which is τ_{gm}. X is not measureable directly. Equation (13.9) has seen widespread use in reverse modeling and has served well in that capacity. Thus, although there is a persistent problem of principle, in practice (13.9) and the measured τ_{gm} are workable.

What of the initial step as marked on Figure 13.23; how does it come about? Until velocity saturation occurs at the drain end of the channel, there is no charge storage by pure delay anywhere in the device. When the electron velocity at the drain end of the channel reaches saturation, so $v(L_G) = v_{\text{sat}}$, the first thing that happens is that L_{II} forms, where L_{II} is the velocity-saturated part of the channel in the shadow of the gate. Beyond it, electrons pass through the depletion extension X_0, which exists when $V_{DS} = 0\text{V}$, that is, $V_{D'G} = V_{S'G}$. If we make the reasonable assumption that both the energy and momentum relaxation times are small relative to the transit times of interest, as soon as the electrons pass into X_0, they will relax to a velocity less than v_{sat}. On this basis, when τ_{gm} first appears it arises as the time taken for the carriers to pass through region L_{II}. As the drain voltage is increased, however, over progressively more of both the channel and the depletion extension the electric field becomes large enough to sustain velocity saturation. Thus, the delay increases owing to L_{II} expanding and to carriers now passing through the first part of X_0 where the field is large enough to sustain v_{sat}. As the process continues, it forms a steep rise in τ_{gm} until further increases in drain voltage are absorbed as an increase in the space-charge-layer extension X, by which time the expansion of L_{II} has slowed dramatically. Qualitatively, therefore, the initial step is due to

$$L_{\text{II}} + X_0$$

where L_{II} is symbolic for growth in L_{II} until it reaches a near enough stable value. So,

$$\tau_{\text{STEP}} \simeq \frac{\left(L_{\text{II}} + X_0\right)}{v_{\text{sat}}}$$

At face value,

$$\text{if } \tau_{\text{STEP}} = 1 \text{ ps then} \left(L_{\text{II}} + X_0\right) = 0.1\ \mu\text{m}$$

and

$$\text{if } \tau_{\text{STEP}} = 3 \text{ ps then} \left(L_{\text{II}} + X_0\right) = 0.3\ \mu\text{m}$$

and so on.

In the context of the extracted functions $\tau_{gm}(V_{GS},V_{DS})$ to be found throughout the book and the devices from which they came, values for L_{II} of 0.1 µm to 0.3 µm seem reasonable. They are more questionable if instead we had written

$$\tau_{\text{STEP}} \simeq \frac{\left(L_{\text{II}} + X_0\right)}{2v_{\text{sat}}}$$

in conformity with (13.9) and the reasoning that led to it, which gives values for $(L_{\text{II}} + X_0)$ twice as large. However, τ_{STEP} is embodied in all the values of τ_{gm} in (13.9), a relationship which, in practice, has proved dependable. Thus, which is the better of the two equations for τ_{STEP} remains an open question.

Whichever way we look at it, L_{II} as well as X contributes to τ_{gm}. For the reason given after (13.9), we will henceforth assume that L_{II} makes only a small contribution relative to X so (13.9) applies as a reasonable approximation.

Because the principle of charge storage by pure delay (rather than capacitance) is an unfamiliar concept, it is worth looking at from a slightly different angle to see what (13.8) entails. Whatever different perspective can be conjured, even if only slight, is useful to understanding the limitations of trying to model delay by capacitance in a nonlinear circuit model for the HFET.

Consider a pulse of channel current as in Figure 13.27.

At any time t after the application of the excitation, the charge stored in region II-III of the depletion edge is

$$Q_{\text{II-III}} = \int_0^t \left(I(t) - I(t-\tau)\right) dt \tag{13.10}$$

This relationship applies so long as τ is less than the time at which $C(t)$ reaches its maximum in Figure 13.26.

Generally, (13.10) is not a simple time-varying function, depending upon the form of the excitation or signal. For the simple case of Figure 13.27, however,

$$Q_{\text{II-III}}(\tau) = I\tau \tag{13.11}$$

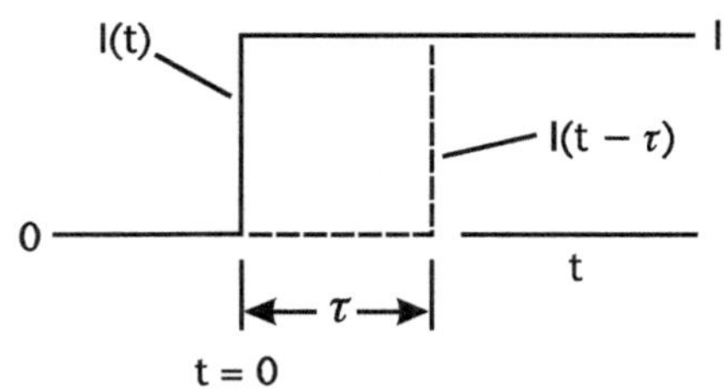

Figure 13.27 Current pulse and delayed current pulse.

By differentiating with respect to V_{GC}, we end up with a quantity having the dimensions of capacitance:

$$\frac{\partial Q_{\text{II-III}}}{\partial V_{GC}} = \frac{\partial I}{\partial V_{GC}}\tau \simeq g_{m0}\tau \tag{13.12}$$

which holds only if τ is a weak function of V_{GC}, which sometimes it is but not always. Thus, we have to be guarded in invoking (13.8).

If thinking in terms of capacitance is habitual, the foregoing equations, especially (13.8), may be useful. Nevertheless, the limitations of the relationship should be borne in mind by looking back over the simplifying assumptions. The imperative is that charge is stored by a mechanism of pure delay operating on current.

We next turn to the information about the device structure that can be gleaned from the measured dependence $\tau_{gm}(V_{GS}, V_{DS})$. (To avoid complicating the matter no distinction is drawn between V_{DS} and $V_{D'G}$; instead, it is supposed that trends exhibited by τ_{gm} with the one on the horizontal axis are representative of trends with the other.)

Referring to (13.9), τ_{gm} is synonymous with X. Therefore, at the point P in Figure 13.28, both X and $V_{D'G}$ have clearly defined values.

From (13.3), it follows that, at point P, Nd is a constant, or

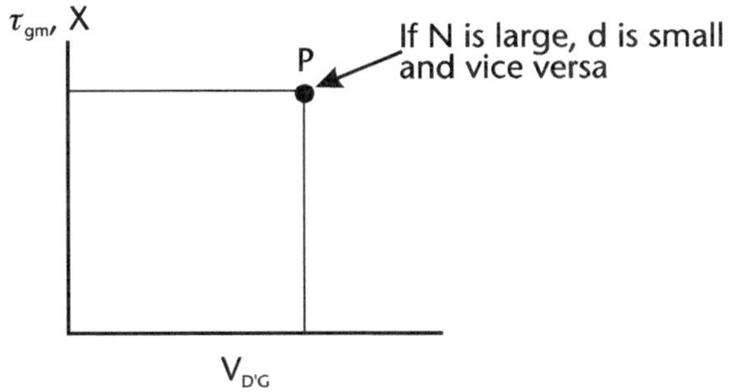

Figure 13.28 The τ_{gm}-drain voltage plane.

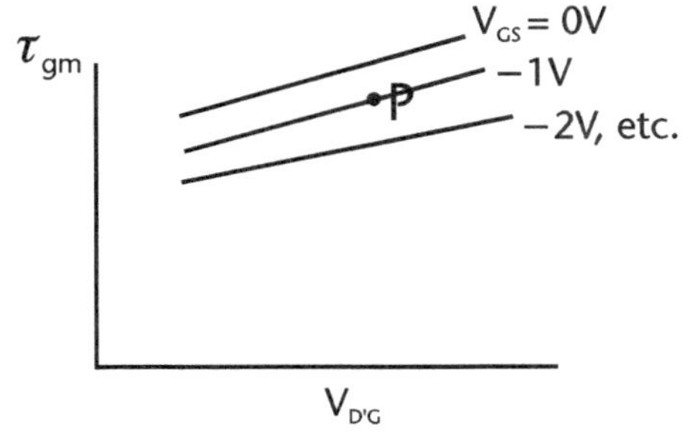

Figure 13.29 Ordering of curves of τ_{gm} with respect to V_{GS} for uniform N.

$$d \propto \frac{1}{N} \tag{13.13}$$

hence the arrowed comment in the figure.

Now, for uniform doping density N, from Figure 6.15 in [1], we have a dependence like Figure 13.29.

In light of (13.13), for $N_2 < N_1$, the example point P will require a larger value of d, that is, it will occur at a more negative value of V_{GS}. The implication is twofold:

- A tail to the doping profile causes the curves to bunch up, as per the sketch of Figure 13.30 and the practical example of Figure 13.31. Ion-implanted channel layers are one example.
- Depending upon how far d reaches into the tail, or lower-doped region, the ordering of the curves may reverse compared with Figure 13.29. This extreme accounts for the highest curve in Figure 13.31 being the closest to pinch-off. The sketch of Figure 13.32 highlights the issue.

In summary, the way that τ_{gm} is ordered with respect to V_{GS} can yield information about the doping profile.

A surprising amount of intelligence can be gathered from τ_{gm} on what kind of device a set of measured S-parameters has come from. Also surprising, unless a special effort has been made to look for it, is the degree of correlation which exists between some of the elements in the fitted standard model. Figure 13.33 is one example.

Not only is the correlation plausible, but also it can be interpreted in terms of device structural factors, as follows.

It will be shown in Section 15.3.3 that a relationship for the gate-source branch transcapacitance is

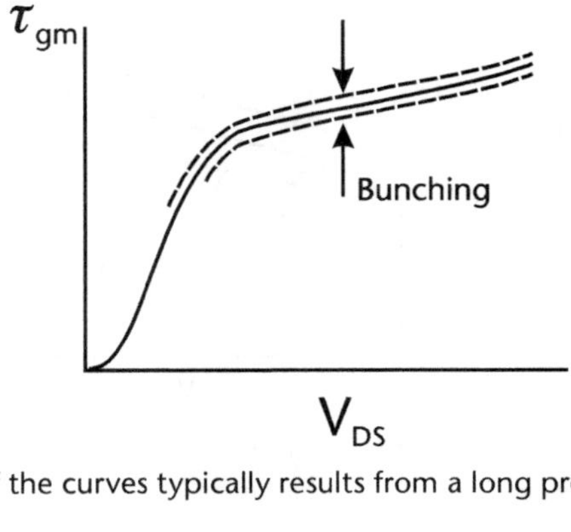

Figure 13.30 Bunching of the curves typically results from a long profile tail.

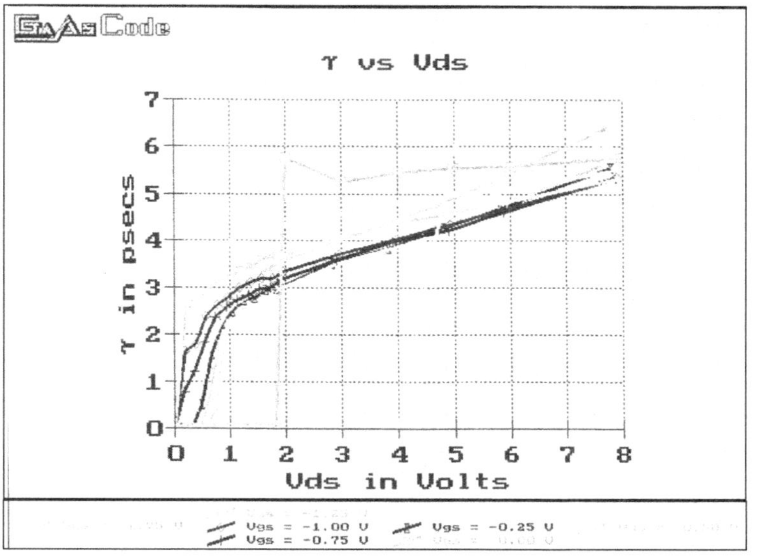

Figure 13.31 A practical example of curve bunching.

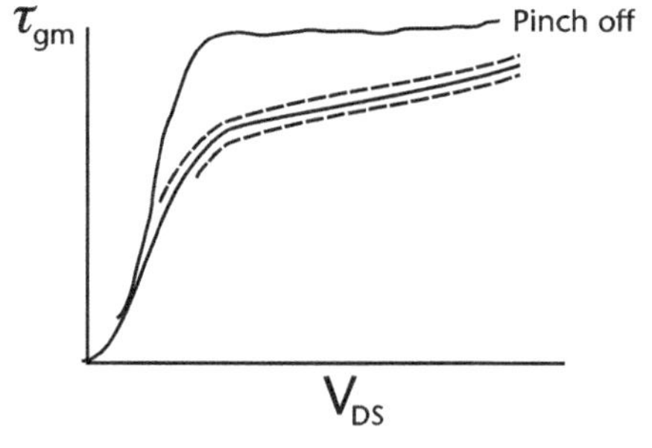

Figure 13.32 A profile tail naturally leads to the highest curve being for V_{GS} near pinch-off.

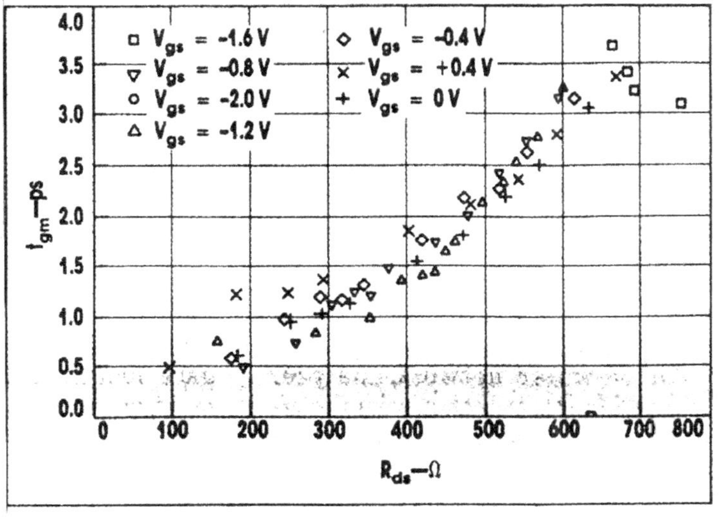

Figure 13.33 Practical correlation between τ_{gm} and R_{ds}.

$$C_M \simeq \frac{\epsilon Z_G}{\left(1 + \frac{2X}{L_G}\right)}$$

and that $g_0 = 1/R_{ds}$ is related to C_M by

$$g_0 = \frac{C_M v_{sat}}{L_G}$$

From these two equations:

$$R_{ds} = \frac{L_G}{\epsilon v_{sat} Z_G} + \frac{2X}{\epsilon v_{sat} Z_G}$$

After substituting from (13.9),

$$R_{ds} = \frac{L_G}{\epsilon v_{sat} Z_G} + \frac{4\tau_{gm}}{\epsilon Z_G}$$

The first term on the right side is the intercept on the horizontal axis; the second term is the slope.

When there is scatter in empirical data, the principled way to proceed is subject it to regression analysis, leading to a most probable value with error bounds. Such has not been tackled here, only a line drawn subjectively through the data points. Starting with the second term for this GaAs device,

$$Z_G \simeq 200\ \mu\text{m}$$

From the intercept (with $v_{sat} = 10^7 \text{cms}^{-1}$):

$$L_G \simeq 0.35\ \mu\text{m}$$

No firm data were available from the device manufacturer to compare with these reverse-modeled numbers. Quick trials with the method soon show that the detailed values recovered are sensitive to the particular line that is drawn through the data points, so the foregoing results have to be seen as just one possibility among several. Knowing the manufacturer's gate width would be a big help in retrieving a value for the gate length with greater certainty.

In conclusion, τ_{gm} and X are fundamental to understanding and predicting how FET devices work.

Notwithstanding the foreseeable challenge of implementing a voltage-dependent time delay in a harmonic balance simulator, any effort to ensure the experimental determination of τ_{gm} is as good as can be is time well spent. Besides, there is no comparable difficulty with a time-domain simulator (the reformed model that is the subject at hand was implemented in both). To that end, a special version of the fitting routine used in all this work, GaAsCode's package FETLink, was developed to optimize the extraction of the delay function $\tau_{gm}(V_{GS},V_{DS})$. This variant was also offered as a commercial product under the identifier FETLink Large Signal (FLLS).

References

[1] Ladbrooke, P. H., *MMIC Design: GaAs FETs and HEMTs*, Norwood, MA: Artech House, 1989.

[2] Ladbrooke, P. H., and J. E. Carroll, "Dielectric Relaxation as a Limit on Transistor Switching Speed," *Electron. Lett.*, Vol. 32, 16, August 1, 1996, pp. 1511–1513.

[3] Ruch, J. G., "Electron Dynamics in Short-Channel Field-Effect Transistors," *IEEE Transactions on Electronic Devices*, Vol. ED-19, 1972, pp. 652–654.

[4] Ladbrooke, P. H., "Some Possible Consequences of a Transverse Doping Gradient in Bulk-Effect Oscillators," *J. Appl. Phys.*, Vol. 3, March 1970, pp. 437–442.

[5] Brennan, K., and K. Hess, "A Theory of Enhanced Impact Ionization Due to the Gate Field and Mobility Degradation in the Inversion Layer of MOSFETs," *IEEE Electron Device Letters*, Vol. EDL-7, No. 2, February 1986, pp. 86–88.

[6] Ladbrooke, P. H., "Reverse Modelling of GaAs MESFETs and HEMTs," *GEC Journal of Research*, Vol. 6, 1988, pp. 1–9.

[7] Roblin, P., and H. Rohdin, *High-Speed Heterostructure Devices*, Fig. 14.5, Cambridge, U.K.: Cambridge University Press, 2002.

[8] Cameron, N. Private communication, University of Glasgow, 1993.

[9] *Silicon Carbide (SiC): Transport Properties in High Electric Fields*, NSM Archive, ww.iofe.ru.

[10] Shul, R. J., and J. Barker, *High Field Effects of GaN HEMTs*, Sandia Report SAND2004-4404, September 2004.

[11] Morton, C. G., C. M. Snowden, and M. J. Howes, "A New Quasi-Two Dimensional HEMT Model," in *Simulation of Semiconductor Devices and Processes*, H. Ryssel and R. Pichler, (eds.), New York: Springer-Verlag, 1995, pp. 352–355.

[12] Unpublished work carried out at Plessey Research (Caswell) Ltd., 1978–1979.

CHAPTER 14

Current and Charge Conservation

14.1 Channel Current

For the most part, the prime concern is to understand what happens above the knee in the $I(V)$ characteristics, the objective being to develop approximations for the key elements in the equivalent circuit model. With the three regions identified in Figure 13.22, a region-by-region examination of the processes governing electron movement proceeds with arguably the easiest, region II–III.

In region II–III, electrons are traveling at their saturated drift velocity v_{sat}, as in Figure 13.2. It is the channel opening $(W - d)$ at L_{II} which determines the current collected by the drain. Also, charge exchange with the depletion edge at d is accomplished by electrons traveling at v_{sat}, as already covered in Section 13.8. Relative to current entering, current leaving region II–III is subject to delay by τ_{gm}, with implications discussed in Section 14.4.

One way or another, region II–III also governs the slope of the $I(V)$ characterisitics above the knee, equivalently the output conductance g_0. When *MMIC Design: GaAs FETs and HEMTs* [1] was written, scattering of hot electrons into the substrate was thought to be the mechanism responsible, whereas the evidence gathered since favors d-lift. Nonetheless, deconfinement of hot electrons is involved in the manner of Figure 14.1.

The view given in [1] is that substrate current is in addition to channel current. An alternative view, a two-step argument, is:

1. The electrons become energetic enough to scatter out of the channel only when they have traveled maybe 0.1 μm through region II.

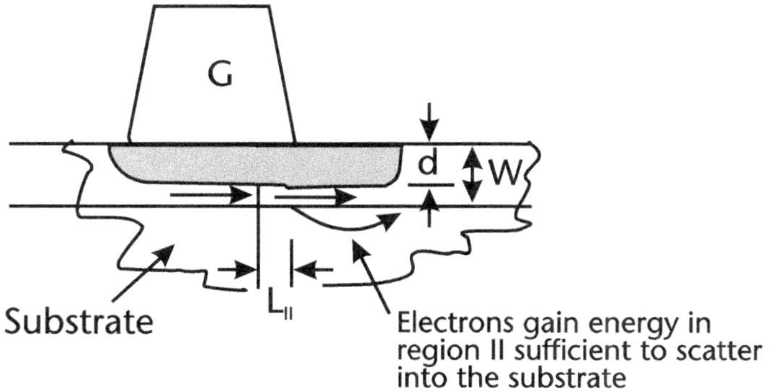

Figure 14.1 Deconfinement of hot electrons.

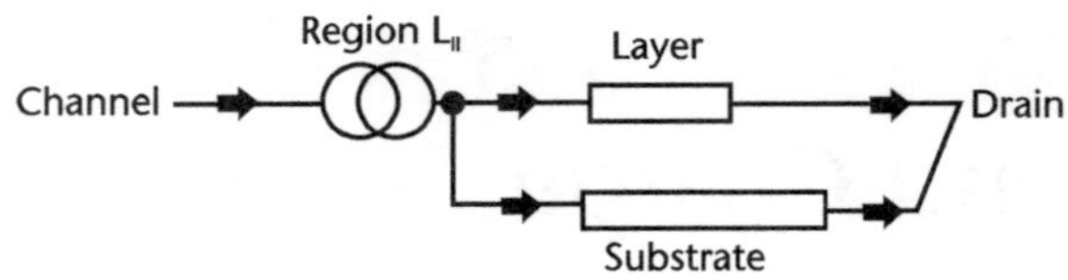

Figure 14.2 Equivalent circuit for the current paths in Figure 14.1.

2. By then, the magnitude of the current has been predetermined by the channel opening at region II.

Logically, then, although hot electrons may indeed scatter into the substrate, they do so only after the current has been predetermined. Consequently, substrate current does not create a parallel path that adds extra current over and above that passing through the channel but merely offers an alternative path along which part of the predetermined current can flow. A circuit equivalent is shown in Figure 14.2.

With this view, the current collected by the drain and the associated output conductance are determined solely by d-lift and not by the mechanism put forward in [1].

Next, we examine the intricacies of region I.

Charge movement in region I is difficult to pin down into a clean one-size-fits-all mechanism (clues as to how the mode changes are to be found in macro-cell simulator results). After considering the shifting regimes, an approximation is proposed that works adequately for foretelling the trends in model elements, notably the region I branch capacitance C_{IB}.

If we take charging to mean electrons entering the depletion edge (as, for example, when the gate is made more positive) and discharging to mean electrons leaving it (typically when the gate is made more negative), the charging and discharging paths are different. Electrons to charge the depletion generally come from the source, as in Figure 14.3, whereas at least some of the electrons released by the depletion expanding to a greater depth d are swept by the channel field towards the drain.

Thus, the specter is raised that the mechanisms governing charging and discharging may not be the same.

To the extent that Figure 13.11 is representative of the x-directed field in the channel when velocity saturation just sets in at the drain end of the gate, the field in region I will be low. Therefore, the mobility will be high and the simple theory suggests that charging of the depletion edge takes place by dielectric relaxation, modified by geometrical factors associated with the channel [1]. However, for the

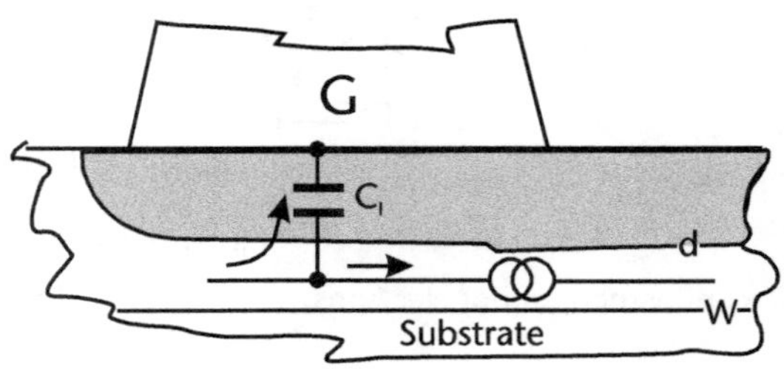

Figure 14.3 Charging of region I as represented by the capacitance C_I.

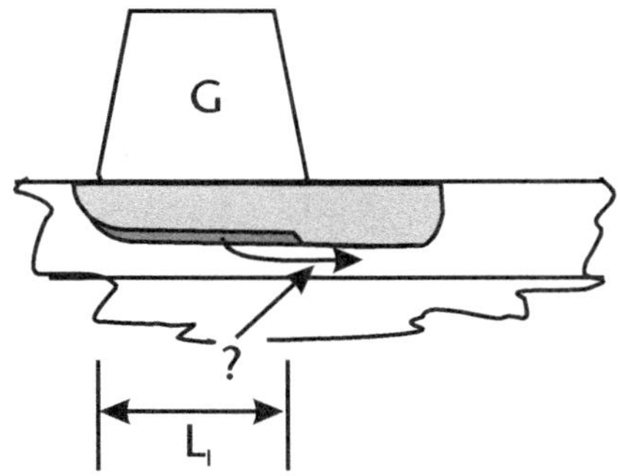

Figure 14.4 Are some of the electrons released from the depletion edge removed from the channel at v_{sat}?

inverse case of discharging, exactly what happens to the released electrons is unclear and we have to resort to speculation.

Imagining an infinitely fast but small negative pulse to be applied to the gate, before the depletion region can react the potential in the channel at the virtual source S′ will instantaneously go negative. In response, the current through the source-gate region decreases, thereby creating capacity for the electrons from the depletion edge nearby to be absorbed. To the extent that this picture is valid, the discharge point is S′, at least for the released electrons nearby, but it is unlikely to account for all the electrons released throughout region I. For those closer to the L_I–L_{II} boundary, the path shown in Figure 14.4 is the more probable, in which event the rate at which some of the newly released electrons can be swept out of the channel is restricted by v_{sat}.

Unarguably, the depletion edge is a charge storage element, but is it a capacitance? Given that a lumped component capacitor is charged and discharged by the same path, we are faced with a charge storage element that is a departure from the capacitor as we know it. The capacitance of region I, C_I in Figure 14.3, may not be charged and discharged in the same way. When it comes to representing the physical processes that occur in a FET by an equivalent circuit, however, we have no option but to adopt lumped capacitances as charge storage elements. All we can do is make sure that the connection points are the best compromise for modeling the charging and discharging paths that the lumped-component circuit allows.

Keeping in mind the aim of predicting functional trends of the elements in the reformed equivalent circuit model, we have to answer for the field (and hence electron transport) conditions in region I.

As the drain voltage is increased, E_x in region I becomes progressively larger: the electron mobility degrades so dielectric relaxation may eventually cease to be the prime mechanism of charge exchange. Instead, the regime enters a hybrid phase between dielectric relaxation and velocity saturation.

What is the limit to this process? To answer this question, we have to look for guidance at the empirical and theoretical results of Section 4.2.2 where it was shown that the turnover in C_{gc} of Figure 4.4 is due to pronounced nonlinearity in R_s, in turn brought about by high electric fields penetrating into the gate-source region. The only place these fields could have come from is the depletion extension via the channel. If large x-directed fields thereby exist in region I, the electron velocity there must be in the upper reaches of Figure 13.2, approaching v_{sat} (and at least the charging and discharging mechanisms are then in accord with one another). In any event, as noted elsewhere the boundary between regions I and II is not sharp but ill-defined, a gradual transition.

So where does this leave us? It leaves us with no hard and fast guideline to meet the objective stated at the beginning of this section. Consequently, where necessary, we have to be restrictive to make progress with formulating equations for the forms of the equivalent circuit elements, to which end we assume our device to be operating far enough above the knee for the electron velocity in region I to be v_{sat}, or nearly so. Yet again, although the foundation appears somewhat shaky, the assumption works well enough in practice.

14.2 Transreactance Current

Referring to Figure 14.5 (which is a redrawn version of Figure 6.9 in [1]) when the gate is made more negative while $V_{D'G}$ is held constant, the primary response is an increase in the depth of depletion d. There is a secondary response: the depletion extension in the gate-drain space shrinks. For the extension X to have shrunken there has to have been an inflow of electrons to infill the space that was devoid of them.

Similarly, had the polarity of the V_{GS} change been reversed, the initial outline of the space-charge layer would be as per the dotted contour and the solid line would be the end condition, in which event electrons would have been displaced from the now-expanded X region.

In either case, there is a net change in the electron charge in the gate-drain region resulting from a voltage change in the gate-source region. In circuit terms, there exists a capacitive reactance in the gate-drain branch of the equivalent circuit which is driven by the voltage applied to the gate-source branch. The stimulus is V_{GS} and the response is a reactive current flow in the branch of the network that is orthogonal to that bearing the stimulus, hence the accurate tag of orthogonal branch current. "Transreactance current" is more descriptive, being in the same vein as transconductance current. Inclusion of this transreactance, termed "transcapacitance" by Root and Hughes [2], is essential if a circuit model is to be faithful to the physics. Understandably, "transcapacitance" is the label that has stuck.

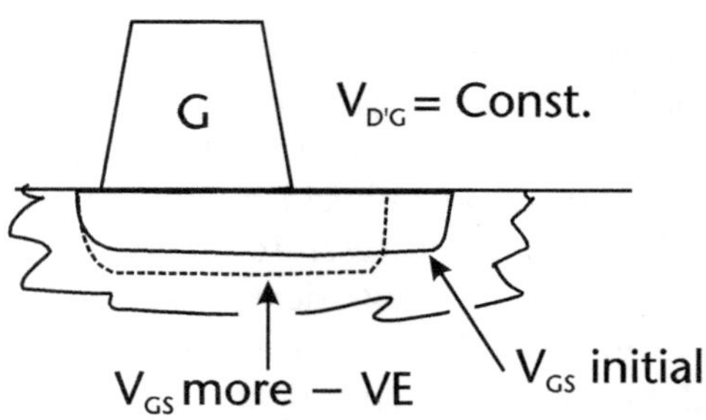

Figure 14.5 Sketch to illustrate the changes in d and X in response to a change in V_{GS} with $V_{D'G}$ held constant. (*After:* [1].)

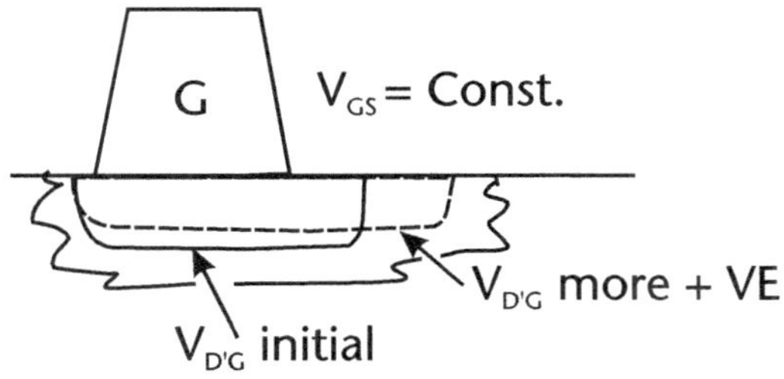

Figure 14.6 Illustrating the changes to *d* and *X* when V_{GS} is held constant and $V_{D'G}$ is changed.

There is an exactly parallel phenomenon which occurs when the drain voltage is changed while the gate-source voltage is maintained constant (Figure 14.6).

When $V_{D'G}$ is increased positively, the primary response is an increase in X; the secondary response is a decrease in d, the phenomenon analyzed as d-lift in Section 13.7 (and, naturally, the converse applies if the sense of all changes is reversed). Once again, there is a current response in one region, or branch, following a voltage stimulus in the other.

The upshot is that, when revising the FET and HEMT model, two transcapacitances must be included to represent the two orthogonal branch charge changes that occur in a real device: one in the gate-drain branch, and one in the gate-source branch.

14.3 Charge Conservation

Charge conservation is a vexed issue: some circuit simulators prefer the nonlinear voltage-dependent charge storage properties of a FET to be specified as branch charges or nodal charges rather than capacitance. Charge functions must be conservative; if they are not, an erroneous accumulation of charge will occur somewhere with the inevitable outcome of invalid results if not outright convergence failure. Furthermore, there must be at least two of them, typically gate charge and drain charge. As already mentioned in Chapter 10, various schemes have been proposed for forming gate-charge and drain-charge functions that are conservative [3, 4].

The dilemma is that the practical device behaves differently: there is only one conservative nodal charge and that is the gate charge. The conservative gate charge, consisting of electrons, is balanced by an equal but opposite charge in the depletion region made up of positive ions. It is a single, cohesive, spatial distribution that cannot be subdivided into, for example, separate source and drain charges. Our task now is to substantiate this assertion from a physics point of view, yet again by looking at how the device actually works.

Charge conservation as applied to a quantity $Q(V_1,V_2...)$ means, at the end of a closed cycle in which $(V_1,V_2...)$ vary but eventually return to their starting values, that Q will return to exactly the same value as it had to begin with. Expressed another way, if all the changes δQ that Q experiences throughout the cycle are summed the result will be zero. Applying this principle to an example representative of practice allows us to see if it holds, where it holds and, importantly, if it does not hold, how and where it fails.

The tools needed to construct an example are Figure 13.1 and caption, together with (13.3) and (13.4), repeated here for easy reference:

$$V_{D'G} \simeq \frac{qNd}{2\epsilon}\left(X + \frac{X^2}{L_G}\right) \qquad V_{GC} \simeq -\frac{qNd^2}{2\epsilon}\left(1 + \frac{2X}{L_G}\right)$$

Note the reinstatement of the negative sign in the expression for V_{GC} to keep everything correct.

We opt for a material and structural specification, bare as it is, appropriate to a GaAs power FET, which in operation produces a space-charge layer extension about twice the gate length. Such exaggerated behavior better illustrates the issues at stake. So:

$$L_G = 0.5\ \mu\text{m};\ N = 5 \times 10^{22}\ \text{m}^{-3} = 5 \times 10^{16}\ \text{cm}^{-3}$$

with a voltage cycle as in Table 14.1.

For each case, the equations for V_{GC} and $V_{D'G}$ can be solved simultaneously to find the values of d and X.

Sparing the mathematical detail (which is not difficult but somewhat drawn out), the answers are as shown in Table 14.2.

A diagram for each case appears in Figure 14.7.

The next question is how to process these results in a form that will tell us what we want: is charge conserved, or not, in a way suitable for a circuit simulator?

Suppose that we divide the space-charge layer into two separate regions as in Figure 14.8.

Self-evidently, in many ways this is a logical partition. The distinguishing feature is the split into a block of charge Q_I in the shadow of the gate, and a block of charge Q_{II-III} immediately identifiable as the space-charge layer extension (here we are assuming region II in Figure 13.1 to be small relative to L_G).

For each case in Figure 14.7, the two charges can be calculated as in Table 14.3. Also shown are the changes from row to row. To save carrying awkward numbers, the tabulated charges are simply the product of the tabulated values of d and L_G and d and X, multiplied by 100.

Of major significance is that the sum of the changes undergone by both Q_I and Q_{II-III} is zero. That is, both Q_I and Q_{II-III} are conservative. On the face of it therefore, herewith a structured basis for implementing a conservative charge-based model with the requisite two branch or nodal charges. Unfortunately, it is impractical:

Table 14.1 Short Cycle of Voltage Changes for the Illustrative Example

	V_{GC}	$V_{D'G}$
Start	−1	3
Then	−2	3
Then	−2	10
Then	−1	10
Finish	−1	3

All values of V_{GC} and $V_{D'G}$ are in volts.

Table 14.2 Solutions for Size of the Depletion Layer

V_{GC}	$V_{D'G}$	d	X
−1	3	0.1	0.45
−2	3	0.16	0.33
−2	10	0.09	1.04
−1	10	0.074	1.17
−1	3	0.1	0.45

All values of d and X are in micrometers.

without knowing d and X there is no way of deciding how to divide up the total charge Q in the depletion region, even if Q itself could be determined. Nonetheless, the particular partition of Figure 14.8 is of theoretical importance inasmuch as it establishes the fact that, in principle (but not practice), a conservative charge formulation is possible.

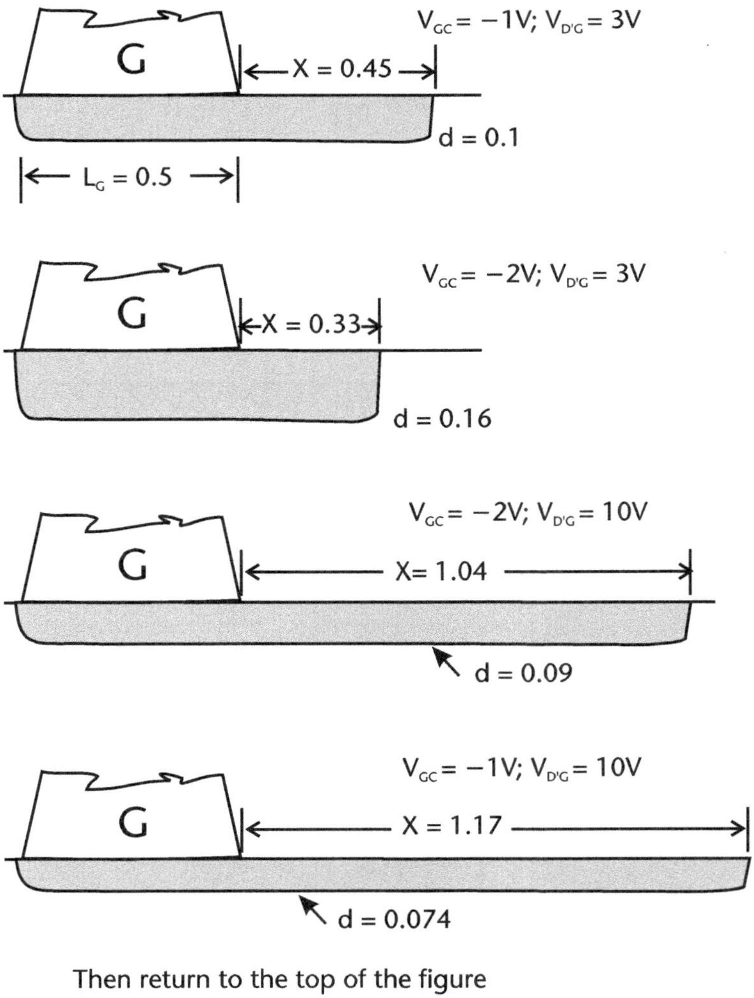

Figure 14.7 Depletion dimensions for a cycle of voltages in the numerical example.

Table 14.3 Values for the Charge Contained Within the Two Areas Delineated in Figure 14.8

V_{GC}	$V_{D'G}$	d	X	Q_I	$Q_{II\text{–}III}$	δQ_I	$\delta Q_{II\text{–}III}$
−1	3	0.1	0.45	5	4.5		
−2	3	0.16	0.33	8	5.28	+3	+0.78
−2	10	0.09	1.04	4.5	9.36	−3.5	+4.08
−1	10	0.074	1.17	3.7	8.66	−0.8	−0.7
−1	3	0.1	0.45	5	4.5	+1.3	−4.16
						$\Sigma = 0$	$\Sigma = 0$

Charge units are arbitrary (see text).

We now turn to a different division that follows closely the way in which charge changes occur in the d and X edges of an operating FET (Figure 14.9).

In the figure, δQ_d represents the change in charge in the d edge of the depletion region, while δQ_X signifies the change in the X edge. All the information needed to study these changes as the device responds to the voltage cycle in Table 14.1 is already available from Figure 14.7. Calculating the changes requires a little more thought than previously.

To make clear how to compute the charge changes from case to case, a separate drawing like Figure 14.10 for each change event helps. Figure 14.10 pertains to the first change event, that is, $V_{GC} = -1\text{V} \rightarrow -2\text{V}$ with $V_{D'G}$unchanged at 3V.

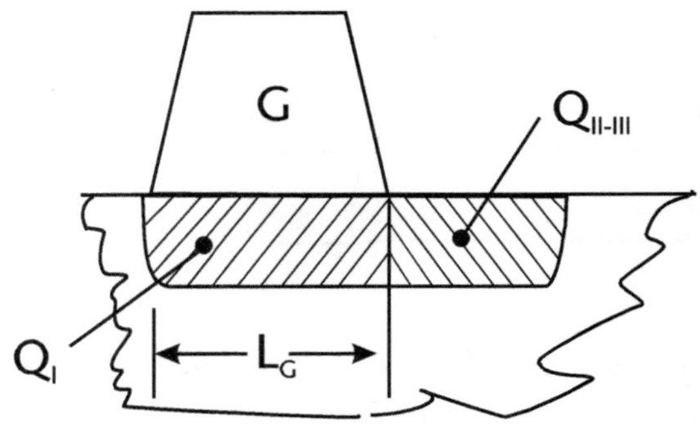

Figure 14.8 Division of the depletion region with geometrically fixed boundaries.

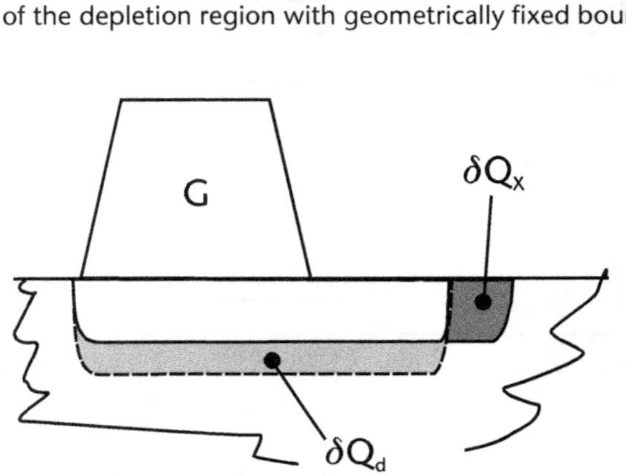

Figure 14.9 A different two-regional split of charge.

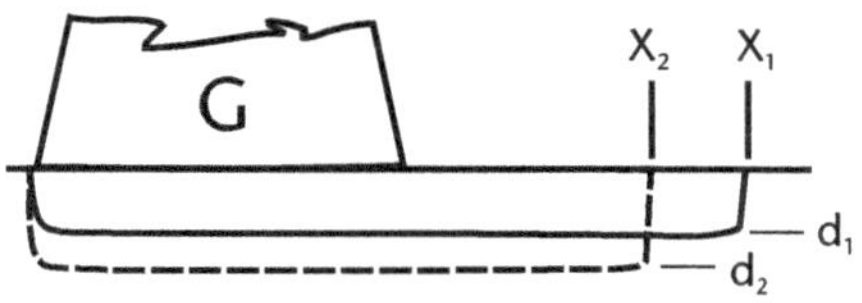

Figure 14.10 Sketch of the first change to the space-charge layer to help to calculate the changes δQ_d and δQ_X.

The continuous line represents the initial contour of the depletion, the dotted line represents the new contour.

For this, the first case, the positive charge gained in the d edge is

$$\delta Q_d = (d_2 - d_1)(L_G + X_1) - (d_2 - d_1)(X_1 - X_2)$$
$$\delta Q_d = (d_2 - d_1)(L_G + X_2)$$

while the charge lost in the X edge is

$$\delta Q_X = (X_1 - X_2)d_1$$

Using the same arbitrary units for charge as before, the results are as given in Table 14.4. Each of the other case-by-case changes can be calculated similarly.

What we now see is that, for each of the two individual charges taken separately, the sum of the changes in each does not come to zero. Therefore, neither is, nor offers the prospect of becoming, a conservative charge function. Note, however, that $(\delta Q_d + \delta Q_X)$ is zero, meaning that charge in the depletion region overall is conserved (i.e., the gate charge is conserved).

As we chose, arbitrarily, a charge gain to mean an increase in positive charge, we could say that, with the cycle of events analyzed, positive charge is pumped from the source to the drain. Had we run the voltage sequence in reverse, the conclusion would have been that the electron charge is pumped from the source to the drain.

The charge change approach involving δQ_d and δQ_X recognizes moving boundaries as defining the two separate regions in which movement of charge occurs. Being concerned with charge changes in response to voltage changes is, by nature,

Table 14.4 Charge Changes from Case to Case in Figure 14.7 Calculated According to the Division of Figure 14.9

V_{GC}	$V_{D'G}$	d	X	δQ_d	δQ_X
−1	3	0.1	0.45		
−2	3	0.16	0.33	+4.98	−1.2
−2	10	0.09	1.04	−5.81	+6.39
−1	10	0.074	1.17	−2.464	+0.962
−1	3	0.1	0.45	+2.47	−5.238
			Σ =	−0.824	+0.824

A + sign signifies an increase in positive charge.

a capacitance-oriented approach and is therefore closely aligned with what can be determined about a device empirically. It shows how charge is shuffled between the two edges at d and X in such a manner that, at the end of a closed cycle of voltage changes, there can be a net gain of charge in one region exactly balanced by a net loss in the other. In turn, that means, first, the individual regions are not charge conservative; second, the total charge in the depletion region, and therefore the charge on the gate, is the one and only charge that is conservative. Imbalance reflects the reality that the depletion region can be charged through one node (say, the source S) but discharged through the other (i.e., the drain D).

The $\delta Q_d, \delta Q_X$ picture is pretty much what happens in practice. Consequently, to be true to life, a model has to recognize only a single conservative nodal charge, the gate charge. Any attempt to conjure a second nodal (or branch) charge that is conservative axiomatically conflicts with the aim of preserving compliance with the way FETs and HEMTs actually work. Hence, any time-domain or harmonic balance simulator that requires a model specified in terms of conservative nodal (or branch) charges forces upon us a nonphysical configuration. Instead, what we want is a nonlinear solver that uses only capacitances of the form $C_{ij}(V_{GC}, V_{D'G})$ or, more familiarly, $C_{ij}(V_{GS}, V_{DS})$, and leaves the modeling engineer to ensure that such capacitances form a conservatively linked set in the sense that they are all derivatives of a common charge, or physically identifiable parts of that charge, the depletion charge Q.

Section 10.4 deals with the general conditions that capacitive elements in the reformed equivalent circuit model should satisfy.

14.4 Charge Storage by Pure Delay τ

Inescapably, charge storage by pure delay has to be confronted, as unfamiliar as the concept might be. Examining the problem from as many angles as possible is worthwhile in the expectation that one or another will make clear how it is different from charge storage by capacitance and why the second is no substitute for the first. Some of the contrasts have already been shown in Section 13.8. Here, we try from a complementary viewpoint. While still basing most of the argument in the time domain (τ is, after all, a time quantity) relating the findings to the frequency domain is obligatory for those whose preference is to think in terms of waves.

In a network containing lumped components that introduce into the response $R(t)$ some sort of delay relative to the input or stimulus $S(t)$, $R(t)$ becomes finite as soon as $S(t)$ is applied. Also, $R(t)$ has a shape or form that is changed from $S(t)$. The change may be small or gross, but it is different nevertheless. Neither of these two properties is a characteristic of a pure delay element, introducing a delay τ. Instead:

- $R(t)$ remains zero until the time t attains the value τ.
- At time $t = \tau$ and thereafter, $R(t)$ takes the form of an exact replica of $S(t)$ without distortion.

A simple example helps visualize the difference. Figure 14.11 shows, on the left, a schematic of a network containing a single capacitance-resistance (CR) delay and,

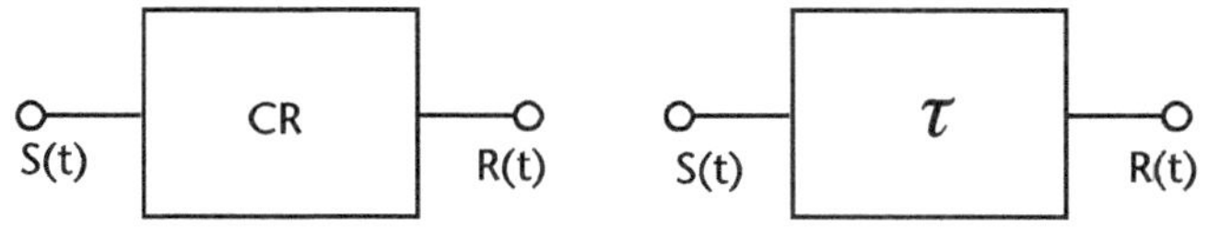

Figure 14.11 A CR delay network and a network with pure delay.

on the right, a schematic for a pure delay element. A step function stimulus as in Figure 14.12 produces the responses sketched in the lower part of the figure. They are quite dissimilar; moreover, the way that the charge is stored, which is centrally what interests us, is very different as well (Figure 14.13).

The simple example is just the beginning. When a device is driven hard, both $S(t)$ and $R(t)$ are large amplitude and usually nonlinear. (We adopt the shorthand "large-signal" to indicate the condition even though the waveforms may be far removed from signals.) Denoting a general large-signal current by $I(t)$, we are concerned with the charge being stored by the difference between $I(t)$ and its delayed version $I(t - \tau)$; thus:

$$\delta I(t) = I(t) - I(t - \tau) \tag{14.1}$$

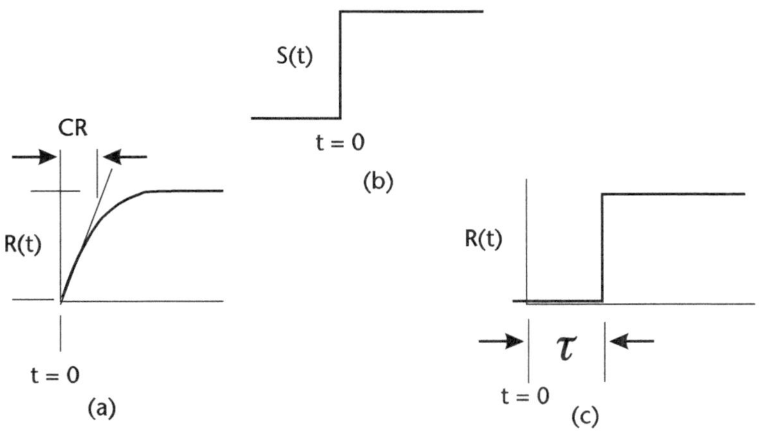

Figure 14.12 Illustrating the different responses of the delay networks in Figure 14.11.

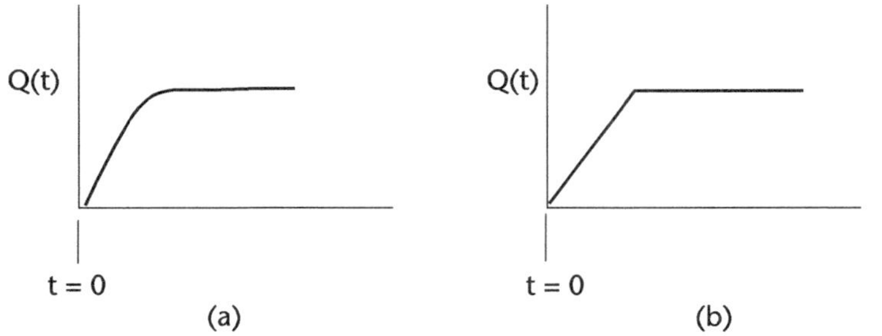

Figure 14.13 Time-dependent charge storage in the two networks of Figure 14.11.

In terms of charge,

$$I(t) = \frac{\partial Q(t)}{\partial t} \qquad I(t-\tau) = \frac{\partial Q(t-\tau)}{\partial t}$$

However,

$$\frac{\partial Q(t-\tau)}{\partial t} = \frac{\partial Q(t-\tau)}{\partial(t-\tau)}\frac{\partial(t-\tau)}{\partial t} = \frac{\partial Q(t)}{\partial(t)}\frac{\partial(t-\tau)}{\partial t}$$

So

$$\delta I(t) = \frac{\partial Q(t)}{\partial(t)}\left[1 - \frac{\partial(t-\tau)}{\partial t}\right]$$

or

$$\delta I(t) = I(t)\left[1 - \frac{\partial(t-\tau)}{\partial t}\right] \tag{14.2}$$

Equation (14.2) is just an unusual way of rewriting (14.1) for the charging current in the large-signal case. The reason for it is to contrast with the equivalent expression in the frequency domain in the small-signal limit.

If, instead of being large-signal, the excitation and the stimulus were small-signal and sinusoidal, namely,

$$i(t) = i_0 e^{j\omega t} = j\omega Q(t)$$

then

$$\delta i(t) = i(t) - i(t-\tau) = j\omega Q\left(1 - e^{-j\omega\tau}\right)$$

In the limit of $\omega\tau \ll 1$,

$$\delta i(t) \simeq j\omega Q(1 - 1 + j\omega\tau)$$

or

$$\delta i(t) \simeq i(1 - 1 + j\omega\tau)$$
$$\delta i(t) \simeq i(j\omega\tau) \tag{14.3}$$

We now have (14.3) to compare with (14.2), but, before we do, if $i = g_{m0}v$ (where v is a small-signal voltage), then

$$\delta i(t) \simeq i(j\omega\tau) = j\omega g_{m0}\tau v$$

and we come back to the idea that the delay can be represented by a capacitance

$$C = g_{m0}\tau \tag{14.4}$$

Note that when in the FET or HEMT this relationship is applied to region II–III as defined in Figure 13.22, it may be written

$$C_{\text{II-III}} = g_{m0}\tau_{gm}$$

Returning to the main argument, the key comparison is between (14.2) and (14.3):

$$\delta I(t) = I(t)\left[1 - \frac{\partial(t-\tau)}{\partial t}\right]$$

$$\delta i(t) \approx i(1 - 1 + j\omega\tau)$$

Trying to assert that these two are equivalent is tantamount to saying

$$\frac{\partial(t-\tau)}{\partial t} \approx (1 - j\omega\tau)$$

Imposing a small-signal sinusoidal condition on the left side requires a stretch of the imagination to arrive at equivalence (($t - \tau$) has a slope of unity and intersects the vertical axis at $-\tau$). The left side is the correct form for large-signal analysis and cannot be supplanted by the right side. Put another way, using (14.4), or any capacitance formulation that embodies it, even implicitly, is invalid for large-signal, nonlinear modeling. C_{gc} in the standard model is the prime example.

While the foregoing is but a restatement of arguments already made, it is confirmation that a large-signal model must be based upon (14.1). Given that, other deductions of a general nature follow.

Looking to the wider implications of Figure 14.12, not only will the amplitude response be different between the two cases, but in the frequency domain so will the phase. Phase has an importance that in some applications has to be teased out from a raft of performance specifications, with amplification of a digital bit stream being one example. In the time domain, a nonperiodic stream of bits amounts to a train of pulses with the interpulse period being random (or pseudo-random). In the frequency domain, the pulse train can be decomposed into sinusoidal waves such that when added together, reversing the process, the waves reconstitute the pulse pattern. If the phases of the constituent waves become shifted during amplification, the result of the summation may look quite unlike the desired train of pulses. Thus, phase integrity is in some applications non-negotiable, which means that the model itself should not be a source of error.

As always, a numerical example helps to gauge the need. If the total delay through a FET is, say, 7 ps, then at the seemingly low frequency of 2.4 GHz, the phase shift is 6°; for the second harmonic, it is 12°, and it is 18° at the third harmonic. From this bare example, phase errors arising from numerical errors in the delay can be quickly quantified at any frequency (e.g., a 2 ps error leads to a phase error of 19° at 26.5 GHz).

When $I(t)$ is an arbitrary function of time, trying to depict how charge is stored becomes messy, especially when τ is a function of the voltages, $\tau(V_{GS},V_{DS})$ For the straightforward case of a periodic waveform and a fixed delay τ, Figure 14.14 shows

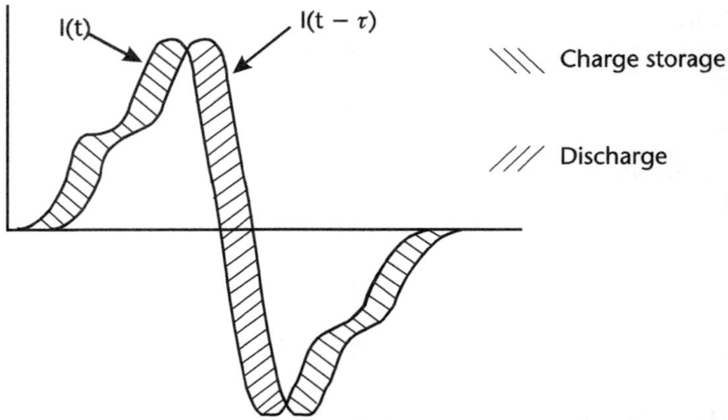

Figure 14.14 Storage and release of charge throughout the cycle of a periodic current subject to pure delay.

how charge is stored and released cycle by cycle. From this simple picture, what happens in a FET with a delay $\tau(V_{GS},V_{DS})$, driven with arbitrary large signals, has to be imagined. Suffice to say, it is a challenging task for a circuit simulator.

Pure delay governs the storage of charge in region II–III of an operating FET or HEMT as discussed in Section 13.8.

14.5 Resistances R_S and R_I

R_S is often treated as the bulk resistance of the source-gate access region with little attention paid to elementary physics governing the region. Yet even from the dc $I(V)$ characteristics, a simple deduction can be made. Taking, for example, Figure 3.5 allied with Figure 13.21, where the practical $I_D(R_S + R_D)$ drop is 1V and if half is due to R_S, then $I_D R_S = 0.5$V. If the length of the access region is 0.5 μm, then the average electric field strength is 10 kVcm^{-1}, by which time (for GaAs devices), R_S has become nonlinear. For this reason alone, when fitting an equivalent circuit model to measured S-parameters, R_S should not automatically be treated as a fixture. In some devices, when driven hard, the electric field in the source-gate space can take on values large enough to render R_S highly nonlinear, as in the PHEMT of Figure 4.16.

We now have to deal with R_I, frequently referred to as the intrinsic channel resistance. Continuing with the theme begun in Section 14.1, region I can always be charged via the left edge of the depletion region, at least until the extreme is reached of the x-directed field becoming so large that R_S becomes nonlinear. If we acknowledge that charging can also take place by electrons already in the channel, R_I in some form is needed in the equivalent circuit model.

Unarguably, the part of the conducting channel in region L_I has bulk resistance that can be readily calculated analytically as an approximation and that is available with greater accuracy from a device simulator such as the result of Figure 4.2. Whether that is a good enough reason to include it in a circuit model has been a topic of debate, it being variously proposed that the R_I needed in the circuit model should be $R_{bulk}/3$, $1/g_{m0}$, $1/3g_{m0}$ or not needed at all. If it is to feature, its purpose has to be clear. To that end, a major consideration is the charging (and discharging) time of the depletion edge in region I: it should be finite, implying that in some form or another R_I should be included.

The short answer is that its value is best left to practical determination. For all that, the finer points of the principle contributing to the empirical value are worth stating, if only for passing interest. The following circuitous argument suggests that charge exchange with the depletion edge is either relatively fast or relatively slow depending upon position under the gate, as surmised in Section 14.1.

We have already seen in Section 14.1 how electrons involved in both charging and discharging the depletion edge at d pass through the channel in region I. It came to light also that the charging and discharging paths, at least for some electrons, are different. Consequently, whatever we choose for R_I cannot theoretically be right for both, so, empirically, we will end up with the best compromise and the least error.

Figure 14.15 is taken from [1]. It is a circuit representation of the space-charge layer and channel in region I, segmented into elementary cells. The simplifying assumption made there was that all cells had the same elemental capacitance ΔC and the same elemental resistance ΔR. Recognizing that the field E_x is not uniform but increases with x suggests the incremental resistances shown in Figure 14.15 are not equal but graded, to which end we now write δR and, to keep in step, δC.

Thus, electrons to charge the depletion edge, coming mainly from the left, pass through resistances of different value from those in the discharge path, mainly to the right.

Consider a cell of length δL at position x. The potential across it is

$$\delta V = E_x(x)\delta L$$

The current through the cell is

$$I = qN(W - d)v(x)Z_G$$

where Z_G is the gate width. Invoking the approximations that we have already tried to justify, flat d and $v \simeq v_{sat}$ throughout region I, the elemental resistance is

$$\delta R = \frac{E_x(x)\tau_{\delta L}}{qN(W - d)Z_G}$$

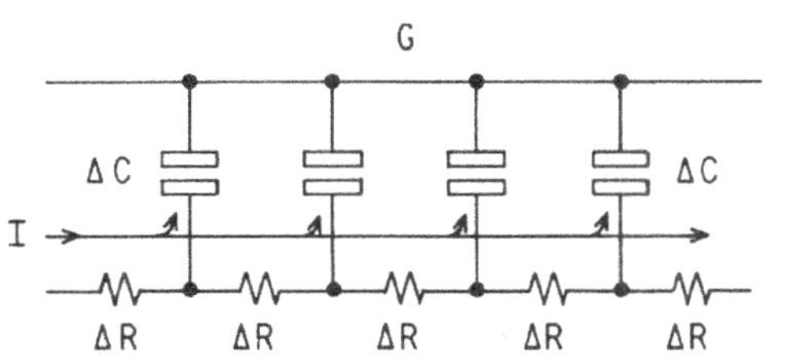

Figure 14.15 Segmentation of region I into elemental cells.

where $\tau_{\delta L}$ is the transit time through δL. Casting the equation thus is a useful way of being reminded that there are fundamental physical processes at work on a short timescale, and that we should be wary of constructing a theory or explanation involving even shorter timescales. In the present context, the limit is the momentum relaxation time τ_m, typically a few tenths of a picosecond.

Multiplying δR by the cell capacitance $\delta C = C_I \delta L / L_I$:

$$\delta C \delta R = \frac{E_x(x)\tau_{\delta L}}{qN(W-d)Z_G} \frac{C_I \delta L}{L_I}$$

According to a simple theory [1], $\delta C \delta R$ is close to the dielectric relaxation time, which for a semiconductor doped to 10^{17} cm^{-3}, for example, comes to $\approx 10^{-14}$s. The problem is that such a number is way smaller than τ_m, it being the case that the proponents of dielectric relaxation theory forgot all about the inertial effects of momentum [5]. Thus, to retain any sense of physical validity, the cell size has to be chosen such that $\delta C \delta R$ is at least a few tenths of a picosecond. Substituting $\delta L = v_{\text{sat}} \tau_{\delta L}$ and rearranging:

$$\frac{\delta C \delta R}{\tau_{\delta L}} = \frac{E_x(x)\tau_{\delta L}}{qN(W-d)Z_G} \frac{C_I v_{\text{sat}}}{L_I}$$

If we take $\tau_{\delta L} = 1$ ps, then to retain physical meaning, the left side must be a fraction of at least a few tenths. By way of specific example:

$C_I = 0.2$ pF

$v_{\text{sat}} = 10^7$ cms^{-1}

$N = 10^{17}$ cm^{-3}

$W - d = 0.05$ µm

$Z_G = 200$ µm

$L_I \simeq L_G = 0.5$ µm

With these numbers

$$\frac{\delta C \delta R}{\tau_{\delta L}} = 2.5 \times 10^{-7} E_x(x)$$

Referring to Table 14.5 with the assumed value $\tau_{\delta L} = 1$ ps, the first entry gives a charge transfer time of $\delta C \delta R = 0.25$ ps, a value that we can just about justify as reasonable as it is comparable with, but not smaller than, τ_m. It implies charge transfer is by dielectric relaxation (although subject to the inertial limit imposed by the momentum relaxation time).

At 40 kVcm^{-1}, the dielectric relaxation time and the transit time through the cell are equal. For higher field strengths, charge transfer in the cell is governed by the transit time. Because the field increases with position along the channel towards the drain, the inference is that at the source end of the gate charge transfer is by dielectric relaxation (although subject to the inertia reservation), but, progressively, the mechanism shifts to charge movement limited by velocity saturation towards

Table 14.5 Charge Transfer Time for R_I Cells (Example Only)

$E_x(x)$	$\delta C \delta R / \tau_{\delta L}$
10 $kVcm^{-1}$	0.25
25 $kVcm^{-1}$	0.625
40 $kVcm^{-1}$	1
100 $kVcm^{-1}$	2.5
200 $kVcm^{-1}$	5

the drain end. Straightforwardly, it is the degradation in the chordal mobility, v/E, with increasing E, which underlies these findings.

The foregoing argument is illustrative, not definitive, and is intended only to give credence to the idea that charging and discharging of the depletion edge is a far-from-simple, uniform process. None of the numbers have special significance; they are simply examples.

Accepting the stratagem of bypassing convoluted theoretical considerations by relying on fitting to microwave data to retrieve $R_I(V_{GS},V_{DS})$, there is still the question of where in the circuit model it should be connected. If charging and discharging of the depletion edge occur by different paths, as appears to be the case, a circuit model cannot represent correctly these physical processes of charge and discharge. A lumped capacitance, in the present context C_I, is charged and discharged by the same path. Thus, whatever connection we choose will necessarily be a compromise.

References

[1] Ladbrooke, P. H., *MMIC Design: GaAs FETs and HEMTs*, Norwood, MA: Artech House, 1989.

[2] Root, D. E., and B. Hughes, "Principles of Nonlinear Active Device Modeling for Circuit Simulation," *32nd IEEE MTT ARFTG Conference Proceedings*, Tempe, AZ, December 1988, pp. 3–24.

[3] Pedro, J. C., et al., *Nonlinear Circuit Simulation and Modeling*, Cambridge, U.K.: Cambridge University Press, 2018.

[4] Aaen, P. H., J. A. Plá, and J. Wood, *Modelling and Characterization of RF and Microwave Power FETs*, Cambridge, U.K.: Cambridge University Press, 2007.

[5] Ladbrooke, P. H., and J. E. Carroll, "Dielectric Relaxation as a Limit on Transistor Switching Speed," *Electron. Lett.*, Vol. 32, No. 16, August 1, 1996, pp. 1511–1513.

CHAPTER 15

Charge Storage

15.1 Revisiting Capacitance

In this chapter we delineate the distinct regions within a FET or HEMT where charge is stored and then look at how each can be accounted for in a circuit model. Unfamiliar concepts arise, like charge that is stored via one path yet discharged through another, charge storage that cannot be modeled by capacitance at all, and negative capacitance. A sound appreciation of each is required to make as good a job as possible of formulating a revised equivalent circuit model. We begin by reviewing some essential principles and definitions having to do with the notion of capacitance.

Two key principles must be observed when deciding whether the storage and release of charge can be represented by capacitance:

1. A physical effect involving the flow of displacement current or reactive current can properly be represented as a lumped capacitance if, and only if, the current flow path for charging and discharging is exactly the same path.
2. The path of the current flow referred to in (1) determines where in the equivalent circuit model the capacitance representing that current flow should be located. If condition 1 does not hold because the paths by which charge is stored and released are different, and if at the same time there is no alternative to a lumped capacitance, we somehow have to decide on the best compromise placement of the capacitor in the model.

From these two principles, decisions can be made as to how many capacitive elements need to be included in the model to represent correctly the dynamic behavior of FETs and HEMTs.

In addition, we need to use consistent terminology to avoid confusion of what constitutes charging and discharging of the depletion layer and what constitutes negative versus positive capacitance. The convention is arbitrary, but we need a clear understanding of statements like:

- Charging of the depletion layer;
- The depletion layer discharges as a result of . . . ;
- A flow of negative capacitive current;
- The capacitance is positive.

Provided that the convention is used consistently and also that the end result is a capacitive current flowing in the same direction in the model as in the physical device, the naming convention is open to choice.

In this book, the phraseology is chosen to be descriptive of what electrons do, rather than descriptive of the space charge conditions that result from such electron movements. Thus, the phrase "charging of the depletion layer" refers to the act of electrons moving into the edges of the depletion layer (even though the result is neutralization of the positive ionic space charge in those regions into which the electrons have now moved).

Correspondingly, the phrase "discharging of the space charge layer" describes removal of electrons from the edges of the depletion layer (even though that process uncovers positive ionic charge there and so increases the total positive charge bounded by the depletion layer).

To help understand the fact that both positive and negative capacitance can arise from such charge movements in FETs and HEMTs, a brief review of the concept of positive capacitance is warranted.

For an ordinary fixed metal plate capacitor, the application of a voltage results in an increase in the negative (electron) charge on the plate to which the negative is applied, and an increase in the positive charge on the plate to which the positive polarity is applied. The sameness of sign of the charge change and the polarity of the voltage on either plate suggests a positive value for the capacitance, a long-established and familiar concept. However, it is a property of simple 1-D electrostatics and does not mean that the same must happen in 2-D systems like FETs and HEMTs.

In FETs and HEMTs, positive capacitance arises in exactly the manner just defined: a positive change in charge appears at the edge of the depletion region at, or closest to, the terminal of the device to which positive potential is applied.

To cite an example of positive capacitance, imagine a 1-D Schottky diode on an N-type layer. A negative voltage applied to the metal results in an expansion of the depletion region in the N-layer. The additional positive ionic charge gives rise to an electric field, which induces a negative charge of extra electrons on the metal Schottky contact. From beginning to end, a negative voltage applied to an electrode has led to negative charge on it: positive capacitance.

Unlike the simple metal plate capacitor or the 1-D Schottky diode, in the 2-D HFET, negative capacitance arises as well. It is a vital concept with which to get to grips if the functioning of FETs and HEMTs is to be understood, and essential to constructing a circuit model compliant with the physics.

According to our definitions, negative capacitance results from the charging of the depletion layer boundary at, or close to, the terminal to which positive polarity of potential is applied. Charging of the depletion edge boundary, as we have mentioned earlier, means that electrons move into that boundary. So, if the number of electrons increases at a boundary closest to the terminal to which a positive potential was applied, we have negative capacitance. Equivalently, negative capacitance arises if the application of a positive potential to one (of two) electrodes results in an increase in positive charge on the other electrode.

There are two places along the depletion edge contour in an active device that give rise to negative capacitance.

The first is the decrease in the depletion depth d, which, in the v_{sat} regime, results from the application of a positive increment in $V_{D'G}$, what we call d-lift. An influx of electrons into the d edge implies, according to the principle of charge balance, a positive change in the charge induced on the gate metal at the gate-semiconductor interface (i.e., the application of a positive potential to the drain results in a positive charge change on the gate, which therefore constitutes a negative capacitance).

The second is the decrease in the depletion layer extension, X, resulting from a negative increment of potential being applied to the gate, what we might call X pull-back (the complement being X push-out). An influx of electrons into the X edge implies, according to the principle of charge balance, a positive change in the charge induced on the gate metal at the gate-semiconductor interface (i.e., the application of a negative potential to the gate results in a component of positive charge change there, which therefore constitutes a negative capacitance).

Both these changes are summarized in Figure 15.1.

By means of diagrams like these, examining the way the space-charge layer changes in shape and size in response to changes in the external voltages highlights the regions of charge change that need to be accounted for in the equivalent circuit model. The aim now is to work out whereabouts in the model the charge storage elements (capacitive or not) should be placed, and to develop a relationship for the functional form each should take. To do so, we need the equations already established for the gate-channel voltage, the gate-drain voltage, and the relationship between δd and δX from the d-lift section (these equations originally appeared in Chapter 13):

$$V_{D'G} \simeq \frac{qNd}{2\epsilon}\left(X + \frac{X^2}{L_G}\right)$$

$$V_{GC} \simeq \frac{qNd^2}{2\epsilon}\left(1 + \frac{2X}{L_G}\right)$$

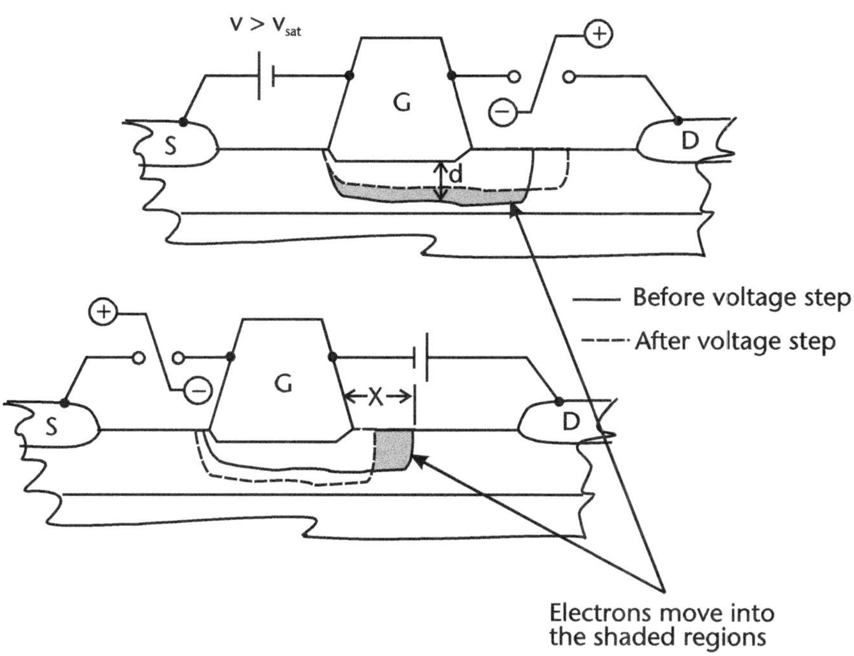

Figure 15.1 Charge changes constituting negative capacitance.

(Note: for consistency with the polarity inferred by the subscript notation, V_{GC} should strictly be written as a negative number. We continue to drop the negative sign for ease and take it as implied.)

$$\delta d = -d\frac{\delta X}{L_G}$$

As an aide-mémoire, the negative sign merely signifies that an increase in X leads to a decrease in d.

For Section 15.3.3, we will also need the moderated version,

$$\delta d = -\alpha d\frac{\delta X}{L_G}$$

where α has a value of less than unity, typically around 0.5.

We begin with the lower of the two diagrams in Figure 15.1, where the gate voltage changes but the drain voltage remains unchanged.

15.2 When V_{GS} Changes

In contrast to the title, which relates to a terminal or extrinsic voltage, analysis proceeds in terms of the intrinsic voltages: the voltages that exist at the virtual electrodes, S′ and D′ in Figure 13.1, belonging to the intrinsic device. Apart from having to begin any analysis with the intrinsic device because that is what dictates the terminal behavior, to stop there bypasses a wealth of tedious algebra and untidy equations. Thus, the development proceeds on the basis of specifying changes in $V_{GS'}$ (which we also call V_{GC}), not V_{GS}.

Additionally, as the elements of interest are functions of both terminal (or virtual terminal) voltages, it is necessary to specify what is to be the state of the second, in this case, the gate-drain voltage $V_{D'G}$. While V_{GC} is varying, $V_{D'G}$ will be held constant.

15.2.1 The Overall Picture

Figure 15.2 shows the changes to the space-charge layer. As a minor point, the diagram of necessity shows the terminal voltages as variously being constant or changing, whereas as already said we are imposing such conditions on the intrinsic device not the extrinsic device. Thus, the voltage dispositions in the figure should be regarded as symbolic, not definitive.

Identified on the figure are three distinct locations where charge change occurs, and in each the mechanism is different. Accordingly, we treat each separately.

15.2.2 Branch Capacitance

Much of what is involved in the charging and discharging of the depletion edge in region I has been addressed in Section 14.5 in connection with R_I. Setting aside for the moment the different paths by which the region is charged and discharged, the

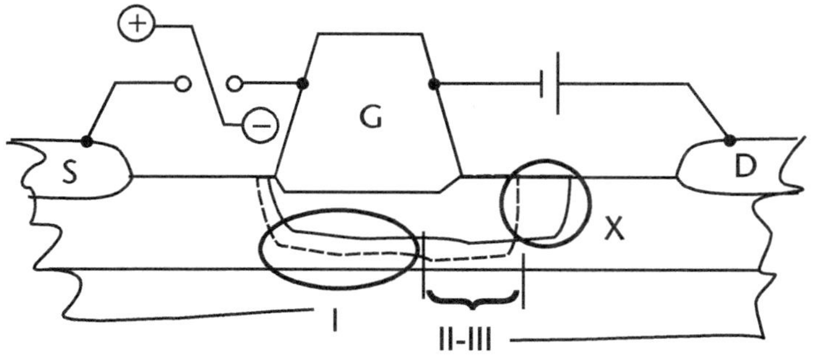

Figure 15.2 Three distinct areas where charge changes.

fact remains that it is located in the same part of the device, the gate-channel part, as that across which the voltage change is applied. Therefore, when it comes to the equivalent circuit, the charge storage element has to be located in the gate-channel branch or, less precisely, in the gate-source branch.

Despite the different paths issue, we have no option but to model region I as a capacitance, a branch capacitance C_{IB}. Because the upper electrode in the physical device through which the charging current passes is the gate, in the equivalent circuit the corresponding connection for the upper electrode of C_{IB} is the gate. That leaves the matter of where the other, or lower, electrode should be connected, rather where the C_{IB}-R_I combination should be connected. It is the one issue that cannot be resolved until all the rest of model is established, as in Part II.

Treating region I as a parallel plate capacitor,

$$C_{IB} = \frac{\epsilon L_I Z_G}{d}$$

or, taking it L_{II} that is small,

$$C_{IB} \simeq \frac{\epsilon L_G Z_G}{d} \tag{15.1}$$

Given (6.18) of [1],

$$g_{m0} \simeq \frac{\epsilon v_{sat} Z_G}{d}$$

there results

$$C_{IB} \simeq \frac{g_{m0} L_G}{v_{sat}} \tag{15.2}$$

Note that the g_{m0} in (15.2) is the intrinsic, not extrinsic, transconductance.

This is the first of the many approximate expressions that we are seeking against which to judge the efficacy of freely fitted elements when extracting the model from measurements. It tells us that C_{IB} and g_{m0} should be very similar functions.

15.2.3 Transcapacitance

For the case of $V_{D'G}$ constant and varying V_{GC}, the region encircled as X on Figure 15.2 contributes transcapacitance. Importantly:

- In contrast to a 1-D Schottky contact where greater depletion results from a negative voltage applied to the gate (classed as positive capacitance), at the X boundary in the 2-D FET of Figure 15.2, there is now less positive charge, meaning the X-change constitutes a negative capacitance.
- Charge movements at X occur in the gate-drain region, so the corresponding negative capacitance should be positioned in the gate-drain branch of the equivalent circuit.
- Because the charge movements at X take place in the branch of the equivalent circuit that is orthogonal to that where the voltage was applied, the element is a transcapacitance.
- Resisting the temptation for endless subscripts, the symbol C_N is distinguishing enough.

To decide upon the functional form of C_N, Figure 15.3 is needed.

To arrive at an expression for C_N requires a little mathematics (for those uninterested in the development, the result is given in (15.7).

$$Q_X = qNdXZ_G \tag{15.3}$$

$$C_N \overset{\text{def}}{=} \frac{\partial Q_X}{\partial V_{GC}}\Big|V_{D'G} = \frac{\partial Q_X}{\partial d}\frac{\partial d}{\partial V_{GC}} + \frac{\partial Q_X}{\partial X}\frac{\partial X}{\partial V_{GC}} \tag{15.4}$$

where $| \, V_{D'G}$ signifies $V_{D'G}$ is held constant.

Differentiating (13.4):

$$\frac{\partial V_{GC}}{\partial X} = \frac{qNd^2}{\epsilon L_G} \tag{15.5}$$

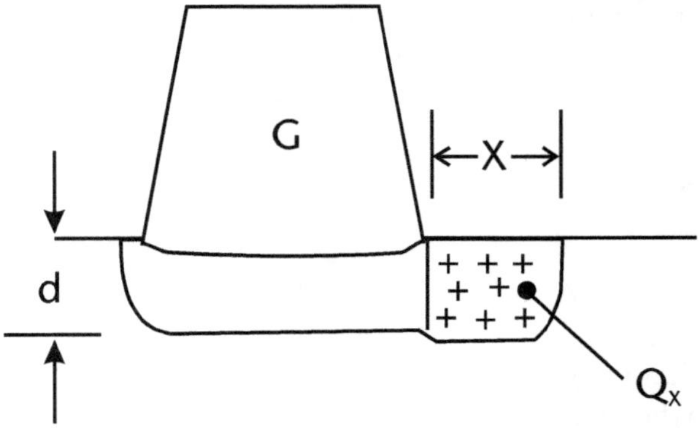

Figure 15.3 Transcapacitance C_N results from changes in Q_X.

so

$$C_N = qNXZ_G \frac{\partial d}{\partial V_{GC}} + \frac{\epsilon L_G Z_G}{d} \tag{15.6}$$

The differential on the right side is constrained by the condition that $V_{D'G}$ is constant, as follows.

From (13.3):

$$\begin{aligned} \delta V_{D'G} &= \frac{\partial V_{D'G}}{\partial d}\delta d + \frac{\partial V_{D'G}}{\partial X}\delta X \\ &= \frac{qN}{2\epsilon}\left(X + \frac{X^2}{L_G}\right)\delta d + \frac{qNd}{2\epsilon}\left(1 + \frac{2X}{L_G}\right)\delta X \end{aligned}$$

Thus, for $\delta V_{D'G} = 0$,

$$\frac{\delta X}{X} = -\frac{\left(1 + \dfrac{X}{L_G}\right)}{\left(1 + \dfrac{2X}{L_G}\right)}\frac{\delta d}{d}$$

The minus sign means X decreases when d increases.

After rearranging, we can write:

$$\frac{\delta d}{\delta V_{GC}} \rightarrow \frac{\partial d}{\partial V_{GC}} = -\frac{\left(1 + \dfrac{2X}{L_G}\right)}{\left(1 + \dfrac{X}{L_G}\right)}\frac{d}{X}\frac{\partial X}{\partial V_{GC}}$$

Using (15.5) to replace the differential term on the right side:

$$\frac{\partial d}{\partial V_{GC}} = -\frac{\left(1 + \dfrac{2X}{L_G}\right)}{\left(1 + \dfrac{X}{L_G}\right)}\frac{\epsilon L_G}{qNdX}$$

Substituting in (15.6),

$$C_N = -\frac{\epsilon L_G Z_G}{d}\left\{\frac{1 + \dfrac{2X}{L_G}}{1 + \dfrac{X}{L_G}} - 1\right\}$$

$$C_N = -\frac{C_{IB}}{1 + \dfrac{L_G}{X}} \tag{15.7}$$

Herewith an interesting but not altogether surprising result: C_N should bear a close resemblance to C_{IB}. The space-charge layer extension is related to the transconductance delay by (13.9):

$$X = 2v_{\text{sat}}\tau_{gm}$$

and, from Figure 3.28 or Figure 13.31. in the saturation regime where the foregoing analysis applies, τ_{gm} is but a moderate function of V_{DS}.

By way of numerical example, for the 0.3 μm HFET of Figures 4.18 through 4.21, Figure 15.4 is a plot of the extracted delay.

From the figure, $\tau_{gm} \simeq 2.5$ ps. Therefore:

$$X \simeq 0.5\ \mu\text{m},\ \frac{L_G}{X} \simeq 0.6$$

$$C_N = -0.6C_{IB}$$

which gives some idea of the magnitude to be expected.

Assembling the findings of this section and those of Section 15.2.2, the contributions to the equivalent circuit so far are shown in Figure 15.5.

15.2.4 Branch Charge Storage by Pure Delay

In the region marked II–III in Figure 15.2, current emerges at the drain end of the depletion region delayed by τ_{gm} relative to the channel current which entered at the left end. An instantaneous change in the current entering the region ripples through $L_{\text{II–III}}$ at velocity v_{sat}, so charge is stored progressively in the depletion edge in the

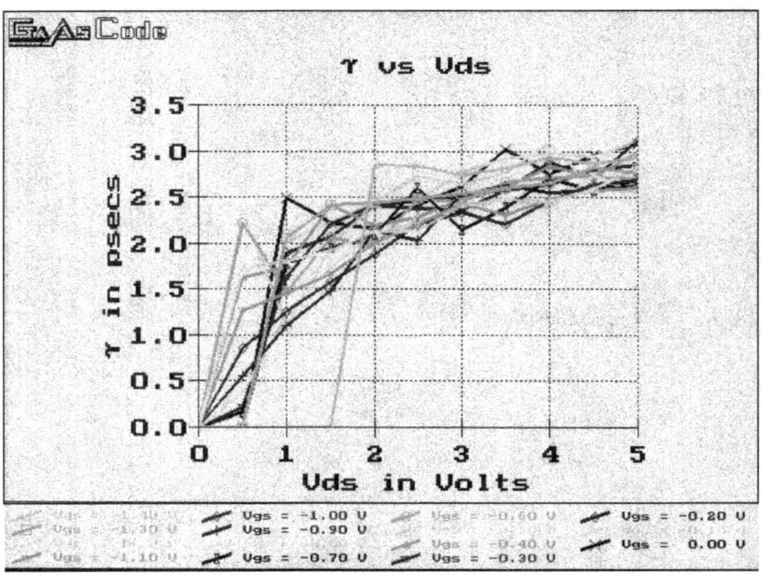

MATCOM 0.3μm FET

Figure 15.4 Extracted transconductance delay for a 0.3 μm FET.

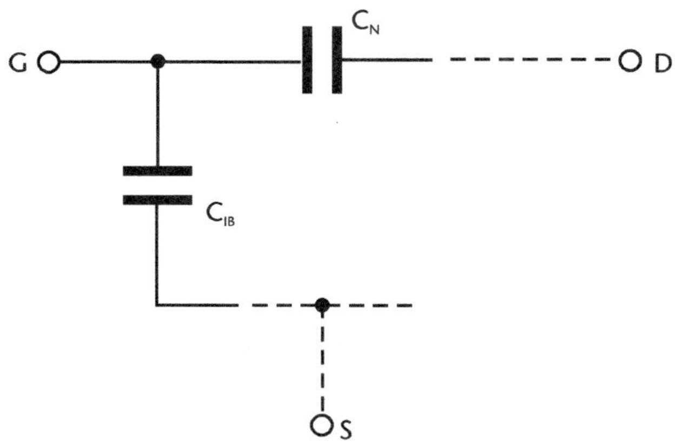

Figure 15.5 The beginnings of the reformed model: branch capacitance C_{IB} and transcapacitance C_N.

time it takes the change to travel from one end of $L_{\text{II–III}}$ to the other, which is τ_{gm}. Section 14.4 dealt with charge storage by pure delay, both in the large-signal case and in the small-signal sinusoidal limit.

How charge is stored in region II–III under large-signal nonlinear conditions is the foundation for reformulating the equivalent circuit model. The rationale, which is rather lengthy, has been given in Section 14.4. In the present section, we give the much shorter account for the small-signal case while stressing that the reasoning which follows, and the result to which it leads, are not valid for large-signal, nonlinear use.

Customized to region II–III, (14.4) becomes

$$C_{\text{II–III}} = g_{m0}\tau_{gm}$$

which has the dimensions of capacitance, but that is as far as it goes; it is not a capacitor. Irrespective, given that the small-signal channel current when the transconductance generator is driven by a small-signal voltage δV_{GC} is $g_{m0}\delta V_{GC}$, we have it that current $g_{m0}\delta V_{GC}$ stores charge $C_{\text{II–III}}\delta V_{GC}$ in a time τ_{gm}.

Furthermore, as the charge is stored in the same part of the device as that across which the stimulus $\delta V_{GC}$was applied, it qualifies as a branch charge. Thus, to the extent that $C_{\text{II–III}}$ can be viewed as a capacitance of some sort, it belongs in the gate-channel (or gate-source) branch of the equivalent circuit.

As the objective is to build a nonlinear circuit model, the foregoing cannot feature in the nascent equivalent circuit of Figure 15.5 because, to emphasize the point, it is strictly a small-signal concept. Its relevance to, and where it fits in, a small-signal model is considered in Section 15.4.

15.3 When V_{DS} Changes

As in Section 15.2, analysis is undertaken with the intrinsic voltages, the voltages that exist at the virtual electrodes, S′ and D′ (Figure 13.1) of the intrinsic device. The intrinsic device still dictates the terminal behavior, and ending the analysis at its boundaries avoids having to engage in tedious algebra and untidy equations. Thus, the development proceeds on the basis of changes in $V_{D'G}$ not V_{DS}.

Additionally, as the elements of interest are functions of both terminal (or virtual terminal) voltages, it is necessary to specify what is to be the state of the second, in this case the gate-channel voltage V_{GC}. While $V_{D'G}$ is varying, V_{GC} will be held constant.

15.3.1 The Overall Picture

Figure 15.6 is the starting point for analysis. As in Section 15.2, the diagram shows the terminal voltages as being constant (in the case of V_{GS}) or changing (in the case of V_{DS}) even though we are imposing such conditions on the intrinsic device not the extrinsic device. Thus, once again, the voltage dispositions on the figure should be regarded as symbolic, not definitive.

When $V_{D'G}$ is the changing voltage, the primary response is an increase in the space-charge layer extension and the secondary response is charge change in the channel.

The primary charge change at X is an increase in positive charge. Correspondingly we accord it a positive + sign; it is a positive branch capacitance. Conversely, the secondary charge changes in regions I and II–III involve an increase in negative charge. An increase in stored negative charge constitutes a negative type of capacitance, not in an absolute sense but in the sense that it reduces the overall or net capacitance. Thus, when it comes to analyzing the increase in the negative charge in regions I and II–III, the resultant equations have to be accorded a negative sign.

Following the terminology of the previous section, to indicate that a voltage is maintained constant, in this case V_{GC}, the marker $| V_{GC}$ is used in the succeeding sections.

15.3.2 Branch Capacitance

We denote the branch capacitance in the gate-drain branch by the symbol C_{gdB}:

$$C_{gdB} \stackrel{\text{def}}{=} \frac{\partial Q_X}{\partial V_{D'G}} \Big| V_{GC} = \frac{\partial Q_X}{\partial X} \frac{\partial X}{\partial V_{D'G}}$$

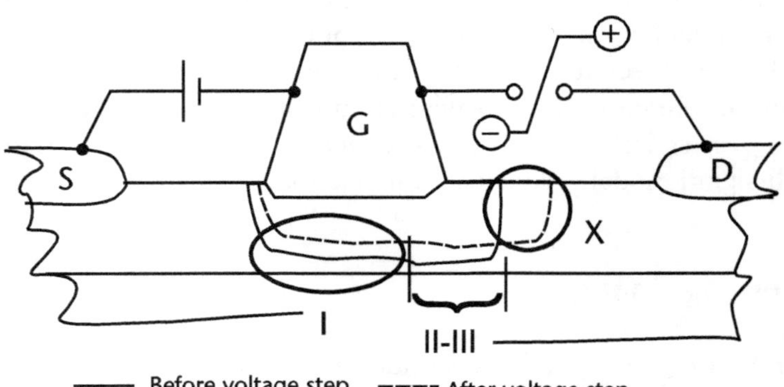

Figure 15.6 The primary charge changes at X, and the secondary charge changes at I and II–III.

where (see Figure 15.3)

$$Q_X = qNdXZ_G$$

Differentiating (13.3):

$$\frac{\partial V_{D'G}}{\partial X} = \frac{qNd}{2\epsilon}\left(1+\frac{2X}{L_G}\right)$$

Hence,

$$C_{gdB} = \frac{2\epsilon Z_G}{\left(1+\dfrac{2X}{L_G}\right)} \tag{15.8}$$

This equation was given in [1] as (6.15) where it was simply regarded as C_{gd} with no distinction drawn between the net gate-drain capacitance in the standard model (the case treated in [1]) and the gate-drain branch capacitance (as here). There is more on the subject in Section 15.4.

Although the foregoing quantitative development is restricted to the v_{sat} regime of operation, it is of interest to look qualitatively at what happens in the ohmic or $v < v_{sat}$ regime.

When $V_{D'G}$ is incremented positively, the depletion region changes shape in the general manner sketched in Figure 15.7. Importantly, the d edge is pulled down (i.e., d increases).

Electrons are drawn out of the d-edge (as well as the X-edge) by the positive potential applied to the drain, and the path for current continuity is the gate-drain loop.

If the change to $V_{D'G}$ is reversed, electrons in the gate-drain region are attracted back into the d-edge and the X-edge by the positive ionic charge that was exposed in the preceding (discharging) step. The current flow path for this recharging is the same as it was for the discharging event, so a capacitance can be used to represent these reactive current flows in a model. As the current is driven by $V_{D'G}$, a branch capacitance is required. We can regard this as a below-the-knee component of the branch capacitance C_{gdB} already identified in (15.8).

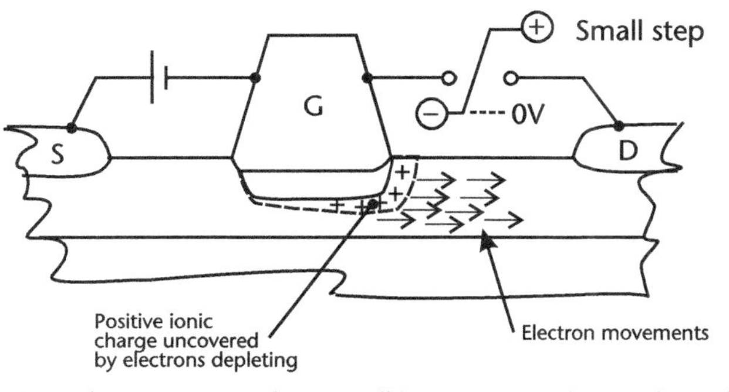

Figure 15.7 Depletion edge movement for a small increment in drain voltage from 0V.

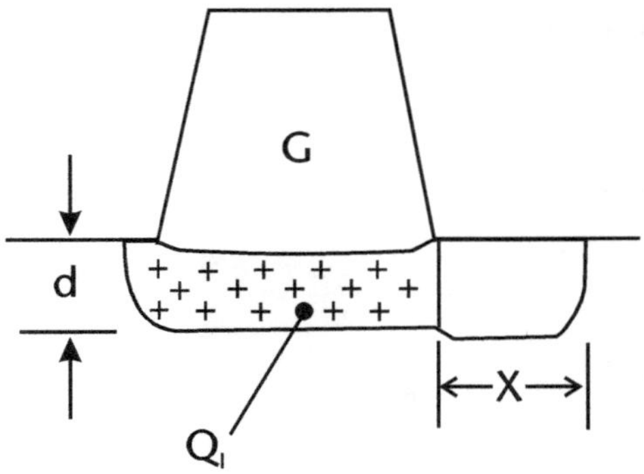

Figure 15.8 Area of charge in region I involved in the transcapacitance C_M.

15.3.3 Transcapacitance

To find an expression for the transcapacitance, here given the symbol C_M to avoid too many subscripts, Figure 15.8 is the starting point.

In carrying out the analysis we have to ensure that the condition V_{GC} = CONSTANT is met. Recalling (13.4),

$$V_{GC} \simeq \frac{qNd^2}{2\epsilon}\left(1 + \frac{2X}{L_G}\right)$$

it follows that there must be a relationship between the changes in d and the changes in X to preserve $\delta V_{GC} = 0$. The constraining relationship was worked out in (13.5) as

$$\delta d = -d\frac{\delta X}{L_G}$$

We recall that the negative sign merely means an increase in X leads to a decrease in d. Figure 15.9 summarizes all the relevant changes.

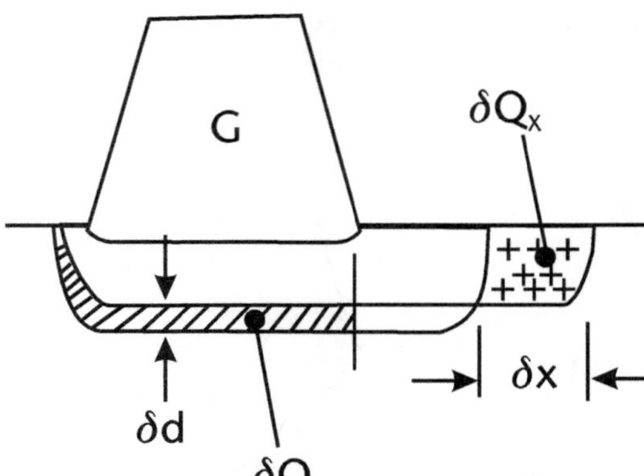

Figure 15.9 Identifying the changes relevant to finding an expression for the transcapacitance C_M.

By definition,

$$C_M \stackrel{\text{def}}{=} \frac{\partial Q_I}{\partial V_{D'G}} \Big| V_{GC} \tag{15.9}$$

where

$$Q_I = qNdL_G Z_G$$

$$\delta Q_I = qNL_G Z_G \delta d$$

Substituting from (13.5):

$$\delta Q_I = -qNdZ_G \delta X$$

We now have a problem. It is that $\delta Q_I = -\delta Q_X$ so, if we were to continue with the calculation of C_M according to (15.9) we would end up with $C_M = -C_{gdB}$, hardly a sensible result. Somewhere, there is a false assumption.

The fault lies in the simplification that the entire flux of E_{yX} field lines originating from within X exerts a lifting field on the depletion edge. In reality, not all the field lines are wholly y-directed throughout the whole depth d; some are part x-directed. x-directed components of external (or fringing) field introduced into the depletion do not contribute to d-lift. On these grounds, (13.5) overestimates the magnitude of d-lift.

One way of redressing the overestimation is to introduce into (13.5) a factor α where $\alpha < 1$ (which becomes (13.6)):

$$\delta d = -\alpha d \frac{\delta X}{L_G}$$

In reality, α will vary right across the range of V_{GC} (or V_{GS}), being larger near pinch-off and smaller near the open channel. Accounting for such variation is beyond the scope of the present endeavor, so a contrived, fixed, value has to do, a value that produces a ratio C_M/C_{gdB} reminiscent of practice. A value typically around 0.5 fits the bill. Thus:

$$C_M \simeq -\alpha C_{gdB} \qquad \alpha \approx 0.5 \tag{15.10}$$

Any misgivings aside, the guideline that we now have is that the transcapacitance C_M should look much like the branch capacitance C_{gdB} only smaller.

Written out in full:

$$C_M = \frac{2\epsilon\alpha Z_G}{1 + \dfrac{2X}{L_G}} \tag{15.11}$$

or, with $\alpha \approx 0.5$:

$$C_M \simeq \frac{\epsilon Z_G}{1 + \frac{2X}{L_G}} \qquad (15.12)$$

We now find something unexpected. From the end of Section 13.7:

$$g_0 = \frac{\delta I_D}{\delta V_{D'G}} = \frac{2\epsilon \alpha v_{sat} Z_G}{\left(L_G + 2X\right)}$$

Allying with (15.11),

$$C_M = \frac{g_0 L_G}{v_{sat}} \qquad (15.13)$$

which is an exact parallel of (15.2).

When at last we come to implementation of the revised model in practice, (15.13) provides yet another measure of whether the extracted elements are believable: C_M and g_0 should look exactly like one another.

Finally, putting together the findings of this section together with those of Section 15.3.2, the additions to the reformed circuit model are as in Figure 15.10.

15.3.4 Orthogonal Branch Charge Storage by Pure Delay

This heading conveys two facts:

1. Charge is stored in that part of the device which is not the part across which the voltage was applied.
2. As the mechanism of storage is by pure delay, it cannot be said to be a transcapacitance.

It is important not to lose sight of the fact that charge storage by pure delay in the large-signal case and in the small-signal sinusoidal limit are quite different

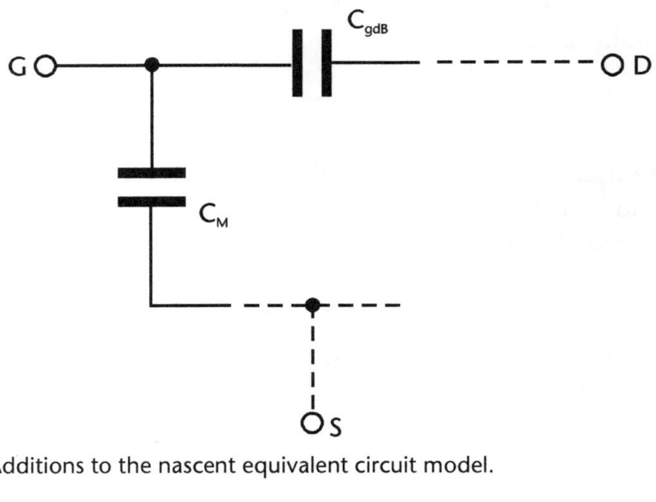

Figure 15.10 Additions to the nascent equivalent circuit model.

in general and very different in detail, as addressed in Section 14.4. Here, we give the much shorter account for the small-signal case while stressing that the reasoning which follows, and the result to which it leads, are not valid for large-signal, nonlinear use.

When the gate-drain voltage changes by an increment $\delta V_{D'G}$, the channel current changes by an amount

$$\delta I_{CH} = g_0 \delta V_{D'G}$$

In the region marked II–III in Figure 15.6, current emerges at the drain end of the depletion region delayed by τ_{gm} relative to the channel current which entered at the left end. A change in the current entering the region ripples through $L_{\text{II–III}}$ at velocity v_{sat}, so the charge is stored progressively in the depletion edge in the time it takes the change to travel from one end of $L_{\text{II–III}}$ to the other, which is τ_{gm}. Thus, in a time equal to τ_{gm}, current $\delta I_{CH} = g_0 \delta V_{D'G}$ stores charge $g_0 \tau_{gm} \delta V_{D'G}$.

This result is an exact parallel of what we found for the complementary case of changing V_{GC} in Section 15.2.4.

Like the term $g_{m0}\tau_{gm}$, $g_0\tau_{gm}$ has dimensions of capacitance but it cannot be represented by a lumped capacitor in the model. Unlike $g_{m0}\tau_{gm}$, which is the branch element, $g_0\tau_{gm}$ is an orthogonal branch element. What they have in common is that both belong in the gate-channel branch of the circuit model.

As confusing as all this sounds, they are facts with which we have to live and which, hopefully, will be a little less confusing after a reading of Section 15.4.

15.4 One Last Visit

This section summarizes some decisive findings.

15.4.1 Reconciliation of the Main Capacitances

Gathering the results of Section 15.2, in response to a change δV_{GC} in the gate-channel voltage, the total charge in the depletion region changes by

$$\delta Q = \left(C_{\text{IB}} + g_{m0}\tau_{gm} - C_N\right)\delta V_{GC} \tag{15.14}$$

The generality of the first and last terms in the brackets is without limit; the middle term restricts the validity of the whole to the small-signal sinusoidal limit.

In the standard model, we would write

$$\delta Q = C_{gc}\delta V_{GC} \tag{15.15}$$

so the suggestion is that

$$C_{gc} = C_{\text{IB}} + g_{m0}\tau_{gm} - C_N \tag{15.16}$$

It may be useful to those familiar with [1] to relate the composition of C_{gc}, (15.16), to what has gone before. The antecedent to (15.16) was

$$C_{gc} = C_{IB} + g_{m0}\tau_{gm} - C_{gd}$$

which follows from Figure 6.9 and (6.21), (6.18), and (6.10) of [1]. The equation immediately preceding (8.1) of [1] is another form of the same relationship. At that time, C_{gd} was taken as the transcapacitance term because the true nature of the transcapacitance, what we are now calling C_N, had yet to be worked out. The correct version is (15.16).

Gathering together the results of Section 15.3, in response to a change $\delta V_{D'G}$ in the gate-drain voltage the total charge in the depletion region changes by

$$\delta Q = \left(C_{gdB} - g_0\tau_{gm} - C_M\right)\delta V_{D'G} \tag{15.17}$$

The generality of the first and last terms in the brackets is without limit; the middle term restricts the validity of the whole to the small-signal sinusoidal limit.

In the standard model, we would write

$$\delta Q = C_{gd}\delta V_{D'G} \tag{15.18}$$

so the suggestion is that

$$C_{gd} = C_{gdB} - g_0\tau_{gm} - C_M \tag{15.19}$$

Here, the wording has been chosen carefully: "suggestion" means precisely what it says, it does not necessarily mean that a fit to multibias S-parameters will yield an empirical function for C_{gd} equivalent to the right side of (15.19). What a fitting routine fits for C_{gd} depends upon the fitter, as discussed in the next section.

15.4.2 Wherefore C_{ds}?

At the end of Section 5.1, the question of what C_{ds} might be from a circuit point of view was left unresolved. Among the equivalent circuit fraternity, most proponents include it. That said, it is hard to find a consistent account of what it actually is or from what it derives.

A clearer view emerges if we resort once more to the physical picture. Figure 15.11 is a redrawn version of Figure 15.6, but now showing the circuit-related elements deduced in Section 15.3.

Bearing in mind that, due to the restriction imposed by characterizing the orthogonal branch storage of charge as $(-g_0\tau_{gm})$ we are talking about a small-signal representation, in those terms the gate-source is shorted. Despite the short, the increase in electron charge in the channel induces a charge of the same magnitude but opposite sign on the gate. With the gate-source shorted, there is no straightforward way such a charge change can be represented in a circuit consisting of just capacitors (namely, the standard model). The only alternatives when fitting the standard model therefore are:

- The change in charge is not represented at all.
- It appears in some place or fashion in the equivalent circuit almost by default.

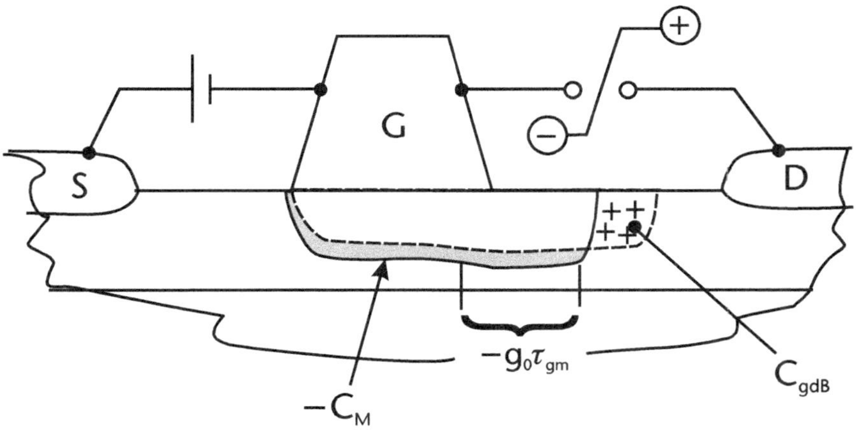

Figure 15.11 Focusing on $-C_M$ and $-g_0\tau_{gm}$.

Adopting conjecture as the path of least criticism, the guess is that the outcome depends upon both the topology of the standard model and the fitter, the routine used to fit the model to the measured S-parameters. In the fitter used throughout this work [2], it turns out that:

1. $(C_M + g_0\tau_{gm})$ is fitted as C_{ds}

which means that

2. C_{gd} is fitted as C_{gdB}.

To emphasize the point, neither of these outcomes is enforced, it is just the way it is. Both require explanation.

A plausible explanation for (1) starts with looking down the channel from the drain towards the source in Figure 15.12. There is a charge change in the edge of the depletion layer governed by $(C_M + g_0\tau_{gm})$. It is written as negative consistent with being a charge of electrons. It induces a positive charge on the gate, and on the source too as they are connected.

Logically, if these changes are to be replicated by the equivalent circuit, a capacitance in the drain-source is the next best place given that the gate-channel branch is shorted. That being so, with the source now bearing a charge change which is positive, the original negative charge change at the depletion edge in the channel becomes associated with the drain electrode, the electrode to which the positive step change was applied in Figure 15.12. Thereby we have a negative charge change at the electrode to which a positive voltage was applied, meaning that in the drain-source branch of the equivalent circuit, $(C_M + g_0\tau_{gm})$ is still ranked as a negative capacitance. It accounts nicely for the steep negative-going excursions in C_{ds} towards V_{DS} = 0V as in Figures 3.26 and 4.11. For reasons not understood, however, the fitter fits C_{ds} with a positive offset. In practical circumstances such as Figure 3.26 a positive offset could be accounted for by geometric (pad) capacitance, but not in fundamental principle as in Figure 4.11 where there is no geometrical capacitance.

With $(C_M + g_0\tau_{gm})$ now accounted for by C_{ds}, C_{gd} cannot be given by (15.19) because the last two terms are already represented. That leaves only C_{gdB} and hence the outcome at (2).

In the context of the present work, although the fact had been long known that the fitted function for C_{gd} appeared by form and numerical size to match the analytical version of (6.15) of [1], it was something of an enigma. The problem was that, had the fitted C_{gd} been given by (15.19), it would be under half the size. Accounting properly for the discrepancy had to await the developments described in this book, which came much later.

Not knowing how other fitting routines are structured, we can only speculate whether (1) and (2) above apply, either fully or approximately, to the $C_{ds}$and C_{gd} elements emerging from those fitters. Judging by the variation in published forms, there is no consensus. Therefore, the following is not a hard and fast rule but just the best interpretation we can offer:

$$C_{ds} = \left(C_M + g_0\tau_{gm}\right)$$

Finally, bringing in the result from Section 15.5,

$$C_{ds} = \left(C_M + g_0\tau_{gm}\right) = g_0\tau_T$$

so it is no surprise that this capacitance should finish up in parallel with the output conductance g_0 in the commonplace model.

15.4.3 The True Nature of the Standard Model

We now reach a crucial point in understanding what the standard model actually is:

1. The right sides of (15.14) and (15.17) represent correctly the changes to the charge stored in the depletion region once stored. Therefore, the left sides represent the charge change once the change has occurred.
2. Therefore, in the static sense, the standard model is correct.
3. In (15.14), the current by which the first two terms change on the right side flows in the gate-channel (or gate-source) part of the device. The current by which the last term changes flows in the gate-drain region of the device. Such is the dynamic reality (with which the reformed model will comply). In sharp contrast, (15.15) has all the current flowing in the gate-channel (or gate-source) branch of the circuit model.
4. Therefore, in the context of (3), the standard model is dynamically incorrect.
5. In (15.17), the first term on the right side is a change involving current flowing in the gate-drain part of the device. The second and last terms are changes due to current in the gate-channel part of the device. Again, whereas the reformed model will comply with this dynamic reality, the standard model lacks the same certainty. At best, it has the current associated with the last two terms flowing in the drain-source branch of the equivalent circuit, which is not the same as the gate-channel branch.
6. Therefore, in the context of (5), the standard model once again is dynamically incorrect.

In summary:

- The standard model is only a static model; it is not a dynamic model.
- Dynamically, the standard model has the wrong reactive currents flowing in the wrong branches. Under small-signal conditions, this fault leads directly to the necessity for some of the fitted elements (e.g., R_s, R_d, L_s, L_d) to take on apparently nonphysical values when freely fitting to measured S-parameters (alternatively, for poor fits to result when the minor elements are fixed at physically sensible values).
- Even as a static representation, the standard model consists of (some) elements that are valid strictly in the small-signal limit only.

If the foregoing assessment is correct, it is little wonder that the standard model fails the fidelity-to-practice test in many circuit applications, and no amount of improved fits to measured $I(V)$ characteristics, the historical thrust of much of Chapter 2, is ever going to fix the problem. The word "static" may fail to convey the seriousness of the limitations, partly because it is close to "quasi-static," which is frequently taken to mean "good enough for RF use." Beyond static, which of the following adjectives would cement the idea that the standard model is fundamentally invalid for dynamic use? Inert? Dormant? Perhaps that is overplaying emphasis where none is needed.

15.5 Enter the Transit Time

To round off Part III on the physics of HFET devices, the transit time τ_T [1] appears in a revealing symmetry. It concerns the sum of the branch charge storage mechanisms in the gate-channel branch and, separately, the sum of the transcapacitance-type storage mechanisms in the same branch.

Denoting by the symbol Σ_C^T the sum of the transcapacitance-type storage elements in the gate-channel part of the device, from Section 15.3:

$$\Sigma_C^T = C_M + g_0\tau_{gm} = \frac{g_0 L_G}{v_{\text{sat}}} + g_0\frac{X}{2v_{\text{sat}}}$$

where we have invoked (13.9) for τ_{gm}.

$$\Sigma_C^T = \frac{g_0}{v_{\text{sat}}}\left(L_G + \frac{X}{2}\right) = g_0\tau_T$$

Now, adding the branch charge storage elements in the same gate-channel part of the device, from Section 15.2:

$$\Sigma_C^B = C_{IB} + g_{m0}\tau_{gm} = \frac{g_{m0}L_G}{v_{\text{sat}}} + g_{m0}\frac{X}{2v_{\text{sat}}}$$

$$\Sigma_C^B = \frac{g_{m0}}{v_{\text{sat}}}\left(L_G + \frac{X}{2}\right) = g_{m0}\tau_T$$

In conclusion, in the gate-channel branch: the total branch capacitance is $g_{m0}\tau_T$ and the total transcapacitance is $g_0\tau_T$.

Many hours have been spent puzzling over whether these elegantly symmetrical results are nothing more than a property of the physics methodology, or whether they are much more than that: a genuine property of physical devices.

References

[1] Ladbrooke, P. H., *MMIC Design: GaAs FETs and HEMTs*, Norwood, MA: Artech House, 1989.

[2] GaAsCode's Fitter FETLink.

CHAPTER 16

Macro-Cell Simulators

16.1 Introduction

In brief, what we call a macro-cell solver is a simplified type of physics-based device simulator that tries to find a practically useful compromise between rigor and speed of execution.

To technologists and circuit designers, a physics-based FET simulator can easily be dismissed as the plaything of blue-sky academics. Yet a device simulator of the right type in the right hands is a boon to the practical effort. A little history sets the scene.

After the years of involvement recounted in this book, a particular thought-provoking outcome remained: offering physics-based HFET modeling software commercially was a flop, but applying it in contract engineering proved quite the opposite. Why might this have been so when the alternative of finding answers to problems of device performance and reproducibility, that of trial and error, was in many ways so wasteful?

Undoubtedly, part of the answer lies with the simulators themselves. At one extreme there are methods of numerical solution based on finite difference, finite box, or finite element schemes. Simulators of this type can take years to write, so the issue of time alone can render them irrelevant: by the time answers are forthcoming, they are answers to yesterday's problem. In addition, among the materials growers, fabrication technologists and circuit designers abounded a suspicion there that the writers of simulation software were aloof, too far removed from the cut and thrust of making things, and therefore not building physics of practical significance into their simulators. Creating such simulators was left to university academics and students, perforce not closely linked to the hard and sometimes messy business having to produce integrated circuits and high-performance devices for a living. Lastly, having the simulator generate RF data at multiple bias points was a hurdle too high for many, so the very essence of what is important to an RF engineer was not addressed. All in all, from the point of view of the device and circuit manufacturer, full-blown simulators spelled commitment of large resources for little gain.

At the other extreme were simple analytic models of the Shockley gradual-channel type, capable of providing fast answers that all too often were inapplicable or outright wrong.

In between the extremes existed a sprinkling of approaches that can be generally classified as quasi-two-dimensional or quasi-2-D. Often missing was enough detail of the theory and assumptions to convey a proper understanding of how they worked, so the term seemed (and still seems) to mean different things to different people. In their favor, such models did at least involve workers closer to practical engineering. However, an arguable criticism was that for their results to be close to practice, they needed one or more adjustable parameters to fit what had already been made and measured; in other words, they were retrospective, hardly an asset to a process manager desperately in need of advice to cure the latest failure.

Through making presentations on physics-based software to companies, mostly in North America, it became apparent that the foregoing attitudes were reasonable and understandable, so our claim to have software both fast-executing and predictive was usually met with polite indifference: the audience had heard it all before. Occasionally, there would be a fractious encounter, few in number but difficult nevertheless. They occurred in organizations with their own device physics expert, into whose realm we were unwittingly making an incursion. Such events left the impression of being a hapless intruder to which there was no answer.

Yet there were exceptions, usually people with a wealth of experience but open-minded; they made all the difference.

- Dr. C. E. C. Wood, who as manager of the GaAs MMIC laboratory in the Hirst Research Centre of the GEC Company, Wembley, United Kingdom, sanctioned the creation of a device physics group, trusting that it would serve the needs of the material and technological effort as a whole. It led to the kind of physics-based device model in [1], which, allied to a commercial circuit simulator, gave the impetus to device centering and yield-driven design of MMICs [2, 3].
- Pacific Monolithics was a particularly innovative company on the U.S. West Coast, the brainchild of Alan Podell, who understood the key component often left unquestioned by circuit designers, the variability of the active devices.
- Pieter Abrie and Frank Traut, circuit designers with long experience and ideas always worth listening to.
- Peter Blakey, a device specialist who, despite long-standing innovation in the field, was open to views and methods not of his making.

As described in Chapter 1, physics-based algorithms can be put to work both to predict the dc and RF characteristics of devices in terms of their material and technological makeup (forward modeling) and, in the inverse sense, to recover data of a technological nature by analyzing measured S-parameters (reverse modeling). Apart from the openness of the few willing to engage with forward modeling, there were others more given to the possibility of what fast fitters and reverse modeling could do. Prominent among the early adopters was Charlie Huang of Anadigics. Jim Turner, of Plessey Research Caswell, was a proponent of it all.

Although commercial take-up of forward modeling software was minimal, over time it proved its worth in contract engineering. It meant that, in some commercial enterprises at least, there was a disposition to become involved with it even though they did not own it. Perhaps, then, the key was not the knowledge the software embodied but knowing how to use it, which (presumably) the originator would be well equipped to do.

What emerges from the historical account is a picture of an underutilized asset, but one whose time has passed to make the impact that it had the potential to make. For all that, the technique may not yet have reached the stage of complete irrelevance, as the insights provided in Section 4.2.2 bear witness. Consequently, a short description of fast, full-range FET and HEMT simulators is warranted in a book on nonlinear design.

16.2 Simulator Requirements

To fulfill the needs of nonlinear design, in no particular order a simulator must:

1. Take in a comprehensive description of the material and technological structure of the device, including layer composition, doping profile, gate cross-section, interelectrode spacing, recess configuration, and surface charge (assumed fixed because to treat it as voltage-dependent involves too many imponderables).
2. Be full-range, meaning from $V_{DS} = 0V$ upwards, and from pinch-off to forward bias.
3. Be predictive, not requiring adjustable parameters or something similar.
4. Predict the RF $i_D(v_{GS},v_{DS})$ characteristics. Given that slow-rate phenomena (notably, the response of surface states D_{ss} and bulk deep levels N_T to changes in bias) are hard to quantify, the simulator perforce will calculate $i_D(v_{GS},v_{DS})$ as determined by the preset values of D_{ss} and N_T anyway. There is no way of knowing whether the assumed values of D_{ss} and N_T match the values set by the particular bias in the practical device. The sensitivity of the device to these indeterminate factors can be assessed by assuming different values and repeating the calculation.
5. Forecast the large-signal properties of the device either in the form of terminal characteristics or as equivalent circuit elements.
6. For each bias point, run in a few tens of seconds or less on a personal computer to provide answers quickly enough to be of immediate use to technologists and circuit designers.

Although none of the fast-executing simulators used in this work incorporated self-heating, to do so should be entirely possible given that the principal determinants are known: the device dimensions, the thermal conductivity, and, from the electrical solution, whereabouts in the device the main volume of heat generation is located.

On the user's side, expectation should be tempered by weighing uncertainties stemming from approximations in the physics (made to gain speed of execution) against uncertainties in the technological data fed into the simulator. In short, the balance of expectation should err on the side of results representative of practice: trends should be forecast well enough to diagnose oddities in FET behavior (as in Section 4.2.2) and to provide sound guidance for modifying the device structure to meet new performance needs.

The issue of execution time bears thinking about. On the face of it, 10 seconds seems very short, but it is not the time taken to complete a single simulation that matters, it is the time taken to produce the answer sought. An example illustrates the point.

Suppose the problem is to see how the nonlinear RF characteristics of a device depend upon the depth and lateral extent of the gate recess, and that 10 different recess configurations are to be tried. Calculations at a minimum of 100 bias points should amount to an adequate determination of the nonlinear characteristics. If each bias point takes 10 seconds, the time required to generate the answer required is getting on for 3 hours. For comparison, the PHEMT results in Section 4.2.2 involved calculation at 600 points.

16.3 Macro-Cell Solvers

16.3.1 The Macro-Cell Idea

Compound semiconductor FETs and HEMTs are innately 2-D devices and highly nonlinear. Devising a scheme for modeling them at a fundamental physical level is not trivial on any basis, let alone one requiring a solution in a few tens of seconds.

Over many years, a method evolved utilizing a coarse mesh of cells that we refer to as macro cells, as in Figure 16.1. The rationale behind this approach is best conveyed by contrasting it with the traditional finite difference or finite element scheme.

Micro-mesh simulators of the finite difference and finite element type employ a fixed mesh with exactly the same set of equations, the set to be solved, throughout the device. Only when iteration ends in a solution does the simulator recognize that some regions contain mobile electrons, some are depleted of electrons, in some parts of the device the carriers move at their scatter-limited velocity and so on. Every time the simulator is run, the general outcome is the same; the technique simply discards knowledge that it has already generated. Looking at the problem critically in this way prompts the idea that, by not discarding such knowledge, a more efficient scheme should be possible. It centers on inverting the entire ethos from finding which physical process applies in each cell of a fixed mesh to finding

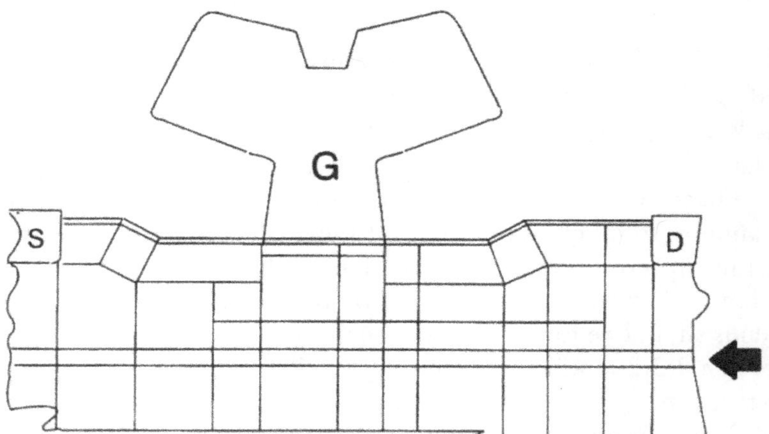

Figure 16.1 Example division of an HFET into macro cells. The device is a bulk GaAs FET with an AlGaAs buffer layer (arrowed).

how large a cell can be while still containing a single physical process. In other words, the mesh now has to be adjustable. Like so many ideas, the adjustable mesh turned out not to be new: it was the centerpiece of the finite box method [4]. In any event, the macro-cell construct was conceived as a logical extension of long-standing analytical models where there are large regions devoid of mobile carriers, others where the electric field strength is low, some where it is high, and so forth.

In summary, we start with an irregular array of interlocking cells. Interior to each a single physical process dominates. Given the terminal voltages, finding a solution then consists of satisfying current and electric flux continuity at the cell boundaries by adjusting the cell sizes, eventually (i.e., meaning after 10 seconds) yielding the magnitude of the physical entity bounded by the cell, for example the number of electrons, together with the spatial extent over which each physical process, entity, or mechanism (e.g., depletion) extends.

Having progressed thus far, we next have to answer for the physics to be lodged in the cells, all the while keeping in sight the big picture, the overriding determinants of device behavior. There are three:

- *d:* the depth of depletion;
- *X:* the space-charge layer extension into the gate-drain space;
- *v:* the electron velocity.

The aim of the solver then is to find values for these three. From them, all else follows. Except for minor additions, the cell physics needed to find them is nothing more than that given in Part III of this book. It is only by integrating these physical principles together in a numerical solver of some sort that a simultaneous solution for them all can be found in the below-the-knee region of operation.

MMIC Design: GaAs FETs and HEMTs demonstrated that, in the v_{sat} regime, analytic equations can be written that describe the bias-dependent small-signal properties of FETs well enough to be practically useful. In large part, it is the fact that the electron velocity, which governs how much drain current will flow, is a constant that makes an analytic solution possible: in effect, it removes a variable, $v(E)$, from the array of variables needing to be found as part of the solution. Nonlinear problems must encompass the below-the-knee region where $v(E)$ plays an interactive part prohibiting an analytic approach. Consequently, some form of simulator is required to manage the interplay between all the variables, finally to achieve a resolution of the competing trends that each presents.

16.3.2 Construction

For those motivated to develop a macro-cell simulator, the itemized procedure below might be useful as a guide:

1. Put physical understanding first and the use of equations second. Starting with equations often leads to simplifications being made (to render the equations tractable), which in the process may risk losing some highly essential physics.

2. Discard the inessential and focus on the essential. "Inessential" means physical processes that have far less effect on the device's behavior than do uncertainties in technology and device structure. Diffusion is one example.
3. Work out what the shape of the depletion region must be on physical grounds and what determines its size.
4. Decide upon the likely consequences of the fact that the transit time is comparable with the momentum relaxation time, a fundamental mechanism that, together with the energy relaxation time and intervalley scattering time, governs electron transport in the material in which the device is constructed.
5. Determine, on physical grounds, how electrons are transported in the x-direction in the presence of y-directed fields.
6. Decide what confines the charge carriers and where in the device this happens.
7. Fathom out the consequences of electrons becoming so hot that they may deconfine.
8. Work out which regions are dominant in determining the device behavior and construct the macro-mesh accordingly.
9. Understand where the electric field arises and where it terminates, a key factor.
10. Finally, work out which equations, or parts thereof, should apply to which cells of the macro-mesh and only then begin to follow what equations tell you.

16.3.3 Choosing the Cells

The cell boundaries are decided principally by three factors:

1. The cross-sectional makeup of the device consisting of the materials, doping profile, recess size, and shape; the cells need not be rectangular boxes.
2. The physical mechanisms dominant region by region, with prominent regions being L_{I}, L_{II}, and X.
3. Given that for predicting microwave performance it is convenient to work with an equivalent circuit, wherever possible the mesh also needs to set boundaries to enclose each circuit element (e.g., R_S, R_I, R_D).

Note that surface charge must be included if adequate representation of practical devices is the objective. Surface charge syphons off the electric field arising from positive charge within the space-charge layer extension X, thereby decreasing the net density of positive charge within X. Because the gate-drain potential $V_{D'G}$ depends upon the product of the net charge density within the extension X and the value of X itself, X is pushed out towards the drain to compensate. The decrease in the net positive charge density also results in a decrease in the y-directed field E_{yX} terminating on the gate, which, in turn, decreases d-lift, thereby flattening the $I(V)$ curves. In principle, surface charge could be made time-dependent to account for the memory effect to which it gives rise, a speculative possibility not tried.

In summary, the opening premise is to have the mathematics serve the physics. Once the dominant 2-D physical process operating within each cell is prescribed, mathematics can be applied to quantify the physics cell by cell, treating each cell as a small boundary-value problem matching with its neighbors. If the procedure is inverted by starting with the mathematics, simplifications made to render the mathematical problem tractable may mean that essential physics is lost. Mathematics must be seen as the servant of the physics, not the dictator.

16.3.4 Below-the-Knee Realism

Given the importance of the region of the $I(V)$ characteristics below the knee, the predictions made by a macro-mesh solver have to replicate the practical $I(V)$ relationship, not only in shape but magnitude as well. Figure 4.17 is a fair example. Predominantly, in this regime, any change in the depth of depletion d in response to increases in V_{DS} is so small as to have a barely noticeable effect on the channel opening. As discussed in Section 13.8 in connection with the step τ_{STEP} in the delay, increases in drain voltage are absorbed initially as an expansion of region II. It is only when the space-charge layer extension X increases that the field gives rise to d-lift. Hence, the shape of $I(V)$ in the below-the-knee region is dominated by two factors: the velocity-field characteristic and the $I_D(R_S + R_D)$ voltage drop (see Figure 13.21).

16.3.5 Deconfinement of Hot Electrons

Over large parts of the operational range, electrons become energetic enough to scatter out of the conducting channel proper. The question is whether in the act of so doing they add a parallel path for current that is eventually collected by the drain. It is a question to be answered as it has a bearing on how far into the substrate the macro-mesh needs to extend. The relevant passages are in Section 14.1.

16.4 The PHEMT Macro-Cell Solver

In the case of the HEMT-PHEMT, no compromise is possible with the core part: the material layer structure and its quantum-mechanical solution in 1-D. It is a job for a quantum physicist, not a device engineer. As in all aspects of device physics touched upon, a good understanding of practice is a prerequisite. Like the simpler FET solver, a HEMT simulator must run in reasonable time, not tens of minutes.

The core of the PHEMT simulator was a quantum analyzer for heterojunction structures developed by Dr. Duncan Allsopp, then at the University of York. From fundamental principles, this versatile tool calculated the free-carrier distribution within the 2-DEG when subjected to external bias.

Given the core 1-D quantum solver, the task reduces to working out how to construct a 2-D simulator around it. While the principles remain the same as for a FET solver, the detail is slightly different. Beyond the drain edge of the gate there must still be a net positive charge to support the gate-drain potential and, as before, that in turn gives rise to a field which eventually terminates on the gate. Within the coarse network of cells overlaid on the cross-section of the device are now cells in which the prescribed physics is that of the quantum analyzer. At each of these special cells underneath the gate, the 2-D field effectively biases the quantum well structure, so the 1-D quantum solver has to be run afresh for each such cell. The same is true for the bulk channel FET, but there the calculation is much simpler involving, as it does, only depleted and undepleted bulk doped regions. Allowing for several iterations to reach an overall solution, the key is to find the balance between run time and the number of cells.

Solution proceeds by adjusting the boundaries of all cells to find the 2-DEG concentration within the special cells consistent with space-charge and electron transport constraints and the 2-D electric field originating from ionic charge in the gate-drain region. A rapidly executing simulator is the result with sufficient closeness to practice to provide a working connection between a device's electrical behavior and its technological makeup. The results given in Section 4.2.2 are typical.

16.5 Applications and Limitations

For the most part, the applications for fast-executing FET and HEMT simulators are those where no other facility exists to generate the answers required. They include:

- Studying the relationship between the terminal RF properties of a device (properties accessible by measurement) and the internal physical processes, inaccessible by measurement, but which must be put into the form of internal circuit elements for a true large-signal model.
- Unraveling where the shapes or forms of the elements of the commonplace model come from when fitted to measured S-parameters.
- Determining the origin of unwanted behavior such as severe transconductance compression, soft breakdown, and catastrophic failure.
- Aiding device design by examining the effect of different material and structural parameters.
- Extending the yield-driven methods of [1] to nonlinear circuits.

As to limitations, the coarseness of the mesh inevitably leads to sharp transitions and kinks in the predicted element functions, as, for instance, in Figure 4.8 and the more extreme example of Figure 4.5. To circumvent the difficulty of overly long execution times that would follow from a finer mesh, an untried possibility is to condition the functions by smoothing with mathematical functions, notably those used to iron out irregularities in empirical plots derived from measured S-parameters. Coupled with further refinement of the simulator proper as well, the status of the technique might be lifted to better than indicative or representative of practice. In the end, however, as with other approaches, that hope could be dashed by the subthreshold region, due not only to shortcomings in the simulator but also to the practical uncertainty involving the vagaries of doping profile tails, conditions at material interfaces, and so forth.

Lastly, the macro-cell solvers referred to herein were based upon on time-independent equations, meaning that they were not designed for direct incorporation into a nonlinear circuit simulator. However, simulation of nonlinear circuits using devices generated by the solvers was possible via the intermediate step of deriving from the simulator equivalent circuit models of elements as, for instance, in Section 4.2.2.

References

[1] Ladbrooke, P. H., *MMIC Design: GaAs FETs and HEMTs*, Norwood, MA: Artech House, 1989.

[2] Ladbrooke, P. H., A. J. Hill, and J. P. Bridge, "CAD for GaAs IC Manufacture-Ability—Thirteen Issues in FET Physics," *GaAs IC Symposium Digest*, 1991, pp. 175–178.

[3] Bridge, J. P., and P. H. Ladbrooke, "Predicting and Enhancing the Manufacturability of MMICs," *Microwave Engineering Europe*, May 1995, pp. 29–31.

[4] Franz, A. F., et al., "Finite Boxes—A Generalization of the Finite-Difference Method Suitable for Semiconductor Device Simulation," *IEEE Transactions on Electronic Devices*, Vol. ED-30, 9, September 1983, pp. 1070–1082.

References

CHAPTER 17

Conclusions

In Part I, we dissected the commonplace equivalent circuit model to see how, why, and where it falls short for nonlinear use. That is not to detract from its value for small-signal applications where it meets the need extremely well, provided that the elements are allowed to be fitted freely. By the same token, the standard model behaves passably in nonlinear circuits with comparatively undemanding specifications at frequencies that are a fraction of the cutoff frequency of the devices, all of which is known.

In Part III, however, the standard model has been shown to be correct in the static limit, but only in the static limit, thus offering insight into its undependability for more demanding nonlinear circuits. As soon as there are time-varying voltages, the currents that store charge in, or deplete charge from, the space-charge layer occur in the wrong branches of the network. The task of correcting the model therefore becomes one of finding where those charge flows should occur. In retrospect, the answer that emerged from applying the principles of device physics seems obvious and simple, as if the pieces of the puzzle were there just waiting to be assembled. It is the proving that takes the time.

Matters of practice needed to be addressed too: where are the empirical data to come from? Should the resource be large-signal measurements, often asserted correctly to be not the same thing as bias-dependent small-signal measurements? What exactly is the nature of the defects to which the latter give rise? What are the consequences of the flaw for practical circuits? That too had to be answered by device physics, in this instance in the form of a fast executing HEMT simulator. Although some potential consequences of the flaw were made apparent, less obvious was the possibility for doing anything about it, thus prompting another way of looking at the problem.

Multibias S-parameters may have gained a reputation for being inadequate not so much because they discount nonlinear terms of the type $v\delta C$ and $i\delta R$ in the model extraction process, but because the extracted elements exhibit distortions attributable to deficiencies in the standard model topology. Could it be more important to have the right elements in the right place in the model (even accepting the omissions of the $v\delta C$ and $i\delta R$ type) rather than extracting the wrong topology model from true nonlinear data? Certainly, from the methods of this book, it appears that way.

Reformulating the topology in Part II involved first characterizing correctly the conduction current flows, namely, the source current, the version of it delayed by a voltage-dependent factor τ (the drain current) and, crucially, acknowledging that the difference between them gives rise to an unfamiliar concept: charge storage by pure delay. Secondly, determining how many capacitive elements there should be and their placement in the various branches.

By using the correct dynamic $I(V)$ characteristics, we axiomatically abandon reliance on the small-signal transconductance and output conductance and the bias dependence thereof. As far as we can tell, τ is recovered from multibias S-parameters in the form required for large-signal operation. Thus, the large-signal equivalents of the small-signal capacitive-type elements $g_{m0}\tau$ and $g_0\tau$ are properly rendered, leaving only the true capacitive elements C_{IB}, C_N, C_{gdB}, C_M to be accounted for.

The underpinning physical theory is large signal (nothing about it restricts it to small signal) and gives the general nonlinear forms these four capacitances should have. By some quirk of fate, we find C_{IB}, C_N, C_{gdB}, C_M derived from multibias S-parameters generally conform to the expected forms, even though accounting for nonlinear terms of the types $i\delta R$ and $v\delta C$ in the fitting-extraction phase has been left unaddressed. The fact that the model and extractor work rather well therefore must mean either these terms are small or they are less important than getting the topology right. That being so, we have ended up with a practically useful large-signal dynamic model.

For a model to be useful as a design tool, it should be judged not on model merit alone but on combined model and extractor merit. The new topology circuit should be no more difficult to extract than the standard version, but therein lurks a stark reality: fitting any model to the resource data must be done well, yet by default it has been accorded something of a Cinderella status in published model accounts. It needs to change.

Although conducted for the grounded-source configuration, provided that the physical operation of the device has been correctly and wholly embodied, the approach should pass for general circuit application where the source may not be grounded. In the intrinsic FET cell it is the interelectrode voltages that govern the device's behavior, so the source need not be grounded for the model derived from grounded-source measurements to be valid. Any restrictions derive from signal paths exterior to the intrinsic device itself, for instance inductance, resistance, and capacitance of via holes and backplane to other parts of a package structure or circuit.

The experience of simulating practical circuits using the new topology model suggests that there may be a fundamental difficulty with harmonic balance: the Fourier transform when the voltage-dependent delay is a parameter of the model. A time-domain simulator reconfigured to accept the new model presented no such difficulty. On that basis, a good way to use the model for multistage circuits promises to be the time-domain simulation of each stage followed by X-parameters for the whole.

Although all the RF results were obtained with devices in the GaAs and related materials system, albeit with wide variation in technology, FETs and HEMTs are generic devices so the findings should apply equally to GaN and SiC (the second having been partly proven within this book anyway).

A frequent shortcoming among accounts of device models is the unaddressed question of where it does not work. To avoid this pitfall, there are weakness and strengths for the reformulated model.

The weaknesses include:

1. Difficulty in coding the model for typical commercial circuit simulators necessitating workarounds;
2. Slow execution, with occasional convergence failure;
3. No clear solution to handling the complexity of slow-rate or memory phenomena (a difficulty common to all).

The strengths include:

1. True-to-life topology with physically correct elements, so the right currents flow in the right branches;
2. Operating principles derived by treating FETs and HEMTs as a genus not allied to any particular material system, so they should apply to GaN and SiC as well as GaAs;
3. Requires only multibias S-parameter measurements;
4. Unites with a routine for reconstructing the bias-dependent RF $I(V)$ characteristics along which devices work in practical circuits;
5. Supported by physics-derived forms expected for the circuit elements to help assess their correctness at various stages of extracting the model from measurements;
6. Features a single analytic equation that fits all current voltage and extracted element functions with smooth interpolation and extrapolation (the latter naturally within limits);
7. Via a two-time-frame formulation, establishes a basis for simulation of slow-rate effects arising from surface states, bulk deep levels, temperature changes during operation, and the interaction between these various mechanisms;
8. Being underpinned by the physics of device operation, the approach links what the technologist does in manufacturing devices to what electrically the circuit engineer is faced with when using them;
9. Fulfills the requirement for a single nonlinear model valid for a wide variety of circuit applications, obviating the need for versions of a model particular to circuit type.

In addition, as with all circuit models, there is a fundamental inability to determine from terminal measurements the exact functions for the intrinsic elements in the equivalent circuit. The shortcoming is a matter of principle. In practice, the resulting approximation is a matter of degree. Also, fitting-with-extraction is an integral but unheralded part of any model.

In summary, when implemented in full the approach detailed herein should offer close to best agreement with practice available from any modeling scheme. It does not require specialized and expensive test instrumentation but, like all models requires the matter of fitting and extraction to be treated with the seriousness it deserves. The difficulties experienced in implementing the whole in commercial circuit simulators could be alleviated by simulator development. Integration of time-domain and X-parameter simulators has special appeal to circumvent the difficulties with harmonic balance simulation which appear to be fundamental when the model embraces voltage-dependent delay. Short of implementing the whole, some of the techniques may be of benefit in improving the fidelity of circuit models in common use.

Acronyms and Abbreviations

2-DEG	Two-dimensional electron gas
ACP	Adjacent channel power
ACPR	Adjacent channel power ratio
AM-AM	Amplitude modulation-to-amplitude modulation, basically gain compression in an amplifier
AM-PM	Amplitude modulation-to-phase modulation
CDMA	Code division multiple access
dc	Direct current
EEHEMT	A model sold by EEsof Inc.
EM	Electromagnetic
FET	Field-effect transistor
FEMT	An alternative for FET or HEMT
HEMT	High electron mobility transistor
HFET	Heterojunction field-effect transistor (also used generically, like FEMT)
IM3	Third-order intermodulation
M3IC	Millimeter-wave monolithic integrated circuit
MBE	Molecular beam epitaxy
MMIC	Monolithic microwave integrated circuit
MOCVD	Metal organic chemical vapor deposition
MODFET	Modulation-doped FET (another term for HEMT)
NLVNA	Nonlinear vector network analyzer
OFDM	Orthogonal frequency division multiplexing
PHEMT	Pseudomorphic HEMT
RF	Radio frequency
rms	Root mean square
S2P	Two-port scattering parameters
TEGFET	Two-dimensional electron gas FET (another term for HEMT)
VNA	Vector network analyzer

List of Symbols

C	capacitance
C_{ds}	drain-source capacitance
C_{gc}	gate-channel space-charge capacitance
C_{gd}	gate-drain space-charge capacitance
C_{dgB}	gate-drain branch capacitance in the reformed model
C_{gs}	gate-source capacitance
C_{I}	space-charge capacitance of region I
$C_{\mathrm{II–III}}$	space-charge capacitance of regions II and III
$C_{\mathrm{I}B}$	branch capacitance of region I in the reformed model
C_M	gate-channel transcapacitance
C_N	gate-drain transcapacitance
C_0	output capacitance
C_x	part of a subcircuit to form differentials
d	depth of depletion
D	electric flux density
D_{SS}	area density of surface states
D_x	x-directed component of D
D_y	y-directed component of D
D	drain electrode
D'	virtual or internal drain
E	electric field
E_B	breakdown electric field
E_S	electric field at which the carrier velocity saturates
E_X	vector electric field arising from X
E_x	x-directed component of E
E_y	y-directed component of E
E_{yX}	y-directed field arising from X
f	frequency; function or functional dependence
f_T	transition frequency; the frequency at which the current gain of a FEMT falls to unity
g_m	shorthand for g_{m0}

g_{m0}	transconductance
g_{m0B}	transconductance at the bias point
g_0	output conductance
g_{0B}	output conductance at the bias point
$g_{0\mathrm{dc}}$	output conductance at dc
$g_{0\mathrm{RF}}$	RF output conductance
G	gate electrode
G_{ass}	associated gain
i_d	small-signal drain current
i_D	large-signal drain current
i_{DGC}	gate-channel diode current
i_{DGD}	gate-drain diode current
i_g	small-signal gate current
i_S	large-signal source current
$i(t)$	large-signal, time-dependent current
$i(t-\tau)$	delayed large-signal current
$i(v)$	large-signal current-voltage characteristics
I_D	drain current at dc
I_{DB}	drain bias current
I_{DP}	dc drain current at a particular point P
I_{DSS}	drain current with gate shorted to source
I_F	maximum drain current under forward gate bias
I_G	gate current
I_S	source current
$I(t)$	general time-dependent current
$I(t-\tau)$	general delayed current
$\mathrm{I(V)}$, $I(V)$	current-voltage characteristics
L	inductance; length
L_d	drain series inductance
L_g	gate series inductance
L_G	gate length
L_{I}	length of region I in the channel
L_{II}	length of region II in the channel
L_{III}	length of region III in the channel
$L_{\mathrm{II-III}}$	combined length of regions I and II
L_s	series source inductance
N	doping density
$NF_{\min}$	minimum noise figure

N_T	volume density of bulk traps or deep levels
P_0	output power
q	charge on the electron
Q	charge; total charge in the depletion region
Q_d, Q_D	drain-side charge in the depletion layer
Q_G	gate charge
Q_{I}	region I depletion charge
$Q_{\text{II–III}}$	region II-III depletion charge
Q_S	source-side charge in the depletion region
Q_X	charge contained within the space-charge-layer extension
r_0	small-signal output resistance
R	resistance
R_d	small-signal drain series resistance
R_D	large-signal drain series resistance
R_{ds}	small-signal drain-source resistance
R_g	microwave gate series resistance
R_{GD}	large-signal resistance in gate-drain branch
R_i	small-signal intrinsic channel resistance
R_I	large-signal intrinsic channel resistance
R_s	small-signal series source resistance
R_S	large-signal series source resistance
R_x	part of a subcircuit to form differentials
$R(t)$	time-dependent response to a stimulus
s_{ij}	scattering parameters
$S(t)$	time-dependent stimulus
S	source electrode
S′	virtual or internal source
t	time
T	period of a wave
v	electron drift velocity; time-varying voltage
v_d	small-signal drain voltage
v_{DG}	large-signal drain-gate voltage
v_{DS}	large-signal drain-source voltage
v_{DSP}	large-signal drain-source voltage at a particular point
v_g	small-signal gate voltage
v_{GC}	large-signal gate-channel voltage
v_{GS}	large-signal gate-source voltage
v_{GSP}	large-signal gate-source voltage at a point P

v_{ij}	signal voltage between nodes i and j
v_{sat}	saturated drift velocity of carriers
$v(E)$	velocity-field characteristic
V_{CH}	gate-channel voltage in a one-dimensional device
$V_{D'G}$	voltage between virtual drain and the gate
V_{DS}	drain-source voltage
V_{DSB}	drain-source bias; drain-source breakdown voltage
V_{DSP}	value of V_{DS} at a particular point P
V_{GC}	gate-channel voltage
V_{GS}	gate-source voltage
V_{GSB}	gate-source bias voltage
$V_{GS'}$	voltage between gate and virtual source
V_P	pinch-off voltage
$V(x)$	potential in the channel as a function of position
W	channel layer thickness
x	spatial variable
X	space-charge layer extension towards the drain
y	spatial variable
y_{ij}	admittance parameters
Y	general array of admittances
Y	2-port of y-parameters
Y^e	admittance 2-port enclosing a smaller 2-port
Z	general array of impedances
Z_G	gate width
δ	incremental operator
Δ	incremental operator
ϵ	absolute permittivity of semiconductor
τ	delay in drain current
$\tau_{\delta L}$	transit time through an element of the channel of length δL
τ_{gm}	transconductance delay; a small-signal version of τ
τ_m	momentum relaxation time
τ_{ss}	characteristic time of slow states or traps
τ_T	transit time
τ_x	time constant of subcircuit to form differentials
τ_0	value of τ at the bias point
ω	angular frequency

About the Author

Peter H. Ladbrooke has spent a professional lifetime on the physics and application of compound semiconductor FETs, following work on nonuniformities in planar Gunn and Schottky devices for which he was awarded a Ph.D. by Cambridge University. He lives on the outskirts of the old university city with his wife of many years, Josephine.

Index

D

Artech House Microwave Library

Behavioral Modeling and Linearization of RF Power Amplifiers, John Wood

Chipless RFID Reader Architecture, Nemai Chandra Karmakar, Prasanna Kalansuriya, Randika Koswatta, and Rubayet E-Azim

Control Components Using Si, GaAs, and GaN Technologies, Inder J. Bahl

Design of Linear RF Outphasing Power Amplifiers, Xuejun Zhang, Lawrence E. Larson, and Peter M. Asbeck

Design Methodology for RF CMOS Phase Locked Loops, Carlos Quemada, Guillermo Bistué, and Iñigo Adin

Design of CMOS Operational Amplifiers, Rasoul Dehghani

Design of RF and Microwave Amplifiers and Oscillators, Second Edition, Pieter L. D. Abrie

Digital Filter Design Solutions, Jolyon M. De Freitas

Discrete Oscillator Design Linear, Nonlinear, Transient, and Noise Domains, Randall W. Rhea

Distortion in RF Power Amplifiers, Joel Vuolevi and Timo Rahkonen

Distributed Power Amplifiers for RF and Microwave Communications, Narendra Kumar and Andrei Grebennikov

Electric Circuits: A Primer, J. C. Olivier

Electronics for Microwave Backhaul, Vittorio Camarchia, Roberto Quaglia, and Marco Pirola, editors

An Engineer's Guide to Automated Testing of High-Speed Interfaces, Second Edition, José Moreira and Hubert Werkmann

Envelope Tracking Power Amplifiers for Wireless Communications, Zhancang Wang

Essentials of RF and Microwave Grounding, Eric Holzman

Frequency Measurement Technology, Ignacio Llamas-Garro, Marcos Tavares de Melo, and Jung-Mu Kim

FAST: Fast Amplifier Synthesis Tool—Software and User's Guide, Dale D. Henkes

Feedforward Linear Power Amplifiers, Nick Pothecary

Filter Synthesis Using Genesys S/Filter, Randall W. Rhea

Foundations of Oscillator Circuit Design, Guillermo Gonzalez

Frequency Synthesizers: Concept to Product, Alexander Chenakin

Fundamentals of Nonlinear Behavioral Modeling for RF and Microwave Design, John Wood and David E. Root, editors

Generalized Filter Design by Computer Optimization, Djuradj Budimir

Handbook of Dielectric and Thermal Properties of Materials at Microwave Frequencies, Vyacheslav V. Komarov

Handbook of RF, Microwave, and Millimeter-Wave Components, Leonid A. Belov, Sergey M. Smolskiy, and Victor N. Kochemasov

High-Efficiency Load Modulation Power Amplifiers for Wireless Communications, Zhancang Wang

High-Linearity RF Amplifier Design, Peter B. Kenington

High-Speed Circuit Board Signal Integrity, Second Edition, Stephen C. Thierauf

Integrated Microwave Front-Ends with Avionics Applications, Leo G. Maloratsky

Intermodulation Distortion in Microwave and Wireless Circuits, José Carlos Pedro and Nuno Borges Carvalho

Introduction to Modeling HBTs, Matthias Rudolph

An Introduction to Packet Microwave Systems and Technologies, Paolo Volpato

Introduction to RF Design Using EM Simulators, Hiroaki Kogure, Yoshie Kogure, and James C. Rautio

Introduction to RF and Microwave Passive Components, Richard Wallace and Krister Andreasson

Klystrons, Traveling Wave Tubes, Magnetrons, Crossed-Field Amplifiers, and Gyrotrons, A. S. Gilmour, Jr.

Lumped Elements for RF and Microwave Circuits, Inder Bahl

Lumped Element Quadrature Hybrids, David Andrews

Microstrip Lines and Slotlines, Third Edition, Ramesh Garg, Inder Bahl, and Maurizio Bozzi

Microwave Component Mechanics, Harri Eskelinen and Pekka Eskelinen

Microwave Differential Circuit Design Using Mixed-Mode S-Parameters, William R. Eisenstadt, Robert Stengel, and Bruce M. Thompson

Microwave Engineers' Handbook, Two Volumes, Theodore Saad, editor

Microwave Filters, Impedance-Matching Networks, and Coupling Structures, George L. Matthaei, Leo Young, and E. M. T. Jones

Microwave Imaging Methods and Applications, Matteo Pastorino and Andrea Randazzo

Microwave Material Applications: Device Miniaturization and Integration, David B. Cruickshank

Microwave Materials and Fabrication Techniques, Second Edition, Thomas S. Laverghetta

Microwave Materials for Wireless Applications, David B. Cruickshank

Microwave Mixer Technology and Applications, Bert Henderson and Edmar Camargo

Microwave Mixers, Second Edition, Stephen A. Maas

Microwave Network Design Using the Scattering Matrix, Janusz A. Dobrowolski

Microwave Power Amplifier Design with MMIC Modules, Howard Hausman

Microwave Radio Transmission Design Guide, Second Edition, Trevor Manning

Microwave and RF Semiconductor Control Device Modeling, Robert H. Caverly

Microwave Transmission Line Circuits, William T. Joines, W. Devereux Palmer, and Jennifer T. Bernhard

Microwaves and Wireless Simplified, Third Edition, Thomas S. Laverghetta

Modern Microwave Circuits, Noyan Kinayman and M. I. Aksun

Modern Microwave Measurements and Techniques, Second Edition, Thomas S. Laverghetta

Modern RF and Microwave Filter Design, Protap Pramanick and Prakash Bhartia

Neural Networks for RF and Microwave Design, Q. J. Zhang and K. C. Gupta

Noise in Linear and Nonlinear Circuits, Stephen A. Maas

Nonlinear Design: FETs and HEMTs, Peter H. Ladbrooke

Nonlinear Microwave and RF Circuits, Second Edition, Stephen A. Maas

On-Wafer Microwave Measurements and De-Embedding, Errikos Lourandakis

Parameter Extraction and Complex Nonlinear Transistor Models, Günter Kompa

Passive RF Component Technology: Materials, Techniques, and Applications, Guoan Wang and Bo Pan, editors

PCB Design Guide to Via and Trace Currents and Temperatures, Douglas Brooks with Johannes Adam

Practical Analog and Digital Filter Design, Les Thede

Practical Microstrip Design and Applications, Günter Kompa

Practical Microwave Circuits, Stephen Maas

Practical RF Circuit Design for Modern Wireless Systems, Volume I: Passive Circuits and Systems, Les Besser and Rowan Gilmore

Practical RF Circuit Design for Modern Wireless Systems, Volume II: Active Circuits and Systems, Rowan Gilmore and Les Besser

Principles of RF and Microwave Design, Matthew A. Morgan

Production Testing of RF and System-on-a-Chip Devices for Wireless Communications, Keith B. Schaub and Joe Kelly

Q Factor Measurements Using MATLAB , Darko Kajfez

Radio Frequency Integrated Circuit Design, Second Edition, John W. M. Rogers and Calvin Plett

Reflectionless Filters, Matthew A. Morgan

RF Bulk Acoustic Wave Filters for Communications, Ken-ya Hashimoto

RF Circuits and Applications for Practicing Engineers, Mouqun Dong

RF Design Guide: Systems, Circuits, and Equations, Peter Vizmuller

RF Linear Accelerators for Medical and Industrial Applications, Samy Hanna

RF Measurements of Die and Packages, Scott A. Wartenberg

The RF and Microwave Circuit Design Handbook, Stephen A. Maas

RF and Microwave Coupled-Line Circuits, Rajesh Mongia, Inder Bahl, and Prakash Bhartia

RF and Microwave Oscillator Design, Michal Odyniec, editor

RF Power Amplifiers for Wireless Communications, Second Edition, Steve C. Cripps

RF Systems, Components, and Circuits Handbook, Ferril A. Losee

Scattering Parameters in RF and Microwave Circuit Analysis and Design, Janusz A. Dobrowolski

The Six-Port Technique with Microwave and Wireless Applications, Fadhel M. Ghannouchi and Abbas Mohammadi

Solid-State Microwave High-Power Amplifiers, Franco Sechi and Marina Bujatti

Stability Analysis of Nonlinear Microwave Circuits, Almudena Suárez and Raymond Quéré

Substrate Noise Coupling in Analog/RF Circuits, Stephane Bronckers, Geert Van der Plas, Gerd Vandersteen, and Yves Rolain

System-in-Package RF Design and Applications, Michael P. Gaynor

Technologies for RF Systems, Terry Edwards

Terahertz Metrology, Mira Naftaly, editor

Understanding Quartz Crystals and Oscillators, Ramón M. Cerda

Vertical GaN and SiC Power Devices, Kazuhiro Mochizuki

The VNA Applications Handbook, Gregory Bonaguide and Neil Jarvis

*Wired and Wireless Seamless Access Systems for Public Infrastructur*e, Tetsuya Kawanishi

For further information on these and other Artech House titles, including previously considered out-of-print books now available through our In-Print-Forever® (IPF®) program, contact:

Artech House Publishers
685 Canton Street
Norwood, MA 02062
Phone: 781-769-9750
Fax: 781-769-6334
e-mail: artech@artechhouse.com

Artech House Books
16 Sussex Street
London SW1V 4RW UK
Phone: +44 (0)20 7596 8750
Fax: +44 (0)20 7630 0166
e-mail: artech-uk@artechhouse.com

Find us on the World Wide Web at: www.artechhouse.com